Techniques and Concepts
of High-Energy Physics VI

NATO ASI Series

Advanced Science Institutes Series

A series presenting the results of activities sponsored by the NATO Science Committee, which aims at the dissemination of advanced scientific and technological knowledge, with a view to strengthening links between scientific communities.

The series is published by an international board of publishers in conjunction with the NATO Scientific Affairs Division

A	Life Sciences	Plenum Publishing Corporation
B	Physics	New York and London
C	Mathematical and Physical Sciences	Kluwer Academic Publishers
D	Behavioral and Social Sciences	Dordrecht, Boston, and London
E	Applied Sciences	
F	Computer and Systems Sciences	Springer-Verlag
G	Ecological Sciences	Berlin, Heidelberg, New York, London,
H	Cell Biology	Paris, Tokyo, Hong Kong, and Barcelona
I	Global Environmental Change	

Recent Volumes in this Series

Volume 268—The Global Geometry of Turbulence: Impact of Nonlinear Dynamics
edited by Javier Jiménez

Volume 269—Methods and Mechanisms for Producing Ions from Large Molecules
edited by K. G. Standing and Werner Ens

Volume 270—Complexity, Chaos, and Biological Evolution
edited by Erik Mosekilde and Lis Mosekilde

Volume 271—Interaction of Charged Particles with Solids and Surfaces
edited by Alberto Gras-Martí, Herbert M. Urbassek,
Néstor R. Arista, and Fernando Flores

Volume 272—Predictability, Stability, and Chaos in N-Body Dynamical Systems
edited by Archie E. Roy

Volume 273—Light Scattering in Semiconductor Structures and Superlattices
edited by David J. Lockwood and Jeff F. Young

Volume 274—Direct Methods of Solving Crystal Structures
edited by Henk Schenk

Volume 275—Techniques and Concepts of High-Energy Physics VI
edited by Thomas Ferbel

Series B: Physics

Techniques and Concepts of High-Energy Physics VI

Edited by
Thomas Ferbel
University of Rochester
Rochester, New York

Plenum Press
New York and London
Published in cooperation with NATO Scientific Affairs Division

Proceedings of the Sixth NATO Advanced Study Institute on
Techniques and Concepts of High-Energy Physics,
held June 14–25, 1990,
in St. Croix, U.S. Virgin Islands

Library of Congress Cataloging-in-Publication Data

NATO Advanced Study Institute on Techniques and Concepts of High
 -Energy Physics (6th : 1990 : St. Croix, V.I.)
  Techniques and concepts of high-energy physics VI / edited by
Thomas Ferbel.
      p.   cm. -- (NATO ASI series. Series B: Physics ; vol. 275)
  "Proceedings of the Sixth NATO Advanced Study Institute on
Techniques and Concepts of High-Energy Physics, held June 14-25,
1990, in St. Croix, U.S. Virgin Islands"--T.p. verso.
  "Published in cooperation with NATO Scientific Affairs Division."
  Includes bibliographical references and index.
  ISBN 0-306-44043-1
  1. Particles (Nuclear physics)--Congresses. I. Ferbel, Thomas.
II. North Atlantic Treaty Organization. Scientific Affairs
Division. III. Title. IV. Series: NATO ASI series. Series B,
Physics ; v. 275.
QC793.N38 1990
539.7'2--dc20   91-29911
                                                                CIP

ISBN 0-306-44043-1

© 1991 Plenum Press, New York
A Division of Plenum Publishing Corporation
233 Spring Street, New York, N.Y. 10013

All rights reserved

No part of this book may be reproduced, stored in a retrieval system, or transmitted
in any form or by any means, electronic, mechanical, photocopying, microfilming,
recording, or otherwise, without written permission from the Publisher

Printed in the United States of America

LECTURERS

A. Clark	Fermilab, Batavia, Illinois
A. Das	University of Rochester, Rochester, New York
D. Leith	Stanford University, Stanford, California
P. Manfredi	University of Pavia, Pavia, Italy
R. Siemann	Cornell University, Ithaca, New York
R. Wigmans	CERN, Geneva, Switzerland
F. Wilczek	Institute for Advanced Study, Princeton, New Jersey

ADVISORY COMMITTEE

B. Barish	Caltech, Pasadena, California
L. DiLella	CERN, Geneva, Switzerland
C. Fabjan	CERN, Geneva, Switzerland
J. Iliopoulos	Ecole Normale Superieure, Paris, France
M. Jacob	CERN, Geneva, Switzerland
C. Quigg	Fermilab, Batavia, Illinois
A. Sessler	University California at Berkeley, California
P. Soding	DESY, Hamburg, Federal Republic of Germany

DIRECTOR

T. Ferbel	University of Rochester, Rochester, New York

PREFACE

The sixth Advanced Study Institute (ASI) on Techniques and Concepts of High Energy Physics was held at the Club St. Croix, in St. Croix, U.S. Virgin Islands. The ASI brought together a total of 70 participants, from 21 different countries. Despite logistical problems caused by hurricane Hugo, it was a very successful meeting. Hugo's destruction did little to dampen the dedication of the inspiring lecturers and the exceptional enthusiasm of the student body; nevertheless, the immense damage caused to the beautiful island was very saddening indeed.

The primary support for the meeting was again provided by the Scientific Affairs Division of NATO. The ASI was cosponsored by the U.S. Department of Energy, by Fermilab, by the National Science Foundation, and by the University of Rochester. A special contribution from the Oliver S. and Jennie R. Donaldson Charitable Trust provided an important degree of flexibility, as well as support for worthy students from developing countries.

As in the case of the previous ASIs, the scientific program was designed for advanced graduate students and recent PhD recipients in experimental particle physics. The present volume of lectures should complement the material published in the first five ASIs, and prove to be of value to a wider audience of physicists.

It is a pleasure to acknowledge the encouragement and support that I have continued to receive from colleagues and friends in organizing this meeting. I am indebted to the members of my Advisory Committee for their infinite patience and excellent advice. I am grateful to the distinguished lecturers for their enthusiastic participation in the ASI, and, of course for their hard work in preparing the lectures and providing the superb manuscripts for the proceedings. I thank John Bythel of the West Indies Lab for his fascinating description of the geology and marine life of St. Croix, and Albert Lang for talking him into this. I thank Harrison Prosper for organizing the student presentations, and both Harrison and Michel Lefebvre for providing the first draft of Frank Wilczek's lecture notes. I also thank Earle Fowler and P.K. Williams for support from the Department of Energy, David Berley and Marcel Bardon for assistance from the National Science Foundation, as well as John Peoples and Angela Gonzales at Fermilab. At Rochester, I am indebted to Ovide Corriveau, Judy Mack and especially Connie Jones for organizational assistance and typing. I owe thanks to Jack Dodds and to Susan and Tommy Borodemos, the managers of the facilities at Club St. Croix, for their and their staff's hospitality. I wish to acknowledge the generosity of Chris Lirakis and Mrs. Marjorie Atwood of the Donaldson Trust, and support from George Blanar and the LeCroy Research Systems Corporation. Finally, I thank Luis da Cunha of NATO for his cooperation and confidence.

T. Ferbel
Rochester, New York

CONTENTS

Introduction to Gauge Theories and Unification 1
    A. Das

Introductory Notes on Particle Physics and Cosmology 49
    F. Wilczek

Experimental Aspects of Hadron Collider Physics 71
    A. G. Clark

A Perspective on Meson Spectroscopy 197
    D. W. G. S. Leith

Introduction to the Physics of Particle Accelerators........................ 283
    R. Siemann

Calorimetry in High Energy Physics... 325
    R. Wigmans

Noise Limits in Detector Charge Measurements.............................. 381
    P. F. Manfredi and V. Speziali

Silicon Junction Field-Effect Transistors in Low-Noise Circuits:
    Research in Progress and Perspectives................................. 423
    P. F. Manfredi and V. Speziali

Index 437

INTRODUCTION TO GAUGE THEORIES AND UNIFICATION

Ashok Das
Department of Physics and Astronomy
University of Rochester
Rochester, NY 14627

Lecture I

Basic Notations

The theories which we will study in these lectures are supposed to describe fundamental processes at extremely high energies. Consequently, these theories will be relativistic theories invariant under Lorentz transformations. Let me, therefore, begin by establishing some notation which I will be using throughout the lectures.

Let us recall that if we have two vectors $\vec{x}$ and $\vec{y}$ in the three dimensional Euclidean space, their product invariant under rotations is defined to be (we will assume repeated indices to be summed unless otherwise specified.)

$$\vec{x} \cdot \vec{y} = \vec{y} \cdot \vec{x} = x_1 y_1 + x_2 y_2 + x_3 y_3 = x_i y_i \qquad i = 1, 2, 3 \qquad (1.1)$$

This is, of course, the scalar product and from this we obtain the length squared of a given vector $\vec{x}$ as

$$\vec{x}^2 = \vec{x} \cdot \vec{x} = x_1^2 + x_2^2 + x_3^2 = x_i x_i \qquad (1.2)$$

which is also invariant under rotations. (Rotations define the isometry group of the Euclidean space.)

In constrast, in the four dimensional Minkowski space, one can define two kinds of vectors, namely, the covariant and the contravariant vectors denoted respectively by A_μ and A^μ. These are four component objects (also known as four-vectors) with μ taking the values, $\mu = 0, 1, 2, 3$. Furthermore, the covariant and the contravariant vectors are related through the metric of the Minkowski space as

$$A^\mu = \eta^{\mu\nu} A_\nu$$
$$A_\mu = \eta_{\mu\nu} A^\nu \qquad (1.3)$$

where I will assume the second rank metric tensors $\eta^{\mu\nu}$ and $\eta_{\mu\nu}$ to take the diagonal matrix form

$$\eta^{\mu\nu} = \begin{pmatrix} 1 & 0 & 0 & 0 \\ 0 & -1 & 0 & 0 \\ 0 & 0 & -1 & 0 \\ 0 & 0 & 0 & -1 \end{pmatrix} = \eta_{\mu\nu} \qquad (1.4)$$

The metric tensors can be used to raise or lower tensor indices and the choice of the metric in Eq. (1.4) is commonly known as the Bjorken-Drell convention.

It is clear now that if we write the components of A^μ as

$$A^\mu = \left(A^0, \vec{A}\right) \qquad (1.5)$$

then the components of A_μ would take the form

$$A_\mu = \eta_{\mu\nu} A^\nu = \left(A^0, -\vec{A}\right) \qquad (1.6)$$

In a sense, the covariant and the contravariant vectors have opposite transformation properties under a Lorentz transformation so that given two vectors A_μ and B^μ, we can define a scalar product

$$A \cdot B = A_\mu B^\mu = A^\mu B_\mu = \eta^{\mu\nu} A_\mu B_\nu = \eta_{\mu\nu} A^\mu B^\nu = A^0 B^0 - \vec{A} \cdot \vec{B} \qquad (1.7)$$

which will be invariant under Lorentz transformations. The length squared of a vector A_μ in Minkowski space now follows to be

$$A^2 = A_\mu A^\mu = \left(A^0\right)^2 - \vec{A}^2 \qquad (1.8)$$

This is Lorentz invariant but is no longer positive definite as would be true in the Euclidean space.

Let us also recall that space and time coordinates define a four vector in Minkowski space. Thus writing

$$x^\mu = (t, \vec{x}) \qquad (1.9)$$

we obtain

$$x_\mu = (t, -\vec{x}) \qquad (1.10)$$

and

$$x^2 = x_\mu x^\mu = t^2 - \vec{x}^2 \qquad (1.11)$$

which is, of course, the invariant length (we will set $\hbar = 1 = c$ throughout). It is clear now that Minkowski space can be divided into four cones and the physical processes are assumed to take place in the forward light cone (so that causality holds) defined by

$$x^2 = t^2 - \vec{x}^2 \geq 0 \qquad t \geq 0 \qquad (1.12)$$

Just as space and time coordinates define a four-vector, derivatives with respect to these coordinates also define a four-vector. Thus the contragradient is defined to be

$$\frac{\partial}{\partial x_\mu} = \partial^\mu = \left(\frac{\partial}{\partial t}, -\vec{\nabla}\right) \qquad (1.13)$$

from which we obtain

$$\frac{\partial}{\partial x^\mu} = \partial_\mu = \eta_{\mu\nu}\partial^\nu = \left(\frac{\partial}{\partial t}, \vec{\nabla}\right) \qquad (1.14)$$

The generalization of the Laplacian to the Minkowski space is known as the D'Alembertian and is given by

$$\Box = \partial_\mu \partial^\mu = \frac{\partial^2}{\partial t^2} - \vec{\nabla}^2 \qquad (1.15)$$

There is one other kind of four-vectors that we will need for our discussions. These are known as the Dirac matrices and are denoted by γ^μ and γ_μ. They satisfy the anticommutation relation

$$\{\gamma^\mu, \gamma^\nu\} = \gamma^\mu \gamma^\nu + \gamma^\nu \gamma^\mu = 2\eta^{\mu\nu} I \qquad (1.16)$$

where I is the identity matrix. It follows, therefore, that

$$(\gamma^0)^2 = I \qquad (\gamma^1)^2 = (\gamma^2)^2 = (\gamma^3)^2 = -I \qquad (1.17)$$

I would choose the Hermiticity properties of these matrices to be

$$(\gamma^0)^\dagger = \gamma^0 \qquad (\gamma^i)^\dagger = -\gamma^i \qquad (1.18)$$

A particular representation for these 4×4 matrices can be written in terms of 2×2 blocks as

$$\gamma^0 = \begin{pmatrix} I & 0 \\ 0 & -I \end{pmatrix} \qquad \gamma^i = \begin{pmatrix} 0 & \sigma_i \\ -\sigma_i & 0 \end{pmatrix}$$

where σ_i's represent the Pauli matrices. From the four Dirac matrices we can construct a nontrival scalar matrix

$$\gamma_5 = i\gamma^0 \gamma^1 \gamma^2 \gamma^3 \qquad (1.20)$$

which satisfies

$$\gamma_5^\dagger = \gamma_5 \qquad (\gamma_5)^2 = I \qquad (1.21)$$

As we will see later, γ_5 describes the chirality or the handedness of a massless Dirac spinor.

Scalar Field Theory

With this introduction, let us look at the simplest of the field theories, namely, the free, massive, real scalar field theory. The Lagrangian (or more appropriately, the Lagrangian density) has the form

$$\mathcal{L} = \frac{1}{2} \partial_\mu \phi \partial^\mu \phi - \frac{m^2}{2} \phi^2 \tag{1.22}$$

where $\phi(x) = \phi(t, \vec{x})$ is Hermitian and is known as a spin zero field or a scalar field because it transforms like a scalar under a Lorentz transformation. Most of the physical theories are at most quadratic in the derivatives. In this case, the Euler-Lagrange equations take the form

$$\partial_\mu \frac{\partial \mathcal{L}}{\partial \partial_\mu \phi} - \frac{\partial \mathcal{L}}{\partial \phi} = 0 \tag{1.23}$$

From Eq. (1.22), therefore, we see the dynamical equations to have the form

$$\left(\Box + m^2\right) \phi = 0 \tag{1.24}$$

This is a generalization of the wave equation, known as the Klein-Gordon equation, whose solutions are plane waves of the form

$$\phi(x) \sim e^{\pm ik \cdot x}$$

with $\tag{1.25}$

$$k^2 = k_\mu k^\mu = m^2$$

The field $\phi(x)$ can describe neutral spin zero particles. In physical processes, however, particles are not completely free - rather they are interacting. Thus a more realistic theory would be one which describes a scalar field interacting with an external source. The Lagrangian, in this case, has the form

$$\mathcal{L} = \frac{1}{2} \partial_\mu \phi \partial^\mu \phi - \frac{m^2}{2} \phi^2 + j\phi \tag{1.26}$$

where $j(x)$ represents an external source and the Euler-Lagrange equation, in this case, takes the form

$$\left(\Box + m^2\right) \phi(x) = j(x) \tag{1.27}$$

The solution to this equation can be obtained from the Greens function for the problem which satisfies

$$\left(\Box_x + m^2\right) G(x - y) = -\delta^{(4)}(x - y) \tag{1.28}$$

In terms of $G(x - y)$ then, we can write

$$\phi(x) = \phi^{(0)}(x) - \int d^4y \, G(x - y) j(y) \tag{1.29}$$

where $\phi^{(0)}(x)$ is any solution of the homogeneous equation (1.24). The formal solution of Eq. (1.29) is useful only if we know the explicit form of $G(x - y)$. Note that in momentum space Eq. (1.28) has the form

$$\left(-k^2 + m^2\right) G(k) = -1$$

$$\text{or,} \quad G(k) = \frac{1}{k^2 - m^2} \tag{1.30}$$

so that we can write

$$G(x - y) = \frac{1}{(2\pi)^4} \int d^4k\, e^{-ik\cdot(x-y)} G(k) \tag{1.31}$$

$$= \frac{1}{(2\pi)^4} \int d^4k\, \frac{e^{-ik\cdot(x-y)}}{k^2 - m^2 - i\epsilon}$$

Here the infinitesimal parameter ϵ is added to the denominator in order to obtain the retarded Greens function. Thus we see that a crucial ingredient in studying any physical system is the Greens function which is also known as the propagator. Note that so far in our discussion we have not brought in the quantum nature of the theory. This can be done simply by noting that from the Lagrangian in Eq. (1.22) or (1.26), we can define a momentum canonically conjugate to the $\phi(x)$ as

$$\Pi(x) = \frac{\partial \mathcal{L}}{\partial \dot{\phi}(x)} = \dot{\phi}(x) \tag{1.32}$$

The quantization rules now follow to be

$$\left[\phi(x), \phi(x')\right]_{t=t'} = 0 = \left[\Pi(x), \Pi(x')\right]_{t=t'}$$

$$\left[\phi(x), \Pi(x')\right]_{t=t'} = i\delta^{(3)}(x - x') \tag{1.33}$$

The fields $\phi(x)$ and $\Pi(x)$ can now be expanded in terms of creation and annihilation operators and we can build up a Hilbert space for the quantum system.

Self-Interacting Scalar Field Theory

Just as a scalar field can interact with an external source, it can also interact with itself. Thus let us choose the following Lagrangian as a model of a self-interacting scalar field theory (also known as the ϕ^4 theory).

$$\mathcal{L} = \frac{1}{2} \partial_\mu \phi \partial^\mu \phi - \frac{m^2}{2} \phi^2 - \frac{\lambda}{4!} \phi^4 \qquad \lambda > 0 \tag{1.34}$$

where λ represents the strength of self interaction or the coupling constant. From this Lagrangian, we can construct the Hamiltonian as

$$H = \Pi \dot{\phi} - \mathcal{L}$$

$$= \frac{1}{2} \Pi^2 + \frac{1}{2} \vec{\nabla}\phi \cdot \vec{\nabla}\phi + \frac{m^2}{2} \phi^2 + \frac{\lambda}{4!} \phi^4 \tag{1.35}$$

It is clear now that classically the field configuration for which the energy would be a minimum has the form

$$\phi_c(x) = \text{constant} = 0 \tag{1.36}$$

Quantum mechanically, we say that the ground state or the vacuum state is one where

$$<0|\phi(x)|0> = 0 \tag{1.37}$$

For a constant field configuration, the minimum of the energy can be simply obtained by noting that in such a case

$$H = V(\phi) = -\mathcal{L}_{\text{int}} = \frac{m^2}{2}\phi^2 + \frac{\lambda}{4!}\phi^4 \tag{1.38}$$

from which we obtain

$$\frac{\partial V(\phi)}{\partial \phi} = 0 \quad \text{for} \quad \phi = \phi_c = 0 \tag{1.39}$$

As we will see later, these observations will be useful in studying the phenomenon of spontaneous symmetry breaking.

In a laboratory experiment, we would like to study the scattering involving particles. The scattering amplitudes can be calculated using the Feynman rules following from the theory, in Eq. (1.34). The theory, as we have seen, has a propagator and a set of interaction vertices. In the present case, we have

$$iG(p) = \frac{i}{p^2 - m^2}$$

$$-i\Gamma^{(4)}(p_1, p_2, p_3, p_4) = -i\left.\frac{\partial^4 V}{\partial \phi^4}\right|_{\phi=0}$$

$$= -i\lambda \delta^{(4)}(p_1 + p_2 + p_3 + p_4) \tag{1.40}$$

Any physical scattering process such as

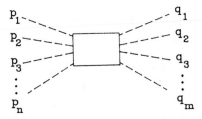

n particles $\longrightarrow$ m particles

can now be constructed and computed using the propagator and the interaction vertices. Thus for example, 3 particles ⟶ 3 particles, in this theory has the lowest order graph given by

and has the value

$$(-i\lambda) \frac{i}{(p_1 + p_2 + p_3)^2 - m^2} (-i\lambda) = -\frac{i\lambda^2}{(p_1 + p_2 + p_3)^2 - m^2} \quad (1.41)$$

The diagram describing the scattering process above is a simple one and such diagrams are known as tree diagrams. However, scattering can take place through complicated diagrams also. For example, in the lowest order in the ϕ^4 theory a particle can scatter by emitting a pair of particles which would annihilate each other. The Feynman diagram corresponding to this would look like

Such a diagram involves an internal loop representing the creation and annihilation of a pair of particles and is known as a loop diagram. In fact, it is called a one-loop diagram since the number of loops involved is one. Use of the Feynman rules now gives this scattering amplitude to be

$$-i\lambda \int \frac{d^4k}{(2\pi)^4} \frac{i}{k^2 - m^2} = \frac{\lambda}{(2\pi)^4} \int d^4k \frac{1}{k^2 - m^2} \quad (1.42)$$

The difference from the tree diagram is now obvious in that we have an integration over a momentum variable. This merely reflects the fact that the process involving the pair creation and annihilation is a virtual process and can occur with any momentum. This integral can be evaluated in many ways. The simplest is to go to the Euclidean space by letting

$$k_0 \longrightarrow ik_4$$

$$k^2 = k_0^2 - \vec{k}^2 \longrightarrow -k_4^2 - \vec{k}^2 = -k_E^2$$

(1.43)

so that the integral takes the form

$$\frac{i\lambda}{(2\pi)^4} \int d^4 k_E \frac{1}{-(k_E^2 + m^2)} = -\frac{i\lambda}{(2\pi)^4} \int k_E^3 dk_E d\Omega \frac{1}{k_E^2 + m^2}$$

$$= -\frac{i\lambda}{(2\pi)^4} \cdot 2\pi^2 \int_0^\infty \frac{1}{2} dk_E^2 \frac{k_E^2}{k_E^2 + m^2}$$

(1.44)

Clearly, the integral in Eq. (1.44) diverges and one way to define the integral is to cut off the integral at some large value of k_E^2. Thus defined this way, Eq. (1.44) becomes

$$-\frac{i\lambda}{16\pi^2} \int_0^{\Lambda^2} dk_E^2 \frac{k_E^2 + m^2 - m^2}{k_E^2 + m^2}$$

$$= -\frac{i\lambda}{16\pi^2} \int_0^{\Lambda^2} dk_E^2 \left(1 - \frac{m^2}{k_E^2 + m^2}\right)$$

(1.45)

$$= -\frac{i\lambda}{16\pi^2} \left(\Lambda^2 - m^2 \ln\left(\frac{\Lambda^2 + m^2}{m^2}\right)\right)$$

The true value of the integral (1.44) is, of course, obtained in the limit $\Lambda \to \infty$ and it diverges. But doing it this way brings out the nature of the divergence. This example also brings home another difference between the tree and the loop diagrams, namely, the loop diagrams diverge and, consequently, need to be regularized.

Lecture II

Dimensional Regularization

As we saw in the last lecture, there are inherent divergences in a quantum field theory which need to be regularized. There are many possible ways of regularizing a theory. For example, in the earlier calculation, we used a cut off to regularize the amplitude. But we could have chosen one of many other available regularization schemes such as the Pauli-Villars regularization or the point splitting regularization or the dimensional regularization or the higher derivative regularization and so on. Given a system, one chooses a regularization scheme which respects all the symmetry properties of the theory. In the case of gauge theories, the regularization that works well (it respects gauge invariance) and has become the standard regularization is dimensional regularization which I will describe next.

Let us now study the ϕ^4 theory, which we have analyzed in some detail, not in four dimensions but rather in n dimensions where $n = 4 - \epsilon$ with ϵ an infinitesimal parameter. The action defined as

$$S = \int d^n x \, \mathcal{L}$$

(2.1)

is a scalar in units of $\hbar = c = 1$ so that the canonical dimension of $\mathcal{L}$ follows to be

$$[\mathcal{L}] = n \tag{2.2}$$

Note that

$$\mathcal{L} = \frac{1}{2}\partial_\mu \phi \partial^\mu \phi - \frac{m^2}{2}\phi^2 - \frac{\lambda}{4!}\phi^4 \tag{2.3}$$

and since

$$[x^\mu] = -1 \qquad [\partial_\mu] = 1 \tag{2.4}$$

the canonical dimension of ϕ now follows to be

$$[\phi] = \frac{n-2}{2} \tag{2.5}$$

We also obtain

$$[m] = 1 \qquad [\lambda] = 4 - n = \epsilon \tag{2.6}$$

We would, however, like the coupling constant λ to be dimensionless and this can be achieved if we introduce an arbitrary mass scale μ and write the Lagrangian as

$$\mathcal{L} = \frac{1}{2}\partial_\mu \phi \partial^\mu \phi - \frac{m^2}{2}\phi^2 - \frac{\mu^\epsilon}{4!}\lambda\phi^4 \tag{2.7}$$

The coupling constant, λ, now will be dimensionless. The Feynman rules for this theory in n-dimensions take the form

$$iG(p) = \frac{i}{p^2 - m^2}$$

$$-i\Gamma^{(4)}(p_1, p_2, p_3, p_4) = -i\frac{\partial^4 V}{\partial \phi^4}$$

$$= -i\mu^\epsilon \lambda \delta^{(4)}(p_1 + p_2 + p_3 + p_4) \tag{2.8}$$

Let us next go on and calculate all the one loop diagrams in this theory. Remembering that we are in n-dimensions, we obtain

$$= -i\mu^\epsilon \lambda \int \frac{d^n k}{(2\pi)^n} \frac{i}{k^2 - m^2}$$

$$= \mu^\epsilon \lambda \int \frac{d^n k}{(2\pi)^n} \frac{1}{k^2 - m^2} \tag{2.9}$$

The fundamental formula for n-dimensional integrals that is of use to us is

$$I_\alpha = \int \frac{d^n k}{(2\pi)^n} \frac{1}{(k^2 + 2k \cdot p - M^2)^\alpha}$$

$$= (-1)^\alpha \frac{i\pi^{n/2}}{(2\pi)^n} \frac{\Gamma(\alpha - n/2)}{\Gamma(\alpha)} \frac{1}{(p^2 + M^2)^{\alpha - n/2}} \quad (2.10)$$

where

$$\Gamma(\alpha + 1) = \alpha!$$

Differentiating $I_{\alpha-1}$ with respect to p^μ, we can obtain other useful formulae such as

$$-\frac{1}{2(\alpha - 1)} \frac{\partial I_{\alpha-1}}{\partial p^\mu} = \int \frac{d^n k}{(2\pi)^n} \frac{k_\mu}{(k^2 + 2k \cdot p - M^2)^\alpha}$$

$$= (-1)^{\alpha-1} \frac{i\pi^{n/2}}{(2\pi)^n} \frac{\Gamma(\alpha - n/2)}{\Gamma(\alpha)} \frac{p_\mu}{(p^2 + M^2)^{\alpha - n/2}} \quad (2.11)$$

and similarly

$$\int \frac{d^n k}{(2\pi)^n} \frac{k_\mu k_\nu}{(k^2 + 2k \cdot p - M^2)^\alpha}$$

$$= (-1)^\alpha \frac{i\pi^{n/2}}{(2\pi)^n} \frac{1}{\Gamma(\alpha)} \frac{1}{(p^2 + M^2)^{\alpha - n/2}} \Big[p_\mu p_\nu \Gamma(\alpha - n/2)$$

$$- \frac{1}{2} \eta_{\mu\nu} (p^2 + M^2) \Gamma(\alpha - 1 - n/2) \Big] \quad (2.12)$$

Using Eq. (2.10), we can now evaluate the expression in Eq. (2.9) which takes the form

$$= \mu^\epsilon \lambda (-1) \frac{i\pi^{n/2}}{(2\pi)^n} \frac{\Gamma(1 - n/2)}{\Gamma(1)} \frac{1}{(m^2)^{1-n/2}}$$

$$= -i\mu^\epsilon \lambda \frac{\pi^{n/2}}{(2\pi)^n} \Gamma\left(-1 + \frac{\epsilon}{2}\right) (m^2)^{1-\epsilon/2} \quad (2.13)$$

We can now use the gamma function identities

$$\Gamma(n) = \frac{\Gamma(n+1)}{n}$$

and

$$\quad (2.14)$$

$$\Gamma\left(\frac{\epsilon}{2}\right) = \frac{2}{\epsilon} + \text{finite}$$

to simplify the expression in Eq. (2.13). Thus

$$\text{[diagram]} = m^2 \cdot \frac{i\lambda}{16\pi^2} \left(\frac{2}{\epsilon} - \ln \frac{m^2}{\mu^2} + \text{finite} \right) \tag{2.15}$$

Thus we see that in $n = 4 - \epsilon$ dimensions, this Feynman amplitude is well defined. However, as we approach $\epsilon \to 0$, the divergence appears as a pole.

There is one other one loop diagram that we can construct in this theory. Let us calculate a simplified version of this.

$$\text{[diagram]} = \frac{1}{2} (-i\mu^\epsilon \lambda)^2 \int \frac{d^n k}{(2\pi)^n} \left(\frac{i}{k^2 - m^2} \right)^2$$

$$= \frac{1}{2} \mu^{2\epsilon} \lambda^2 \int \frac{d^n k}{(2\pi)^n} \frac{1}{(k^2 - m^2)^2}$$

$$= \frac{1}{2} \mu^{2\epsilon} \lambda^2 (-1)^2 \frac{i\pi^{n/2}}{(2\pi)^n} \frac{\Gamma(2 - n/2)}{\Gamma(2)} \frac{1}{(m^2)^{2-n/2}}$$

$$= \frac{1}{2} \mu^{2\epsilon} \lambda^2 \frac{i\pi^{n/2}}{(2\pi)^n} \Gamma\!\left(\frac{\epsilon}{2}\right) (m^2)^{-\epsilon/2}$$

$$\tag{2.16}$$

$$= i\mu^\epsilon \lambda \cdot \frac{\lambda}{32\pi^2} \left(\frac{2}{\epsilon} - \ln \frac{m^2}{\mu^2} + \text{finite} \right)$$

Here the factor $\frac{1}{2}$ is known as the symmetry factor and arises because the amplitude is symmetric under the interchange of the two internal lines. Once again we see that the divergence appears as a pole and is independent of the external momentum. We can now explicitly work out and show that even for arbitrary external momenta, the Feynman amplitude will have the form

$$\text{[diagram]} = i\mu^\epsilon \lambda \cdot \frac{\lambda}{32\pi^2} \left(\frac{2}{\epsilon} - \ln \frac{m^2}{\mu^2} + \text{finite} \right) \tag{2.17}$$

$$p_1 + p_2 + p_3 + p_4 = 0$$

Adding in the other two channels, we see that at one loop the total four particle scattering amplitude will have the form

$$= i\mu^\epsilon \lambda \cdot \frac{3\lambda}{32\pi^2} \left(\frac{2}{\epsilon} - \ln \frac{m^2}{\mu^2} + \text{finite} \right) \qquad (2.18)$$

The divergence structure of the theory is now completely determined at the one loop level and we see that if we start from the theory in Eq. (2.7), then at one loop, the two point function as well as the four point function develop divergences. On the other hand, let us note that if we had started from the theory

$$\mathcal{L}_0 = \frac{1}{2} \partial_\mu \phi \partial^\mu \phi - \frac{m^2}{2} \phi^2 - \frac{\mu^\epsilon \lambda}{4!} \phi^4 - \frac{m^2}{2} A\phi^2 - \frac{\mu^\epsilon \lambda}{4!} B\phi^4 \qquad (2.19)$$

with

$$A = \frac{\lambda}{8\pi^2 \epsilon}$$

$$B = \frac{3\lambda}{16\pi^2 \epsilon} \qquad (2.20)$$

then we would have additional vertices in the theory given by

$$= -im^2 A = -im^2 \frac{\lambda}{8\pi^2 \epsilon}$$

$$= -i\mu^\epsilon \lambda B \delta^4 (p_1 + p_2 + p_3 + p_4) \qquad (2.21)$$

$$= -i\mu^\epsilon \lambda \cdot \frac{3\lambda}{16\pi^2 \epsilon} \delta^4 (p_1 + p_2 + p_3 + p_4)$$

Note that the constants A and B are really one loop quantities. (this would have been easier to see if we had kept the $\hbar$ terms.) Thus in this new theory, at one loop we will have

$$= -im^2 \cdot \frac{\lambda}{8\pi^2 \epsilon} + im^2 \cdot \frac{\lambda}{16\pi^2}\left(\frac{2}{\epsilon} - \ln\frac{m^2}{\mu^2} + \text{finite}\right)$$

$$= -im^2 \cdot \frac{\lambda}{16\pi^2}\left(\ln\frac{m^2}{\mu^2} + \text{finite}\right) \tag{2.22}$$

$(P_1 + P_2 + P_3 + P_4 = 0)$

$$= -i\mu^\epsilon \lambda \cdot \frac{3\lambda}{16\pi^2\epsilon} + i\mu^\epsilon \lambda \cdot \frac{3\lambda}{32\pi^2}\left(\frac{2}{\epsilon} - \ln\frac{m^2}{\mu^2} + \text{finite}\right)$$

$$= -i\mu^\epsilon \lambda \cdot \frac{3\lambda}{32\pi^2}\left(\ln\frac{m^2}{\mu^2} + \text{finite}\right) \tag{2.23}$$

Thus we see that had we started with the theory $\mathcal{L}_0$, there would be no divergence at least at the one loop level. However, the Lagrangian $\mathcal{L}_0$ would appear to be different from $\mathcal{L}$. But on closer inspection we find that $\mathcal{L}_0$ really has the same form as $\mathcal{L}$ with redefined parameters, namely,

$$\mathcal{L}_0 = \frac{1}{2}\partial_\mu\phi_0\partial^\mu\phi_0 - \frac{m_0^2}{2}\phi_0^2 - \frac{\lambda_0}{4!}\phi_0^4 \tag{2.24}$$

where
$$\phi_0 = \phi$$

$$m_0^2 = m^2(1+A) = m^2\left(1 + \frac{\lambda}{8\pi^2\epsilon}\right) \tag{2.25}$$

$$\lambda_0 = \mu^\epsilon\lambda(1+B) = \mu^\epsilon\lambda\left(1 + \frac{3\lambda}{16\pi^2\epsilon}\right)$$

The terms, depending on the constants A and B, which are added to the original Lagrangian to render the amplitudes finite are known as counterterms and this process of removing divergences is known as renormalization. Although we have explicitly shown this only for one-loop, in any given theory one can add counterterms order by order so that the amplitudes are finite at every loop. Furthermore, a renormalizable theory is one (physical theories are renormalizable.) where the counterterms can be completely absorbed into a redefinition of the fields and the various parameters in the theory.

Let me say this again in a different way. Let us call ϕ_0 to be the bare field and m_0, λ_0 to be the bare parameters. Similarly, let us call ϕ to be the renormalized field and m, λ to be the renormalized parameters. Then if we start from the bare Lagrangian $\mathcal{L}_0$ and calculate the amplitudes in terms of bare parameters m_0 and λ_0 then the process of renormalization guarantees that, in a renormalizable theory, when the bare parameters are expressed in terms of a set of renormalized parameters as in Eq. (2.25), the amplitudes would be finite. Thus if we calculated the four point function up to one-loop from $\mathcal{L}_0$ in terms of m and λ, we would obtain

$$= -i\mu^\epsilon \lambda - i\mu^\epsilon \lambda \cdot \frac{3\lambda}{32\pi^2}\left(\ln\frac{m^2}{\mu^2} + \text{finite}\right)$$

$$= -i\mu^\epsilon \lambda \left(1 + \frac{3\lambda}{32\pi^2}\left(\ln\frac{m^2}{\mu^2} + \text{finite}\right)\right) = \text{finite} \qquad (2.26)$$

The choice of the renormalized parameters is, however, not unique. As is obvious, their definition depends on the mass scale μ. Thus for example, given a bare theory $\mathcal{L}_0$, two different people can choose two sets of renormalized parameters - depending on different mass scales - which would lead to finite scattering amplitudes. The mass scale μ, therefore, would correspond in some sense to the subtraction point or the energy scale at which a process is evaluated. Thus we see that in a renormalizable theory, the renormalized parameters become energy dependent. In the case of coupling constant one fondly says that the renormalized coupling runs with energy or that it becomes a running coupling. This dependence of the renormalized parameters on the mass scale μ leads to the renormalization group equation which basically gives how various quantities would change as μ is changed. For the coupling constant, this evolution up to one-loop can be obtained from Eq. (2.25). Note that whereas the renormalized parameters depend on μ, the bare parameters do not depend on the arbitrary mass scale. Thus from Eq. (2.25) we obtain

$$\mu \frac{\partial \lambda_0}{\partial \mu} = 0$$

or, $\quad \mu \dfrac{\partial}{\partial \mu} \left(\mu^\epsilon \lambda \left(1 + \dfrac{3\lambda}{16\pi^2 \epsilon} \right) \right) = 0$

or, $\quad \mu \dfrac{\partial}{\partial \mu} \left(\left(\lambda + \dfrac{3\lambda^2}{16\pi^2 \epsilon} \right) \left(1 + \epsilon \ln \mu + O\left(\epsilon^2\right) \right) \right) = 0$

or, $\quad \mu \dfrac{\partial}{\partial \mu} \left(\lambda + \dfrac{3\lambda^2}{16\pi^2 \epsilon} + \left(\epsilon \lambda + \dfrac{3\lambda^2}{16\pi^2} \right) \ln \mu + O\left(\epsilon^2\right) \right) = 0$

or, $\quad \mu \dfrac{\partial \lambda}{\partial \mu} \left(1 + \dfrac{3\lambda}{8\pi^2 \epsilon} + \left(\epsilon + \dfrac{3\lambda}{8\pi^2} \right) \ln \mu \right) + \epsilon \lambda + \dfrac{3\lambda^2}{16\pi^2} + O\left(\epsilon^2\right) = 0$

or, $\quad \mu \dfrac{\partial \lambda}{\partial \mu} = -\left(\epsilon \lambda + \dfrac{3\lambda^2}{16\pi^2} \right) \left(1 - \dfrac{3\lambda}{8\pi^2 \epsilon} - \left(\epsilon + \dfrac{3\lambda}{8\pi^2} \right) \ln \mu \right) + O\left(\epsilon^2\right) \quad (2.27)$

Keeping terms up to order λ^2 which is the consistent one-loop case, we obtain in the limit $\epsilon \to 0$

$$\mu \frac{\partial \lambda}{\partial \mu} = -\frac{3\lambda^2}{16\pi^2} + \frac{3\lambda^2}{8\pi^2} = \frac{3\lambda^2}{16\pi^2} = \beta(\lambda) \quad (2.28)$$

This equation can be solved in a straightforward manner.

$$\mu \frac{\partial \lambda}{\partial \mu} = \frac{3\lambda^2}{16\pi^2}$$

or, $\quad \dfrac{d\lambda}{\lambda^2} = \dfrac{3}{16\pi^2} \dfrac{d\mu}{\mu}$

or, $\quad \displaystyle\int_{\lambda(\bar\mu)}^{\lambda(\mu)} \dfrac{d\lambda}{\lambda^2} = \dfrac{3}{16\pi^2} \int_{\bar\mu}^{\mu} \dfrac{d\mu}{\mu}$

or, $\quad -\dfrac{1}{\lambda(\mu)} + \dfrac{1}{\lambda(\bar\mu)} = \dfrac{3}{16\pi^2} \ln \dfrac{\mu}{\bar\mu}$

or, $\quad \lambda(\mu) = \dfrac{\lambda(\bar\mu)}{1 - \dfrac{3\lambda(\bar\mu)}{16\pi^2} \ln \dfrac{\mu}{\bar\mu}} \quad (2.29)$

This, indeed, shows how the coupling constant, $\lambda(\mu)$, changes with the scale or energy. As μ increases relative to $\bar\mu$, $\lambda(\mu)$ grows. The coupling becomes stronger and beyond a certain value perturbation theory breaks down. One says that the ϕ^4 theory is not asymptotically free. I would also like to emphasize here that although our discussion so far has been within the context of the real scalar field, it can be generalized to other theories in a straightforward manner.

Lecture III

Complex Scalar Field Theory

Let us next consider a self-interacting scalar field theory where the field $\phi(x)$ is not real. Such a theory can describe processes involving charged spin zero particles. Thus let,

$$\mathcal{L} = \partial_\mu \phi^\dagger \partial^\mu \phi - m^2 \phi^\dagger \phi - \frac{\lambda}{4}\left(\phi^\dagger \phi\right)^2 \qquad \lambda > 0 \qquad (3.1)$$

Since $\phi(x)$ is complex, we can express it in terms of two real fields $\sigma(x)$ and $\zeta(x)$ as

$$\phi(x) = \frac{1}{\sqrt{2}}\left(\sigma(x) + i\zeta(x)\right)$$

$$\phi^\dagger(x) = \frac{1}{\sqrt{2}}\left(\sigma(x) - i\zeta(x)\right) \qquad (3.2)$$

When expressed in terms of $\sigma(x)$ and $\zeta(x)$, the Lagrangian of Eq. (3.1) takes the form

$$\mathcal{L} = \frac{1}{2}\partial_\mu \sigma \partial^\mu \sigma + \frac{1}{2}\partial_\mu \zeta \partial^\mu \zeta - \frac{m^2}{2}\left(\sigma^2 + \zeta^2\right) - \frac{\lambda}{16}\left(\sigma^2 + \zeta^2\right)^2 \qquad (3.3)$$

Thus we see that a self-interacting complex scalar field theory is equivalent to a theory of two coupled, self-interacting real scalar fields.

The fields $\phi(x)$ and $\phi^\dagger(x)$ in Eq. (3.1) can be taken to be independent variables. Correspondingly, the Euler-Lagrange equations for the system are

$$\partial_\mu \frac{\partial \mathcal{L}}{\partial \partial_\mu \phi^\dagger} - \frac{\partial \mathcal{L}}{\partial \phi^\dagger} = 0$$

or, $\qquad (\Box + m^2)\phi = -\frac{\lambda}{2}\left(\phi^\dagger \phi\right)\phi \qquad (3.4)$

and

$$\partial_\mu \frac{\partial \mathcal{L}}{\partial \partial_\mu \phi} - \frac{\partial \mathcal{L}}{\partial \phi} = 0$$

or, $\qquad (\Box + m^2)\phi^\dagger = -\frac{\lambda}{2}\left(\phi^\dagger \phi\right)\phi^\dagger \qquad (3.5)$

The canonical momenta conjugate to $\phi(x)$ and $\phi^\dagger(x)$ are given respectively by

$$\Pi_\phi(x) = \frac{\partial \mathcal{L}}{\partial \dot\phi(x)} = \dot\phi^\dagger(x)$$

$$\Pi_{\phi^\dagger}(x) = \frac{\partial \mathcal{L}}{\partial \dot\phi^\dagger(x)} = \dot\phi(x) \qquad (3.6)$$

The quantization conditions, therefore, become

$$[\phi(x), \Pi_\phi(x')]_{t=t'} = i\delta^{(3)}(x - x') = [\phi^\dagger(x), \Pi_{\phi^\dagger}(x')]_{t=t'} \qquad (3.7)$$

with all other equal time commutators vanishing.

Noether's Theorem

Let us note that the Lagrangian of Eq. (3.1) is invariant under a phase transformation of the form

$$\phi(x) \to \bar{\phi}(x) = e^{i\alpha}\phi(x)$$
$$\phi^\dagger(x) \to \bar{\phi}^\dagger(x) = e^{-i\alpha}\phi^\dagger(x) \quad (3.8)$$

where α is a space-time independent constant parameter. Thus the phase transformations of Eq. (3.8) are a symmetry of the system described by the Lagrangian in Eq. (3.1). The phase transformations do not involve a change in the space-time coordinates of the fields and hence do not correspond to a space-time symmetry transformation. Rather, such a transformation is known as an internal symmetry transformation. Furthermore, the parameter of transformation, α, is a constant - it is the same at all coordinate points. Thus such a transformation defines a global symmetry trnasformation. Often times, it is more convenient to study the infinitesimal form of a transformation. Thus when the parameter of transformation is infinitesimally small, the transformations in Eq. (3.8) can be written as

$$\delta_\epsilon \phi(x) = \bar{\phi}(x) - \phi(x) = i\epsilon\phi(x)$$
$$\delta_\epsilon \phi^\dagger(x) = \bar{\phi}^\dagger(x) - \phi^\dagger(x) = -i\epsilon\phi^\dagger(x) \quad (3.9)$$

where ϵ is the infinitesimal parameter of transformation.

If the Lagrangian of a system is invariant under a continuous symmetry transformation, Noether's theorem guarantees the existence of a conserved current. In the case of an internal symmetry transformation, the conserved current can be obtained as

$$\epsilon j^\mu(x) = \frac{\partial \mathcal{L}}{\partial \partial_\mu \psi(x)} \delta_\epsilon \psi(x) \quad (3.10)$$

where $\psi(x)$ generically represents all the field variables of the theory. Thus for the Lagrangian in Eq. (3.1), we can determine the current associated with the infinitesimal transformations of Eq. (3.9) to be

$$\epsilon j^\mu(x) = \frac{\partial \mathcal{L}}{\partial \partial_\mu \phi(x)} \delta_\epsilon \phi(x) + \frac{\partial \mathcal{L}}{\partial \partial_\mu \phi^\dagger(x)} \delta_\epsilon \phi^\dagger(x)$$

$$= \partial^\mu \phi^\dagger(x)(i\epsilon\phi(x)) + \partial^\mu \phi(x)(-i\epsilon\phi^\dagger(x))$$

$$= -i\epsilon\left(\phi^\dagger(x)\partial^\mu \phi(x) - \partial^\mu \phi^\dagger(x)\phi(x)\right)$$

or, $\quad j^\mu(x) = -i\left(\phi^\dagger(x)\partial^\mu \phi(x) - \partial^\mu \phi^\dagger(x)\phi(x)\right) = -i\phi^\dagger(x)\overleftrightarrow{\partial^\mu}\phi(x) \quad (3.11)$

Let us note that

$$\partial_\mu j^\mu(x) = -i\partial_\mu\left(\phi^\dagger(x)\partial^\mu\phi(x) - \partial^\mu\phi^\dagger(x)\phi(x)\right)$$

$$= -i\left(\phi^\dagger(x)\Box\phi(x) - \Box\phi^\dagger(x)\phi(x)\right) \tag{3.12}$$

Using the equations of motion in (3.4 - 3.5), we see that

$$\partial_\mu j^\mu(x) = -i\left(\phi^\dagger(x)\left(-m^2\phi(x) - \frac{\lambda}{2}(\phi^\dagger\phi)\phi\right)\right.$$

$$\left. - \left(-m^2\phi^\dagger(x) - \frac{\lambda}{2}(\phi^\dagger\phi)\phi^\dagger\right)\phi(x)\right) \tag{3.13}$$

or, $\quad \partial_\mu j^\mu(x) = 0$

In other words, the current associated with the transformations in Eq. (3.9) is, indeed, conserved.

Given a conserved current, j^μ, we can construct a charge as

$$Q = \int d^3x\, j^0(\vec{x},t) \tag{3.14}$$

which can be shown to be independent of time. Thus, for the complex scalar field theory, we see that there exists a charge operator

$$Q = \int d^3x\, j^0(x) = -i\int d^3x(\phi^\dagger(x)\dot\phi(x) - \dot\phi^\dagger(x)\phi(x))$$

$$= -i\int d^3x\,(\phi^\dagger(x)\Pi_{\phi^\dagger}(x) - \Pi_\phi(x)\phi(x)) \tag{3.15}$$

Here we have used the relations in Eq. (3.6). The algebra of the charge can now be obtained using Eq. (3.7) to be

$$[Q,Q] = 0 \tag{3.16}$$

In other words, the conserved charge associated with the symmetry transformations of Eq. (3.9) is Abelian. This can be identified with the electric charge operator of the theory. Note also that using Eq. (3.7), we can show that

$$\delta_\epsilon \phi(x) = i\epsilon\phi(x) = -i[\phi(x), \epsilon Q]$$

$$\delta_\epsilon \phi^\dagger(x) = -i\epsilon\phi^\dagger(x) = -i[\phi^\dagger(x), \epsilon Q] \tag{3.17}$$

This shows that the charge Q is the generator of the infinitesimal transformations in Eq. (3.9).

Spontaneous Symmetry Breaking

Let us now consider the complex scalar field theory described by the Lagrangian

$$\mathcal{L} = \partial_\mu \phi^\dagger \partial^\mu \phi + m^2 \phi^\dagger \phi - \frac{\lambda}{4} (\phi^\dagger \phi)^2 \qquad \lambda > 0 \qquad (3.18)$$

This Lagrangian differs from the one in Eq. (3.1) in the sign of the quadratic term. It is still invariant under the phase transformations of Eq. (3.8) or the infinitesimal transformations of Eq. (3.9). The conserved charge Q of Eq. (3.15) must, therefore, commute with the Hamiltonian of the system and one would naively expect that the charge operator Q would annihilate the vacuum of the theory. But as we will see next, this is not true.

Let us rewrite the Lagrangian in terms of the real fields $\sigma(x)$ and $\zeta(x)$ of Eq. (3.2).

$$\mathcal{L} = \frac{1}{2} \partial_\mu \sigma \partial^\mu \sigma + \frac{1}{2} \partial_\mu \zeta \partial^\mu \zeta + \frac{m^2}{2} (\sigma^2 + \zeta^2) - \frac{\lambda}{16} (\sigma^2 + \zeta^2)^2 \qquad (3.19)$$

In terms of these fields, the infinitesimal transformations of Eq. (3.9) would take the form

$$\delta_\epsilon \sigma(x) = -\frac{1}{\sqrt{2}} \epsilon \zeta(x) = -i[\sigma(x), \epsilon Q]$$

$$\delta_\epsilon \zeta(x) = \frac{1}{\sqrt{2}} \epsilon \sigma(x) = -i[\zeta(x), \epsilon Q] \qquad (3.20)$$

Let us analyze the ground state of the theory. For constant field configurations, we see that the minimum of the potential

$$V = -\frac{m^2}{2} (\sigma^2 + \zeta^2) + \frac{\lambda}{16} (\sigma^2 + \zeta^2)^2 \qquad (3.21)$$

would give the ground state. Note that the solutions to the minimum equations

$$\frac{\partial V}{\partial \sigma} = -m^2 \sigma + \frac{\lambda}{4} \sigma (\sigma^2 + \zeta^2) = 0$$

$$\frac{\partial V}{\partial \zeta} = -m^2 \zeta + \frac{\lambda}{4} \zeta (\sigma^2 + \zeta^2) = 0 \qquad (3.22)$$

are given by

$$\sigma = 0 = \zeta$$

or $\qquad (3.23)$

$$\sigma^2 + \zeta^2 = \frac{4m^2}{\lambda}$$

It can be verified readily that $\sigma = \zeta = 0$ defines a local maximum whereas the true minimum is given by

$$\sigma^2 + \zeta^2 = \frac{4m^2}{\lambda} \qquad (3.24)$$

For simplicity, we choose the minimum to correspond to

$$\zeta = 0 \qquad \sigma = \frac{2m}{\sqrt{\lambda}} \qquad (3.25)$$

As we have discussed before, in the quantum theory, it corresponds to the fact that the vacuum state satisfies

$$<0|\zeta|0> = 0$$
$$<0|\sigma|0> = \frac{2m}{\sqrt{\lambda}} \qquad (3.26)$$

From Eq. (3.20), on the other hand, we see that this implies

$$<0|\delta_\epsilon \zeta|0> = \frac{1}{\sqrt{2}} \epsilon <0|\sigma|0> = \epsilon \sqrt{\frac{2}{\lambda}} m = -i <0|[\zeta, \epsilon Q]|0> \qquad (3.27)$$

This cannot be satisfied if

$$Q|0> = 0 \qquad (3.28)$$

In other words, we see that in this case, a symmetry of the theory is not a symmetry of the vacuum. In such a case, we say that there is a spontaneous breakdown of the symmetry.

To obtain the consequences of spontaneous symmetry breaking, let us observe that any perturbtion of a system can only be stable around the ground state. Thus we should expand the theory around the stable minimum by redefining

$$\sigma \to \sigma + \frac{2m}{\sqrt{\lambda}} \qquad (3.29)$$

The Lagrangian now becomes

$$\mathcal{L} = \frac{1}{2} \partial_\mu \sigma \partial^\mu \sigma + \frac{1}{2} \partial_\mu \zeta \partial^\mu \zeta + \frac{m^2}{2}\left((\sigma + \frac{2m}{\sqrt{\lambda}})^2 + \zeta^2\right)$$

$$- \frac{\lambda}{16}\left((\sigma + \frac{2m}{\sqrt{\lambda}})^2 + \zeta^2\right)^2$$

$$= \frac{1}{2} \partial_\mu \sigma \partial^\mu \sigma + \frac{1}{2} \partial_\mu \zeta \partial^\mu \zeta - m^2 \sigma^2 + \frac{m^4}{\lambda}$$

$$- \frac{m\sqrt{\lambda}}{2} \sigma(\sigma^2 + \zeta^2) - \frac{\lambda}{16}(\sigma^2 + \zeta^2)^2 \qquad (3.30)$$

The interesting point to note in the above Lagrangian is that there is no mass term for the ζ-field. That is, the ζ-field has become massless. This is known as the Goldstone theorem which roughly says that whenever a continuous global symmetry is spontaneously broken, there must arise massless particles in the theory. The massless field ζ is also known as the Goldstone field.

Dirac Field Theories

Let us next discuss theories which describe particles obeying Fermi-Dirac statistics. The simplest theory is, of course, one which describes a massless spin $\frac{1}{2}$ particle. The Lagrangian has the form

$$\mathcal{L} = i\bar{\psi}(x)\gamma^\mu \partial_\mu \psi(x) \tag{3.31}$$

Here γ^μ's are the 4×4 Dirac matrices we discussed in Lecture I and $\psi(x)$ is a four component spinor field which has the form

$$\psi(x) = \begin{pmatrix} \psi_1(x) \\ \psi_2(x) \\ \psi_3(x) \\ \psi_4(x) \end{pmatrix} \tag{3.32}$$

and

$$\bar{\psi}(x) = \psi^\dagger(x)\gamma^0 \tag{3.33}$$

The Euler-Lagrange equations have the form

$$i\gamma^\mu \partial_\mu \psi(x) = i\partial\!\!\!/\psi(x) = 0$$

$$\hspace{6cm} (\slashed{A} = \gamma^\mu A_\mu) \tag{3.34}$$

$$i\bar{\psi}(x)\overleftarrow{\partial}_\mu \gamma^\mu = i\bar{\psi}(x)\overleftarrow{\slashed{\partial}} = 0$$

The momentum canonically conjugate to $\psi_\alpha(x)$ is

$$\Pi_\alpha(x) = \frac{\partial \mathcal{L}}{\partial \dot{\psi}_\alpha(x)} = i\left(\bar{\psi}(x)\gamma^0\right)_\alpha = i\psi^\dagger_\alpha(x) \qquad \alpha = 1,2,3,4 \tag{3.35}$$

The theory can now be quantized in the following way.

$$\{\psi_\alpha(x), \psi_\beta(x')\}_{t=t'} = \left(\psi_\alpha(x)\psi_\beta(x') + \psi_\beta(x')\psi_\alpha(x)\right)_{t=t'} = 0$$

$$\{\Pi_\alpha(x), \Pi_\beta(x')\}_{t=t'} = 0 \tag{3.36}$$

$$\{\psi_\alpha(x), \Pi_\beta(x')\}_{t=t'} = i\delta_{\alpha\beta}\delta^{(3)}(x-x')$$

We see the basic difference between the bosonic theories and the fermionic theories in that the fermionic theories are quantized with anticommutators. This is connected with the fact that the fermionic theories describe particles obeying Fermi-Dirac statistics.

The Lagrangian in Eq. (3.31) is invariant under the constant phase transformations

$$\psi(x) \to \tilde{\psi}(x) = e^{i\alpha}\psi(x)$$
$$\bar{\psi}(x) \to \tilde{\bar{\psi}}(x) = e^{-i\alpha}\bar{\psi}(x)$$
(3.37)

as well as the chiral phase transformations

$$\psi(x) \to \tilde{\tilde{\psi}}(x) = e^{i\gamma_5\beta}\psi(x)$$
$$\bar{\psi}(x) \to \tilde{\tilde{\bar{\psi}}}(x) = \bar{\psi}(x)e^{i\gamma_5\beta}$$
(3.38)

where α and β are constant parameters independent of any space-time coordinates. Infinitesimally, the transformations take the form

$$\delta_\epsilon \psi(x) = \tilde{\psi}(x) - \psi(x) = i\epsilon\psi(x)$$
$$\delta_\epsilon \bar{\psi}(x) = \tilde{\bar{\psi}}(x) - \bar{\psi}(x) = -i\epsilon\bar{\psi}(x)$$
(3.39)

and

$$\delta_\eta \psi(x) = \tilde{\tilde{\psi}}(x) - \psi(x) = i\eta\gamma_5\psi(x)$$
$$\delta_\eta \bar{\psi}(x) = \tilde{\tilde{\bar{\psi}}}(x) - \bar{\psi}(x) = i\eta\bar{\psi}(x)\gamma_5$$
(3.40)

The conserved currents associated with these symmetry transformations can be constructed from the Noether's theorem (see Eq. (3.10)) and take the form

$$\epsilon j^\mu(x) = \frac{\partial \mathcal{L}}{\partial \partial_\mu \psi(x)} \delta_\epsilon \psi(x) = i\bar{\psi}(x)\gamma^\mu(i\epsilon\psi(x))$$

$$\text{or,} \quad j^\mu(x) = -\bar{\psi}(x)\gamma^\mu\psi(x) \tag{3.41}$$

and

$$\eta j_5^\mu(x) = \frac{\partial \mathcal{L}}{\partial \partial_\mu \psi(x)} \delta_\eta \psi(x) = i\bar{\psi}(x)\gamma^\mu(i\eta\gamma_5\psi(x))$$

$$\text{or,} \quad j_5^\mu(x) = -\bar{\psi}(x)\gamma^\mu\gamma_5\psi(x) \tag{3.42}$$

The corresponding conserved charges are

$$Q = \int d^3x\, j^0(x) = -\int d^3x\, \bar{\psi}(x)\gamma^0\psi(x)$$
$$= -\int d^3x\, \psi^\dagger(x)\psi(x) = i\int d^3x\, \Pi_\alpha(x)\psi_\alpha(x) \tag{3.43}$$

and

$$Q_5 = \int d^3x \, j_5^0(x) = -\int d^3x \, \bar{\psi}(x)\gamma^0\gamma_5\psi(x)$$

$$= -\int d^3x \, \psi^\dagger(x)\gamma_5\psi(x)$$

$$= i\int d^3x \, \Pi_\alpha(x)(\gamma_5)_{\alpha\beta}\psi_\beta(x) \tag{3.44}$$

Using the quantization conditions of Eq. (3.36), the charge algebra can now be shown to satisfy

$$[Q,Q] = 0 = [Q_5, Q_5] = [Q, Q_5] \tag{3.45}$$

Let us note here that although the currents j^μ and j_5^μ in Eqs. (3.41) and (3.42) are conserved classically, quantum mechanical corrections may spoil this. When a current is conserved classically, but quantum mechanically

$$\partial_\mu j^\mu(x) \neq 0 \tag{3.46}$$

we say that there are anomalies in the theory and that the symmetry has become anomalous. Anomalies associated with global symmetries are harmless (in fact, sometimes useful as shown in $\pi^0 \to 2\gamma$) but as we will see, anomalies associated with local symmetries can render the theory inconsistent.

Local Symmetry

Let us next analyze what would happen if we tried to make the phase parameter in Eq. (3.37) to be a local function. Under

$$\psi(x) \to \tilde{\psi}(x) = e^{i\alpha(x)}\psi(x)$$

$$\bar{\psi}(x) \to \tilde{\bar{\psi}}(x) = e^{-i\alpha(x)}\bar{\psi}(x) \tag{3.47}$$

we see that

$$\mathcal{L} \to \tilde{\mathcal{L}} = i\tilde{\bar{\psi}}(x)\gamma^\mu \partial_\mu(\tilde{\psi}(x))$$
$$= i\bar{\psi}(x)\gamma^\mu\partial_\mu\psi(x) - \partial_\mu\alpha(x)\bar{\psi}(x)\gamma^\mu\psi(x) \tag{3.48}$$
$$= \mathcal{L} - \partial_\mu\alpha(x)\bar{\psi}(x)\gamma^\mu\psi(x)$$

In other words, the Lagrangian of Eq. (3.31) is not invariant under a local phase transformation. On the other hand, we note that if we had started from the Lagrangian

$$\bar{\mathcal{L}} = i\bar{\psi}\gamma^\mu(\partial_\mu - ieA_\mu)\psi \tag{3.49}$$

where "e" is a constant, then this would be invariant under

$$\psi(x) \to e^{i\alpha(x)}\psi(x)$$

$$\bar{\psi}(x) \to e^{-i\alpha(x)}\bar{\psi}(x) \quad (3.50)$$

$$A_\mu(x) \to A_\mu(x) + \frac{1}{e}\partial_\mu\alpha(x)$$

Infinitesimally, these take the form

$$\delta_\epsilon\psi(x) = i\epsilon(x)\psi(x)$$

$$\delta_\epsilon\bar{\psi}(x) = -i\epsilon(x)\bar{\psi}(x) \quad (3.51)$$

$$\delta_\epsilon A_\mu(x) = \frac{1}{e}\partial_\mu\epsilon(x)$$

Thus we see that the local symmetry of Eq. (3.31) requires an additional field $A_\mu(x)$. $A_\mu(x)$ is known as a gauge field and the transformations of Eqs. (3.50) and (3.51) are known as gauge transformations.

Let us also note here that the Lagrangian $\bar{\mathcal{L}}$ can be obtained from $\mathcal{L}$ with the replacement

$$\partial_\mu \to \partial_\mu - ieA_\mu$$

$$\text{or,} \quad p_\mu = -i\partial_\mu \to p_\mu - eA_\mu \quad (3.52)$$

This is, of course, the prescription of minimal coupling we are familiar with in trying to couple charged particles to electromagnetic fields. Thus we can identify the gauge field, $A_\mu(x)$, with the photon of the theory. This also suggests that a local symmetry must always be accompanied by physical forces. Conversely, we may try to describe physical forces in terms of theories with local symmetries (or gauge theories).

Lecture IV

<u>Quantum Electrodynamics</u>

We have seen how to couple charged spin $\frac{1}{2}$ particles to the electromagnetic field or the photon field. If we now introduce the dynamics of the photon fields, we would have an interacting theory of say, electrons and photons - otherwise known as quantum electrodynamics. The Lagrangian, in this case, has the form (e can now be thought of as the electromagnetic coupling.)

$$\mathcal{L}_{\text{QED}} = i\bar{\psi}\gamma^\mu(\partial_\mu - ieA_\mu)\psi - \frac{1}{4}F_{\mu\nu}F^{\mu\nu} \quad (4.1)$$

where
$$F_{\mu\nu} = \partial_\mu A_\nu - \partial_\nu A_\mu = -F_{\nu\mu}$$
$$F_{0i} = E_i \qquad (4.2)$$
$$F_{ij} = -\epsilon_{ijk} B_k$$

The Euler-Lagrange equations for this theory are

$$i\gamma^\mu (\partial_\mu - ieA_\mu)\psi = (i\slashed{\partial} + e\slashed{A})\psi = 0 \qquad (4.3)$$

and

$$\partial_\mu F^{\mu\nu} = -e\bar\psi \gamma^\nu \psi = ej^\nu \qquad (4.4)$$

The $\nu = 0$ and $\nu = i$ components of Eq. (4.4) give respectively

$$\vec\nabla \cdot \vec E = ej^0$$

and $\qquad (4.5)$

$$\vec\nabla \times \vec B = \frac{\partial \vec E}{\partial t} + e\vec j$$

Similarly, from the definition of $F_{\mu\nu}$ in Eq. (4.2) we see that it satisfies

$$\partial_\mu F_{\nu\lambda} + \partial_\nu F_{\lambda\mu} + \partial_\lambda F_{\mu\nu} = 0 \qquad (4.6)$$

These set of equations can be shown to be equivalent to the pair of equations

$$\vec\nabla \cdot \vec B = 0$$

and $\qquad (4.7)$

$$\vec\nabla \times \vec E = -\frac{\partial \vec B}{\partial t}$$

We recognize Eqs. (4.5) and (4.7) together as the set of Maxwell's equations and, therefore, we conclude that the additional term in the Lagrangian in Eq. (4.1), indeed, gives the dynamics of the photon fields.

The Lagrangian of Eq. (4.1) can be checked to be invariant under the gauge transformations

$$\delta_\epsilon \psi(x) = i\epsilon(x)\psi(x)$$

$$\delta_\epsilon \bar\psi(x) = -i\epsilon(x)\bar\psi(x) \qquad (4.8)$$

$$\delta_\epsilon A_\mu(x) = \frac{1}{e} \partial_\mu \epsilon(x)$$

This is, of course, a local symmetry. Note that gauge invariance requires the photon to be massless since a mass term would break gauge invariance. From Eq. (4.4) we see that since $F^{\mu\nu}$ is antisymmetric, consistency of the equation would require

25

$$\partial_\nu \partial_\mu F^{\mu\nu} = 0 = e\partial_\nu j^\nu \qquad (4.9)$$

In such a case, therefore, the current must be conserved even quantum mechanically. Any violation of current conservation or any anomaly would render the dynamical equations inconsistent. This is, of course, what we have noted earlier, namely, whereas anomalous global symmetries are harmless, anomalies in local symmetries must be avoided.

The gauge invariances of the QED Lagrangian has both advantages as well as disadvantages. To see the advantages, let us write down the fermionic Feynman rules of the theory.

$$\xrightarrow{p} \quad = iS(p) = \frac{i}{\not{p}}$$

$$(4.10)$$

$$= -i\Gamma_\mu(p_1, p_2, p_3) = ie\gamma_\mu \delta^{(4)}(p_1 + p_2 + p_3)$$

From the structure of the propagator and the vertex, we see that

$$\frac{\partial}{\partial p^\mu} \xrightarrow{p} \;=\; \frac{\partial}{\partial p^\mu} \frac{i}{\not{p}} = \frac{i}{\not{p}} i\gamma_\mu \frac{i}{\not{p}} = \frac{1}{e} \left(\begin{array}{c} \text{vertex diagram} \\ k=0 \end{array} \right)$$

or, $\quad \dfrac{\partial}{\partial p^\mu}(iS(p)) = \dfrac{1}{e}(iS(p))\big(-i\Gamma_\mu(p,-p,0)\big)(iS(p))$

or, $\quad \dfrac{\partial S^{-1}(p)}{\partial p^\mu} = -\dfrac{1}{e}\Gamma_\mu(p,-p,0) \qquad (4.11)$

This relation is quite important in that it relates different scattering amplitudes. It is, in fact, a consequence of the gauge invariance of the system (although our simple derivation does not make it seem so). Although, we have derived this relation for the case when the electrons are massless, the same holds for massive electrons. Furthermore, this relation holds order by order in perturbation theory and plays a crucial role in the renormalization of the theory. Relation (4.11) is also known as the Ward identity of QED.

The difficulties of gauge invariance can be seen from Eq. (4.4).

$$\partial_\mu F^{\mu\nu} = ej^\nu$$

or, $$\partial_\mu(\partial^\mu A^\nu - \partial^\nu A^\mu) = ej^\nu$$

or, $$(\Box \eta^{\mu\nu} - \partial^\mu \partial^\nu) A_\mu = ej^\nu \tag{4.12}$$

The Greens function associated with this equation must satisfy

$$(\Box_x \eta^{\mu\nu} - \partial_x^\mu \partial_x^\nu) G_{\nu\lambda}(x - y) = -\delta_\lambda^\mu \delta^{(4)}(x - y) \tag{4.13}$$

But note that the operator $(\Box_x \eta^{\mu\nu} - \partial_x^\mu \partial_x^\nu)$ is a transverse projection operator in the sense that

$$\partial_{x\mu}(\Box_x \eta^{\mu\nu} - \partial_x^\mu \partial_x^\nu) = \Box_x \partial_x^\nu - \Box_x \partial_x^\nu = 0 \tag{4.14}$$

Since projection operators do not have inverses, the Greens function of Eq. (4.13) does not exist. Consequently, the Cauchy initial value problem cannot be solved uniquely. Classically, we know that in such a case, we have to choose a gauge in which the problem can be solved. The rationale for this, of course, comes from the fact that any observable is gauge invariant and is, therefore, insensitive to a choice of gauge.

In the quantum theory, there is a well-defined procedure (known as the Faddeev-Popov procedure) for doing this. One adds a gauge fixing term to the Lagrangian (corresponding to the choice of a gauge) and a compensating ghost Lagrangian. Thus with a covariant gauge choice, the complete gauge fixed Lagrangian for QED takes the form

$$\begin{aligned}\mathcal{L}_{\text{eff}} &= \mathcal{L}_{\text{QED}} - \frac{1}{2\alpha}(\partial_\mu A^\mu)^2 + \partial^\mu \bar{c} \partial_\mu c \\ &= i\bar{\psi}\gamma^\mu(\partial_\mu - ieA_\mu)\psi - \frac{1}{4}F_{\mu\nu}F^{\mu\nu} - \frac{1}{2\alpha}(\partial_\mu A^\mu)^2 + \partial^\mu \bar{c}\partial_\mu c\end{aligned} \tag{4.15}$$

Here α is an arbitrary constant parameter known as the gauge fixing parameter. c and $\bar{c}$ are known as ghost fields and satisfy anticommutation relations like the fermions. (They are scalars with opposite statistics.) Physically, one can think of the ghost fields as subtracting out two degrees of freedom from the four component photon field to give effectively two physical degrees of freedom (namely, the transverse degrees). The complete theory of QED now has the additional Feynman rules given by

$$= iG_{\mu\nu}(p) = -\frac{i}{p^2}\left(\eta_{\mu\nu} - (1-\alpha)\frac{p_\mu p_\nu}{p^2}\right) \tag{4.16}$$

$$= iG_c(p) = \frac{i}{p^2}$$

The procedure of gauge fixing, while gives well defined calculational rules, has changed the theory also (at least appears to). For example, the theory is no longer gauge invariant and, consequently, it is not clear whether the Ward identities which we derived earlier and which characterize gauge invariance still hold in the full theory. A crucial observation which helps answer this question is that even though $\mathcal{L}_{\text{eff}}$ is not gauge invariant, it is invariant under a symmetry transformation involving the ghost field (also known as the BRST transformation). This can be appreciated by rewriting $\mathcal{L}_{\text{eff}}$ as

$$\mathcal{L}_{\text{eff}} = i\bar\psi\gamma^\mu(\partial_\mu - ieA_\mu)\psi - \frac{1}{4}F_{\mu\nu}F^{\mu\nu} - F\partial_\mu A^\mu + \frac{\alpha}{2}F^2 + \partial^\mu \bar c \partial_\mu c \qquad (4.17)$$

Note that if we eliminate the auxiliary field F from Eq. (4.17), the Lagrangian of Eq. (4.15) is obtained. The Lagrangian (4.17) is invariant under the transformations

$$\delta_\beta A_\mu(x) = \frac{\beta}{e}\partial_\mu c(x)$$

$$\delta_\beta \psi(x) = i\beta c(x)\psi(x)$$

$$\delta_\beta \bar\psi(x) = -i\beta c(x)\bar\psi(x) \qquad (4.18)$$

$$\delta_\beta c(x) = 0$$

$$\delta_\beta \bar c(x) = -\frac{\beta}{e}F(x)$$

$$\delta_\beta F(x) = 0$$

Here β is a space-time independent anticommuting parameter. The invariance of the Lagrangian can be checked by noting that these transformations correspond to a gauge transformation if we identify

$$\alpha(x) = \beta c(x) \qquad (4.19)$$

Since the original Lagrangian is gauge invariant, it follows now that

$$\begin{aligned}\delta_\beta \mathcal{L}_{\text{eff}} &= \delta_\beta\left[-F\partial_\mu A^\mu + \frac{\alpha}{2}F^2 + \partial^\mu \bar c \partial_\mu c\right]\\ &= -F\partial_\mu \delta_\beta A^\mu + \partial^\mu \delta_\beta \bar c \partial_\mu c\\ &= -\frac{\beta}{e}F\partial_\mu\partial^\mu c - \frac{\beta}{e}\partial^\mu F \partial_\mu c\\ &= -\frac{\beta}{e}\partial_\mu(F\partial^\mu c)\end{aligned} \qquad (4.20)$$

Therefore, the action is invariant.

The BRST symmetry of the theory imposes relations between different scattering amplitudes which include the Ward identities we discussed earlier. But more importantly, the symmetry transformations of Eq. (4.18) lead to a conserved charge, Q_{BRST}, through the Noether procedure. This charge has the important property that it is nilpotent, that is,

$$Q^2_{\text{BRST}} = 0 \qquad (Q_{\text{BRST}} \text{ is fermionic.}) \qquad (4.21)$$

This allows us to define the physical states in this theory as those states which are annihilated by Q_{BRST}, namely,

$$Q_{\text{BRST}}|\text{phys}> = 0 \qquad (4.22)$$

Note that since Q_{BRST} is the generator of the BRST transformations, we can write (up to a total derivative)

$$-F\partial_\mu A^\mu + \frac{\alpha}{2} F^2 + \partial^\mu \bar{c} \partial_\mu c$$

$$= \delta\left(-e\left(A_\mu \partial^\mu \bar{c} + \frac{\alpha}{2} F\bar{c}\right)\right) \qquad (4.23)$$

$$= \left\{Q_{\text{BRST}}, -e\left(A_\mu \partial^\mu \bar{c} + \frac{\alpha}{2} F\bar{c}\right)\right\}$$

It follows now that

$$<\text{phys}'|\mathcal{L}_{\text{eff}}|\text{phys}>$$

$$=<\text{phys}'|\mathcal{L}_{\text{QED}} - F\partial_\mu A^\mu + \frac{\alpha}{2} F^2 + \partial^\mu \bar{c}\partial_\mu c|\text{phys}>$$

$$=<\text{phys}'|\mathcal{L}_{\text{QED}} - e\left\{Q_{\text{BRST}}, A^\mu \partial_\mu \bar{c} + \frac{\alpha}{2} F\bar{c}\right\}|\text{phys}> \qquad (4.24)$$

$$=<\text{phys}'|\mathcal{L}_{\text{QED}}|\text{phys}>$$

Here in the last step we have used Eq. (4.22). This shows that even though the gauge fixing procedure may have changed the theory, the effect is not observable in the physical sector. The BRST symmetry also plays a crucial role in proving the perturbative unitarity of the theory.

Higgs Mechanism

Let us reconsider the self-interacting theory of the complex scalar field which displays spontaneous symmetry breaking. However, let us also assume the complex (charged) scalar fields to be interacting with photons. As we have seen, interaction with photons can be introduced through the minimal coupling. Thus the Lagrangian for this theory is given by

$$\mathcal{L} = -\frac{1}{4} F_{\mu\nu} F^{\mu\nu} + (D_\mu \phi)^\dagger (D^\mu \phi) + m^2 \phi^\dagger \phi - \frac{\lambda}{4} (\phi^\dagger \phi)^2 \qquad \lambda > 0 \qquad (4.25)$$

where
$$D_\mu \phi = (\partial_\mu - ieA_\mu)\phi$$
$$(D_\mu \phi)^\dagger = (\partial_\mu + ieA_\mu)\phi^\dagger \qquad (4.26)$$

In terms of the real fields σ and ζ, the Lagrangian of Eq. (4.25) takes the form

$$\mathcal{L} = -\frac{1}{4} F_{\mu\nu} F^{\mu\nu} + \frac{1}{2} \partial_\mu \sigma \partial^\mu \sigma + \frac{1}{2} \partial_\mu \zeta \partial^\mu \zeta$$

$$- e\sigma \overleftrightarrow{\partial_\mu} \zeta A^\mu + \frac{e^2}{2} A_\mu A^\mu (\sigma^2 + \zeta^2) \qquad (4.27)$$

$$+ \frac{m^2}{2}(\sigma^2 + \zeta^2) - \frac{\lambda}{16}(\sigma^2 + \zeta^2)^2$$

This Lagrangian is invariant under the gauge transformations

$$\delta_\epsilon \sigma(x) = -\frac{1}{\sqrt{2}} \epsilon(x)\zeta(x)$$

$$\delta_\epsilon \zeta(x) = \frac{1}{\sqrt{2}} \epsilon(x)\sigma(x) \qquad (4.28)$$

$$\delta_\epsilon A_\mu(x) = \frac{1}{e} \partial_\mu \epsilon(x)$$

However, as we have seen earlier, the ground state of this theory occurs for

$$\zeta = 0$$
$$\sigma = \frac{2m}{\sqrt{\lambda}} \qquad (4.29)$$

For a stable perturbation, the theory must be expanded around this ground state by letting

$$\sigma \to \sigma + \frac{2m}{\sqrt{\lambda}} \qquad (4.30)$$

This leads the Lagrangian in Eq. (4.27) to take the form

$$\mathcal{L} = -\frac{1}{4} F_{\mu\nu} F^{\mu\nu} + \frac{1}{2} \partial_\mu \sigma \partial^\mu \sigma + \frac{1}{2} \partial_\mu \zeta \partial^\mu \zeta$$

$$- \frac{2em}{\sqrt{\lambda}} A^\mu \partial_\mu \zeta - e\sigma \overleftrightarrow{\partial_\mu} \zeta A^\mu + \frac{2m^2 e^2}{\lambda} A_\mu A^\mu$$

$$+ \frac{2me^2}{\sqrt{\lambda}} \sigma A_\mu A^\mu + \frac{e^2}{2} A_\mu A^\mu (\sigma^2 + \zeta^2) + \frac{m^4}{\lambda} - m^2 \sigma^2 \qquad (4.31)$$

$$- \frac{m\sqrt{\lambda}}{2} \sigma(\sigma^2 + \zeta^2) - \frac{\lambda}{16}(\sigma^2 + \zeta^2)^2$$

As we have seen before, there is spontaneous breakdown of symmetry in this theory and in the absence of the photon fields, Goldstone's theorem guarantees the existence of massless particles. However, in the presence of the photon field, there is a gauge invariance which allows us to choose a gauge. In particular, if we choose the gauge (this is also known as the unitary gauge)

$$\zeta = 0 \tag{4.32}$$

Then the Lagrangian of Eq. (4.31) takes the form

$$\begin{aligned}\mathcal{L} = &-\frac{1}{4} F_{\mu\nu} F^{\mu\nu} + \frac{1}{2} \partial_\mu \sigma \partial^\mu \sigma + \frac{2m^2 e^2}{\lambda} A_\mu A^\mu \\ &- m^2 \sigma^2 + \frac{m^4}{\lambda} + \frac{2me^2}{\sqrt{\lambda}} \sigma A_\mu A^\mu + \frac{e^2}{2} \sigma^2 A_\mu A^\mu \\ &- \frac{m\sqrt{\lambda}}{2} \sigma^3 - \frac{\lambda}{16} \sigma^4 \end{aligned} \tag{4.33}$$

We note that in such a case, the massless particle has disappeared and instead the photon field has become massive with a mass given by

$$m_{ph} = \frac{2me}{\sqrt{\lambda}} \tag{4.34}$$

This is known as the Higgs mechanism and we say that the photon has become massive by eating the Goldstone boson ζ. Note that a massive photon has three helicity states as opposed to the two states a massless photon can have and, consequently, the total number of degrees of freedom is unchanged (otherwise unitarity would be violated). The Higgs mechanism is quite useful in generating masses for particles in a physical theory.

Lecture V

Non-Abelian Symmetries

So far we have only considered symmetries where the generator of the symmetry, namely, the charge satisfied a trivial algebra - that is, it commuted with itself. Such symmetries are known as Abelian symmetries. Let us next consider some symmetries where the generators satisfy a nontrivial algebra. Such symmetries are known as non-Abelian symmetries and we are quite familiar with them also. For example, we know that the angular momentum operators generate rotations and satisfy the algebra

$$[J^a, J^b] = i\epsilon^{abc} J^c \qquad a, b, c = 1, 2, 3 \tag{5.1}$$

As we know, this is a non-Abelian algebra corresponding to the group SU(2). We also know that the quantum mechanical operator generating rotations is given by

$$U(\theta) = e^{iJ^a \theta^a} \tag{5.2}$$

where θ^a is the parameter of rotation. Thus, for example, we know that if ψ is a two component spinor corresponding to the $j = 1/2$ representation, then under a rotation

$$\psi \longrightarrow \tilde{\psi} = U_{1/2}(\theta)\psi = e^{\frac{i}{2}\sigma^a \theta^a}\psi \tag{5.3}$$

where σ^a are the Pauli matrices and correspond to the angular momentum operators for this representation (actually, $\frac{1}{2}\sigma^a$ corresponds to the generators). Infinitesimally, the two components of the spinor would rotate as

$$\delta_\epsilon \psi^i = i\left(\frac{1}{2}\sigma^a \epsilon^a\right)^{ij} \psi^j \qquad i,j = 1,2 \tag{5.4}$$

The $j = 1/2$ representation is $2j+1 = 2$ dimensional and is also called the fundamental representation of SU(2).

SU(2) is, of course, the simplest of the non-Abelian symmetries. In general, the algebra corresponding to a higher symmetry group SU(n) consists of $n^2 - 1$ Hermitian generators and satisfies an algebra of the form

$$[T^a, T^b] = if^{abc}T^c \qquad a,b,c = 1,2,\ldots,n^2 - 1 \tag{5.5}$$

where the totally antisymmetric constants, f^{abc}, are known as the structure constants of the group. We can think of the generators T^a as generating rotations in a $(n^2 - 1)$ dimensional internal space. Therefore, we can readily generalize many of the results of SU(2) to the SU(n) case. For example, we note that if ψ is a function belonging to the fundamental representation of SU(n), then it will be a n-component object and under an infinitesimal SU(n) rotation, it would transform as

$$\delta_\epsilon \psi^i = i(T^a \epsilon^a)^{ij} \psi^j \qquad i,j = 1,2,\ldots n \tag{5.6}$$

Here T^a corresponds to the SU(n) generators in the fundamental representation and ϵ^a is the parameter of rotation.

Let us next consider a free fermion theory where the fermion field belongs to the fundamental representation of an internal symmetry group SU(n). We are, of course, quite familiar with many such fermionic systems. We know that the three colored quarks belong to the fundamental representation of the color group SU(3). The up and down quarks belong to the fundamental representation of the isospin group SU(2) and so on. Such a system is, therefore, worth studying. The Lagrangian is given by

$$\mathcal{L} = i\bar{\psi}^i \gamma^\mu \partial_\mu \psi^i \qquad i = 1, 2, \ldots n \qquad (5.7)$$

which is just a sum of n-free fermion Lagrangians. Note that the momenta conjugate to $\psi_\alpha^i(x)$ are

$$\Pi_\alpha^i(x) = \frac{\partial \mathcal{L}}{\partial \dot{\psi}_\alpha^i(x)} = i\psi_\alpha^{i\dagger}(x) \qquad \alpha = 1, 2, 3, 4 \qquad (5.8)$$

so that the quantization rules become

$$\{\psi_\alpha^i(x), \psi_\beta^j(x')\}_{t=t'} = 0 = \{\Pi_\alpha^i(x), \Pi_\beta^j(x')\}_{t=t'}$$
$$\{\psi_\alpha^i(x), \Pi_\beta^j(x')\}_{t=t'} = i\delta^{ij}\delta_{\alpha\beta}\delta^{(3)}(x-x') \qquad (5.9)$$

The Lagrangian of Eq. (5.7) is, of course, invariant under the global U(1) transformation we have discussed earlier, namely,

$$\psi^i(x) \longrightarrow e^{i\alpha}\psi^i(x)$$
$$\bar{\psi}^i(x) \longrightarrow e^{-i\alpha}\bar{\psi}^i(x) \qquad (5.10)$$

where α is a constant scalar parameter. But more importantly, the Lagrangian is also invariant under a global SU(n) rotation which has the infinitesimal form

$$\delta_\epsilon \psi^i(x) = i(T^a \epsilon^a)^{ij} \psi^j$$
$$\delta_\epsilon \bar{\psi}^i(x) = -i\bar{\psi}^j (T^a \epsilon^a)^{ji} \qquad (5.11)$$

Here ϵ^a are constant, infinitesimal parameters of the SU(n) transformation. The invariance can be checked readily as

$$\delta_\epsilon \mathcal{L} = i\delta_\epsilon \bar{\psi}^i \gamma^\mu \partial_\mu \psi^i + i\bar{\psi}^i \gamma^\mu \partial_\mu \delta_\epsilon \psi^i$$

$$= \bar{\psi}^j (T^a \epsilon^a)^{ji} \gamma^\mu \partial_\mu \psi^i - \bar{\psi}^i \gamma^\mu \partial_\mu \left((T^a \epsilon^a)^{ij} \psi^j\right)$$

$$= \bar{\psi}^i (T^a \epsilon^a)^{ij} \gamma^\mu \partial_\mu \psi^j - \bar{\psi}^i (T^a \epsilon^a)^{ij} \gamma^\mu \partial_\mu \psi^j$$

or, $\quad \delta_\epsilon \mathcal{L} = 0 \qquad (5.12)$

The conserved current can now be constructed from the Noether procedure and has the form

$$j^{\mu,a} = -\bar{\psi}^i (T^a)^{ij} \gamma^\mu \psi^j = -\bar{\psi} T^a \gamma^\mu \psi \qquad (5.13)$$

The corresponding conserved charges

$$Q^a = \int d^3x \, j^{0,a}(x) \tag{5.14}$$

can be shown using Eqs. (5.5) and (5.9) to satisfy

$$[Q^a, Q^b] = i f^{abc} Q^c \tag{5.15}$$

which, as we have seen, is the SU(n) algebra (see Eq. (5.5)).

The Lagrangian of Eq. (5.7) is, however, not invariant under the SU(n) transformations of Eq. (5.11) if the parameter ϵ^a are coordinate dependent. As we have seen earlier, invariance under a local transformation necessarily requires the introduction of a gauge field. In the present case, the Lagrangian

$$\mathcal{L} = i\bar{\psi}^i \gamma^\mu \left(\delta^{ij} \partial_\mu - ig A_\mu^a (T^a)^{ij} \right) \psi^j \tag{5.16}$$

can be shown to be invariant under the local gauge transformations

$$\delta_\epsilon \psi^i(x) = i \left(T^a \epsilon^a(x) \right)^{ij} \psi^j(x)$$

$$\delta_\epsilon \bar{\psi}^i(x) = -i \bar{\psi}^j(x) \left(T^a \epsilon^a(x) \right)^{ji} \tag{5.17}$$

$$\delta_\epsilon A_\mu^a(x) = \frac{1}{g} \partial_\mu \epsilon^a(x) + f^{abc} A_\mu^b(x) \epsilon^c(x)$$

The invariance can, in fact, be readily checked as

$$\delta_\epsilon \mathcal{L} = i \delta_\epsilon \bar{\psi}^i \gamma^\mu \left(\delta^{ij} \partial_\mu - ig A_\mu^a (T^a)^{ij} \right) \psi^j$$

$$+ i \bar{\psi}^i \gamma^\mu \left(\delta^{ij} \partial_\mu - ig A_\mu^a (T^a)^{ij} \right) \delta_\epsilon \psi^j$$

$$+ g \bar{\psi}^i \gamma^\mu (T^a)^{ij} \delta_\epsilon A_\mu^a \psi^j(x)$$

$$= \bar{\psi}^k \left(T^b \epsilon^b(x) \right)^{ki} \gamma^\mu \left(\delta^{ij} \partial_\mu - ig A_\mu^a (T^a)^{ij} \right) \psi^j$$

$$- \bar{\psi}^i \gamma^\mu \left(\delta^{ij} \partial_\mu - ig A_\mu^a (T^a)^{ij} \right) \left(T^b \epsilon^b(x) \right)^{jk} \psi^k(x)$$

$$+ g \bar{\psi}^i \gamma^\mu (T^a)^{ij} \left(\frac{1}{g} \partial_\mu \epsilon^a(x) + f^{abc} A_\mu^b(x) \epsilon^c(x) \right) \psi^j(x)$$

$$= ig \bar{\psi}^i \gamma^\mu [T^a, T^b]^{ij} A_\mu^a(x) \epsilon^b(x) \psi^j(x)$$

$$+ g \bar{\psi}^i \gamma^\mu f^{abc} (T^a)^{ij} A_\mu^b(x) \epsilon^c(x) \psi^j(x)$$

Using Eq. (5.5), we now obtain

$$\delta_\epsilon \mathcal{L} = - g\bar\psi^i \gamma^\mu f^{abc}(T^c)^{ij} A_\mu^a(x) \epsilon^b(x) \psi^j(x)$$
$$+ g\bar\psi^i \gamma^\mu f^{abc}(T^a)^{ij} A_\mu^b(x) \epsilon^c(x) \psi^j(x) = 0 \quad (5.18)$$

Several comments are in order here. The parameter g can be thought of as the coupling constant for the SU(n) gauge group. Furthermore, in the present case we note that the gauge fields A_μ^a carry SU(n) quantum numbers and hence SU(n) charge. This behavior is quite distinct from the photon which does not carry electric charge. Furthermore, since the gauge fields couple to any source carrying the corresponding charge and since in the case of SU(n) the gauge fields themselves carry SU(n) charge, it is clear that the gauge fields of SU(n) must couple to themselves - that is, they must have self-interaction in contrast to the case of the photon. In fact, the dynamical Lagrangian for the SU(n) gauge fields invariant under the transformations of Eq. (5.17) can be shown to be

$$\mathcal{L}_{\text{gauge}} = -\frac{1}{4} F_{\mu\nu}^a(x) F^{\mu\nu,a}(x) \qquad a = 1,2,\ldots n^2 - 1$$

where (5.19)

$$F_{\mu\nu}^a = \partial_\mu A_\nu^a - \partial_\nu A_\mu^a + g f^{abc} A_\mu^b A_\nu^c$$

The self-coupling is now obvious. Thus for example, if we are considering Quantum Chromodynamics corresponding to the gauge group SU(3), there would be $3^2 - 1 = 8$ gauge fields or gluons which not only couple to the colored quarks but also to themselves. This, of course, has profound consequences leading to asymptotic freedom.

The complete Lagrangian including the fermions and the dynamics of the gauge fields which is invariant under the transformations of Eq. (5.17) is given by

$$\mathcal{L}_{\text{inv}} = -\frac{1}{4} F_{\mu\nu}^a F^{\mu\nu,a} + i\bar\psi^i \gamma^\mu \left(\delta^{ij} \partial_\mu - ig(T^a)^{ij} A_\mu^a \right) \psi^j(x) \quad (5.20)$$

The gauge invariance, as we have seen, presents problems in quantizing the theory. Therefore, following the method due to Faddeev and Popov, we choose a gauge fixing and a ghost Lagrangian. A covariant choice of the gauge in the present case leads to the complete Lagrangian

$$\mathcal{L} = \mathcal{L}_{\text{inv}} - \frac{1}{2\alpha}(\partial_\mu A^{\mu,a})^2 + \partial^\mu \bar c^a(x) \left(\partial_\mu \delta^{ac} + g f^{abc} A_\mu^b \right) c^c(x) \quad (5.21)$$

where $c^a(x)$ and $\bar c^a(x)$ are the respective ghost and anti-ghost fields. The Feynman rules for this Lagrangian can now be derived.

$$iG^{ab}_{\mu\nu}(p) = -\frac{i\delta^{ab}}{p^2}\left(\eta_{\mu\nu} - (1-\alpha)\frac{p_\mu p_\nu}{p^2}\right)$$

$$iS^{ij}(p) = \frac{i\delta^{ij}}{\slashed{p}}$$

$$iG^{ab}(p) = \frac{i\delta^{ab}}{p^2}$$

$$-i\Gamma^{a,ij}_\mu(p_1,p_2,p_3) = ig\gamma_\mu (T^a)^{ij}\delta^{(4)}(p_1+p_2+p_3)$$

$$-i\Gamma^{abc}_\mu(p_1,p_2,p_3) = gp_{1\mu}f^{abc}\delta^{(4)}(p_1+p_2+p_3)$$

$$-i\Gamma^{abc}_{\mu\nu\lambda}(p_1,p_2,p_3) = -gf^{abc}\Big[(p_1-p_2)_\lambda \eta_{\mu\nu}$$
$$+ (p_2-p_3)_\mu \eta_{\nu\lambda} + (p_3-p_1)_\nu \eta_{\lambda\mu}\Big]$$

(5.22)

$$-i\Gamma^{abcd}_{\mu\nu\lambda\rho}(p_1,p_2,p_3,p_4) = -ig^2\Big[f^{abp}f^{cdp}(\eta_{\mu\lambda}\eta_{\nu\rho} - \eta_{\mu\rho}\eta_{\nu\lambda})$$
$$+ f^{acp}f^{dbp}(\eta_{\mu\rho}\eta_{\nu\lambda} - \eta_{\mu\nu}\eta_{\lambda\rho})$$
$$+ f^{adp}f^{bcp}(\eta_{\mu\nu}\eta_{\lambda\rho} - \eta_{\mu\lambda}\eta_{\nu\rho})\Big]$$

The Feynman rules clearly bring out the feature of pure gauge interactions. With these, one can now calculate any scattering amplitude.

The Lagrangian in Eq. (5.21) is no longer gauge invariant. But as in the case of QED, there is a residual (BRST) symmetry involving the anticommuting ghost fields. Thus the Lagrangian in Eq. (5.21) is invariant under the set of transformations

$$\delta_\beta A_\mu^a = \beta \left(\frac{1}{g} \partial_\mu c^a(x) + f^{abc} A_\mu^b(x) c^c(x) \right)$$

$$\delta_\beta \psi^i = i\beta c^a(x) (T^a)^{ij} \psi^j(x)$$

$$\delta_\beta \bar{\psi}^i = -i\beta c^a(x) \bar{\psi}^j (T^a)^{ji} \quad (5.23)$$

$$\delta_\beta c^a(x) = -\frac{\beta}{2} f^{abc} c^b(x) c^c(x)$$

$$\delta_\beta \bar{c}^a(x) = -\frac{\beta}{\alpha} \left(\partial_\mu A^{\mu,a}(x) \right)$$

Here β is an anticommuting constant parameter and the invariance of the Lagrangian can be checked in a straightforward manner. Once again the BRST symmetry leads to relations between different scattering amplitudes known as Ward identities or Slavnov-Taylor identities. These identities are much more complicated than the ones we encountered in the case of QED but are quite useful in renormalizing the theory. The conserved charge associated with the BRST symmetry in the present case can also be shown to be nilpotent. As in the case of QED, this helps us define a physical Hilbert space. In this space, the theory can again be shown to be independent of the choice of the gauge and the parameter α. Furthermore, perturbative unitarity can also be shown to hold in this space.

As we have seen earlier, renormalization introduces a mass scale μ and that all coupling constants become functions of μ. The μ-dependence of the SU(n) gauge coupling in the present case can be calculated since all the Feynman rules are known. In fact, if we assume that there are n_f fermion fields in the fundamental representation interacting with the SU(n) gauge field, then at one-loop level we find

$$\mu \frac{\partial g}{\partial \mu} = \beta(g) = -\frac{g^3}{16\pi^2} \cdot \frac{1}{3} (11n - 2n_f) \quad (5.24)$$

The first term on the right hand side comes from pure gauge interactions whereas the second term which depends on the number of fermion flavors comes from the fermionic interactions. Note that the two terms contribute with opposing signs. As in Eq. (2.29), we can solve Eq. (5.24) to obtain

$$g^2(\mu) = \frac{g^2(\bar{\mu})}{1 + \frac{g^2(\bar{\mu})}{48\pi^2} (11n - 2n_f) \ln \frac{\mu^2}{\bar{\mu}^2}} \quad (5.25)$$

This shows that if

$$11n - 2n_f > 0 \quad (5.26)$$

Then $g(\mu)$ decreases as μ increases with respect to $\bar{\mu}$. In other words, in such a case the coupling becomes weaker as the energy scale increases. In particular, for infinitely large energy values, the coupling vanishes leading us to conclude that such theories are asymptotically free.

Let us note, in particular, that when $n_f = 0$, namely, when no fermions are present, the scale dependence of the gauge coupling is given by

$$g^2(\mu) = \frac{g^2(\bar{\mu})}{1 + \frac{11ng^2(\bar{\mu})}{48\pi^2} \ln \frac{\mu^2}{\bar{\mu}^2}} \tag{5.27}$$

That is, in a pure non-Abelian gauge theory, the coupling is asymptotically free. It is the presence of fermions and other matter fields that spoils asymptotic freedom. Intuitively, one understands this as saying that fermions and other matter fields lead to a screening effect whereas a non Abelian gauge field leads to antiscreening which is responsible for asymptotic freedom. Note also that since in an asymptotically free theory, the coupling is weak at high energies, perturbative calculations can be trusted only at large energies. At low energies, however, the coupling constant grows and hence perturbation theory breaks down.

Let me conclude by pointing out that Quantum Chromodynamics which is the theory of strong interactions is a gauge theory based on the gauge group SU(3). The quarks which are the fermion fields in this theory come in threee colors and belong to the fundamental representation of SU(3). Thus specializing to $n = 3$ we would obtain all the necessary results for QCD. Since we do not see free quarks in nature, we can say that the color symmetry (SU(3)) is unbroken leading to the fact that observables must be color singlet states. As we have seen, since the coupling becomes stronger in non Abelian gauge theories, it supports this hypothesis that the quarks must be strongly bound. However, a conclusive proof of quark confinement is still lacking.

Lecture VI

Weinberg-Salam-Glashow Theory

The strong force can be described by Quantum Chromodynamics which is a gauge theory based on the gauge group SU(3). As we have seen, we understand the basic features of this theory quite well. Thus let us ignore the strong interactions for a moment and try to understand the gauge structure of the other two fundamental interactions, namely, the weak interaction and the electromagnetic interaction. Let us recall that while leptons interact weakly as well as through electromagnetic interactions, they do not have any strong interaction. Consequently, we can, for simplicity, restrict ourselves to the gauge theory involving only leptons in order to understand the weak and electromagnetic forces.

To begin with, let us recall some facts about fundamental particles. We know that all elementary particles can be classified according to the representations of the weak isospin group, SU(2), which is very similar to the rotation group. (For clarity let me emphasize here that the weak isospin is different from the strong isospin which classifies observed hadrons.) Thus, let us list some of the more familiar particles all of which correspond to the $I = 1/2$ representation of the weak isospin group.

$$I_3 = \tfrac{1}{2} \qquad \begin{pmatrix} \nu_e \\ e \end{pmatrix} \begin{pmatrix} \nu_\mu \\ \mu \end{pmatrix} \begin{pmatrix} \nu_\tau \\ \tau \end{pmatrix} \begin{pmatrix} u \\ d \end{pmatrix} \begin{pmatrix} c \\ s \end{pmatrix} \qquad (6.1)$$
$$I_3 = -\tfrac{1}{2}$$

The particles within a given multiplet are arranged so that the member with a higher $I_3(I_z)$ value has a larger electric charge. It is also known that we can assign to every elementary particle a U(1) quantum number known as the weak hypercharge and denoted by Y such that the electric charge of any given particle can be written as

$$Q = I_3 + \frac{Y}{2} \qquad (6.2)$$

(Once again, a word of caution that the weak hypercharge is different from the strong hypercharge which can be identified with the sum of the baryon number and the strangeness number.) Eq. (6.2) can, in fact, be taken as defining the hypercharge of a given particle. Thus, the hypercharges of some of the particles in Eq. (6.1) are

$$Y_e = -1 = Y_{\nu_e}$$
$$Y_u = \frac{1}{3} = Y_d \qquad (6.3)$$

Note here that the hypercharges of the particles within an isospin multiplet are the same.

Phenomenologically, we know that weak interactions are short ranged and, therefore, if they can be written as a gauge theory, the gauge bosons must be massive. Second, we know that they violate parity maximally and have a V−A structure. To understand this better, let us recall that the electromagnetic current, in the case of QED, has the form (see Eq. (3.41))

$$j_V^\mu = -\bar{\psi}\gamma^\mu\psi \qquad (6.4)$$

This behaves like a vector under a Lorentz transformation as well as under a space reflection and is, therefore, called a vector current. An axial vector current, on the other hand, transforms like a vector under a Lorentz transformation but behaves like a pseudo-vector under a space reflection and has the form (see Eq. (3.42))

$$j_A^\mu = -\bar{\psi}\gamma^\mu\gamma_5\psi \qquad (6.5)$$

A V−A current, as the name suggests, has the structure

$$j_{V-A}^\mu = -\frac{1}{2}\bar{\psi}\gamma^\mu(1-\gamma_5)\psi$$
$$= -\bar{\psi}_L\gamma^\mu\psi_L \qquad (6.6)$$

where we have defined

$$\psi_L = \frac{1}{2}(1-\gamma_5)\psi$$
$$\bar{\psi}_L = \frac{1}{2}\bar{\psi}(1+\gamma_5) \qquad (6.7)$$

The quantity $\frac{1}{2}(1-\gamma_5)$ is a projection operator which merely projects out the left handed component of a fermion field. Thus the V−A structure of weak interactions tantamounts to saying that only the left-handed components of particles participate in weak interactions. Consequently, we can think of weak isospin as a left-handed group.

With all this information, let us construct the simplest theory involving only one family of leptons, namely, the electron family. Let

$$\ell = \begin{pmatrix} \nu_e \\ e \end{pmatrix}_L \tag{6.8}$$

This is an isospin doublet. However, whereas we know that the right handed component of the electron exists, right handed neutrinos are not seen in nature. Consequently, there will only be one right-handed lepton in this case which is the right-handed electron. By definition, this will be an isospin singlet.

$$r = e_R \tag{6.9}$$

All the fermions, of course, carry the hypercharge quantum number. We have already determined the hypercharge of the left-handed particles to be

$$Y_\ell = -1 \tag{6.10}$$

The hypercharge of the right-handed electron can, similarly, be determined to be

$$Y_r = -2 \tag{6.11}$$

Note that all the fermions carry both the isospin as well as the hypercharge quantum numbers. Thus the simplest gauge theory that we can think of constructing is one where both these symmetries are local. We can easily write down a fermionic Lagrangian which is invariant under the isospin (SU$_L$(2)) and hypercharge (U$_Y$(1)) gauge transformations.

$$\mathcal{L}_f = i\bar{\ell}^i \gamma^\mu \left(\delta^{ij} \partial_\mu - \frac{ig'}{2} \delta^{ij} Y_\mu - \frac{ig}{2} (\sigma^a)^{ij} W^a_\mu \right) \ell^j \tag{6.12}$$
$$+ i\bar{r}\gamma^\mu (\partial_\mu - ig' Y_\mu) r$$

Here $i, j = 1, 2$ and $a = 1, 2, 3$. We have introduced the gauge fields W^a_μ and Y_μ corresponding to the isospin and hypercharge transformations. g and g' denote respectively the strengths of the isospin and hypercharge interactions. Note that since the right-handed field does not carry any isospin quantum number, it does not couple to W^a_μ.

The dynamics of the gauge fields can now be introduced in a straightforward manner.

$$\mathcal{L}_{\text{gauge}} = -\frac{1}{4} Y_{\mu\nu} Y^{\mu\nu} - \frac{1}{4} W^a_{\mu\nu} W^{\mu\nu,a} \qquad (6.13)$$

where

$$Y_{\mu\nu} = \partial_\mu Y_\nu - \partial_\nu Y_\mu$$

$$W^a_{\mu\nu} = \partial_\mu W^a_\nu - \partial_\nu W^a_\mu + g\epsilon^{abc} W^b_\mu W^c_\nu \qquad (6.14)$$

Relation (6.14) emphasizes that Y_μ is an Abelian gauge field like the photon field since it corresponds to the group $U_Y(1)$. W^a_μ, on the other hand, is a non-Abelian gauge field corresponding to the gauge group $SU_L(2)$. Thus, together,

$$\mathcal{L} = \mathcal{L}_{\text{gauge}} + \mathcal{L}_f \qquad (6.15)$$

defines an interacting gauge theory of leptons based on the gauge group $SU_L(2) \times U_Y(1)$.

The weak interactions, on the other hand, are short ranged which amounts to the corresponding gauge bosons being massive. We can incorporate this into our theory by adding to our Lagrangian a part depending on scalar fields which, as we have seen, can give masses to the gauge bosons through the Higgs mechanism. Let

$$\phi = \begin{pmatrix} \phi^+ \\ \phi^0 \end{pmatrix} \qquad (6.16)$$

denote an isospin doublet of complex scalar fields with charges 1 and 0. Thus the hypercharge quantum number associated with this multiplet is 1. Let us also denote the Hermitian conjugate of ϕ as

$$\phi^\dagger = \overparen{\phi^- \quad \bar{\phi}^0} \qquad (6.17)$$

The scalar Lagrangian, invariant under $SU_L(2) \times U_Y(1)$ transformations, can now be written as

$$\mathcal{L}_{\text{Higgs}} = \left(\left(\delta^{ij} \partial_\mu + \frac{ig'}{2} \delta^{ij} Y_\mu - \frac{ig}{2} (\sigma^a)^{ij} W^a_\mu \right) \phi^j \right)^\dagger$$

$$\cdot \left(\delta^{ik} \partial^\mu + \frac{ig'}{2} \delta^{ik} Y^\mu - \frac{ig}{2} (\sigma^b)^{ik} W^{\mu,b} \right) \phi^k \qquad (6.18)$$

$$+ m^2 \phi^\dagger \phi - \frac{\lambda}{4} (\phi^\dagger \phi)^2 - h(\bar{r}\phi^\dagger \ell + \bar{\ell}\phi r)$$

This is the usual symmetry breaking Lagrangian for the scalar fields except for the last term representing the interaction between the scalar fields and the fermions in the theory. h is known as the strength of the Yukawa coupling. We can now write the total Lagrangian for the theory to be

$$\mathcal{L}_{\text{TOT}} = \mathcal{L}_{\text{gauge}} + \mathcal{L}_f + \mathcal{L}_{\text{Higgs}} \qquad (6.19)$$

Note that if we now define the combinations

$$W_\mu^\pm = \frac{1}{\sqrt{2}} \left(W_\mu^1 \pm i W_\mu^2 \right) \tag{6.20}$$

then $\mathcal{L}_{\text{Higgs}}$ can be written explicitly in terms of the components as

$$\begin{aligned}\mathcal{L}_{\text{Higgs}} = &\left(\left(\partial_\mu - \frac{ig'}{2} Y_\mu\right) \phi^- + \frac{ig}{2} W_\mu^3 \phi^- + \frac{ig}{\sqrt{2}} W_\mu^- \bar\phi^0 \right) \left(\left(\partial^\mu + \frac{ig'}{2} Y^\mu\right) \phi^+ \right.\\ &\left. - \frac{ig}{2} W^{\mu 3} \phi^+ - \frac{ig}{\sqrt{2}} W^{\mu +} \phi^0 \right) + \left(\left(\partial_\mu - \frac{ig'}{2} Y_\mu\right) \bar\phi^0 \right.\\ &\left. + \frac{ig}{\sqrt{2}} W_\mu^+ \phi^- - \frac{ig}{2} W_\mu^3 \bar\phi^0 \right) \left(\left(\partial^\mu + \frac{ig'}{2} Y^\mu\right) \phi^0 \right.\\ &\left. - \frac{ig}{\sqrt{2}} W^{\mu -} \phi^+ + \frac{ig}{2} W^{\mu 3} \phi^0 \right) + m^2 \left(\phi^- \phi^+ + \bar\phi^0 \phi^0 \right)\\ &- \frac{\lambda}{4} \left(\phi^- \phi^+ + \bar\phi^0 \phi^0 \right)^2 - h \bar e_R \nu_{eL} \phi^- - h \bar e_R e_L \bar\phi^0 \\ &- h \bar\nu_{eL} e_R \phi^+ - h \bar e_L e_R \phi^0 \end{aligned} \tag{6.21}$$

As before, we can now calculate the minimum of the potential. To be consistent with our earlier notation, let me define

$$\begin{aligned} \phi^0 &= \frac{1}{\sqrt{2}} (\sigma + i\zeta) \\ \bar\phi^0 &= \frac{1}{\sqrt{2}} (\sigma - i\zeta) \end{aligned} \tag{6.22}$$

In terms of these, the minimum of the potential can be shown to occur at (see Eq. (4.29))

$$\begin{aligned} \phi^+ &= 0 = \phi^- = \zeta \\ \sigma &= v = \frac{2m}{\sqrt{\lambda}} \end{aligned} \tag{6.23}$$

We can now expand the theory around this classical minimum by letting

$$\sigma \to \sigma + v \tag{6.24}$$

To bring out the essential features of the theory, let us first look at only the quadratic terms in $\mathcal{L}_{\text{TOT}}$ after shifting. The quadratic Lagrangian takes the form

$$\mathcal{L}_{\text{Quad.}} = -\frac{1}{2}\left(\partial_\mu W_\nu^+ - \partial_\nu W_\mu^+\right)\left(\partial^\mu W^{\nu-} - \partial^\nu W^{\mu-}\right)$$

$$-\frac{1}{4}\left(\partial_\mu W_\nu^3 - \partial_\nu W_\mu^3\right)\left(\partial^\mu W^{\nu 3} - \partial^\nu W^{\mu 3}\right)$$

$$-\frac{1}{4}\left(\partial_\mu Y_\nu - \partial_\nu Y_\mu\right)\left(\partial^\mu Y^\nu - \partial^\nu Y^\mu\right) + i\bar{\nu}_{eL}\slashed{\partial}\nu_{eL}$$

$$+ i\bar{e}_L \slashed{\partial} e_L + i\bar{e}_R \slashed{\partial} e_R + \partial_\mu \phi^- \partial^\mu \phi^+ \quad (6.25)$$

$$+ \frac{1}{2}\partial_\mu \sigma \partial^\mu \sigma + \frac{1}{2}\partial_\mu \zeta \partial^\mu \zeta - \frac{igv}{2} W_\mu^+ \partial^\mu \phi^-$$

$$+ \frac{igv}{2} W_\mu^- \partial^\mu \phi^+ + \frac{v}{2}\left(g'Y_\mu + gW_\mu^3\right)\partial^\mu \zeta$$

$$+ \frac{g^2 v^2}{4} W_\mu^+ W^{\mu -} + \frac{v^2}{8}\left(g'Y_\mu + gW_\mu^3\right)\left(g'Y^\mu + gW^{\mu 3}\right)$$

$$- \frac{hv}{\sqrt{2}}\bar{e}_R e_L - \frac{hv}{\sqrt{2}}\bar{e}_L e_R - m^2 \sigma^2$$

Although this looks complicated, it can be simplified by redefining variables in the following way. Let

$$\sin\theta_W = \frac{g'}{\sqrt{g^2 + g'^2}} \qquad (\theta_W = \text{Weinberg angle})$$

$$\cos\theta_W = \frac{g}{\sqrt{g^2 + g'^2}}$$

$$Z_\mu = \frac{1}{\sqrt{g^2 + g'^2}}\left(g'Y_\mu + gW_\mu^3\right) = \sin\theta_W Y_\mu + \cos\theta_W W_\mu^3 \quad (6.26)$$

$$A_\mu = \frac{1}{\sqrt{g^2 + g'^2}}\left(gY_\mu - g'W_\mu^3\right) = \cos\theta_W Y_\mu - \sin\theta_W W_\mu^3$$

Then in terms of these variables, the quadratic Lagrangian takes the form

$$\mathcal{L}_{\text{Quad.}} = -\frac{1}{2}\left(\partial_\mu W_\nu^+ - \partial_\nu W_\mu^+\right)\left(\partial^\mu W^{\nu-} - \partial^\nu W^{\mu-}\right) + \frac{g^2 v^2}{4} W_\mu^+ W^{\mu -}$$

$$-\frac{1}{4}\left(\partial_\mu Z_\nu - \partial_\nu Z_\mu\right)\left(\partial^\mu Z^\nu - \partial^\nu Z^\mu\right) + \frac{(g^2 + g'^2)v^2}{8} Z_\mu Z^\mu$$

$$-\frac{1}{4}\left(\partial_\mu A_\nu - \partial_\nu A_\mu\right)\left(\partial^\mu A^\nu - \partial^\nu A^\mu\right)$$

$$+ i\bar{\nu}_{eL}\slashed{\partial}\nu_{eL} + i\bar{e}_L\slashed{\partial}e_L + i\bar{e}_R\slashed{\partial}e_R \tag{6.27}$$

$$- \frac{hv}{\sqrt{2}} \bar{e}_R e_L - \frac{hv}{\sqrt{2}} \bar{e}_L e_R + \partial_\mu \phi^- \partial^\mu \phi^+$$

$$+ \frac{1}{2} \partial_\mu \zeta \partial^\mu \zeta + \frac{1}{2} \partial_\mu \sigma \partial^\mu \sigma - m^2 \sigma^2$$

$$- \frac{igv}{2} W_\mu^+ \partial^\mu \phi^- + \frac{igv}{2} W_\mu^- \partial^\mu \phi^+$$

$$+ \frac{v\sqrt{g^2 + g'^2}}{2} Z_\mu \partial^\mu \zeta$$

We note here that three of the four gauge fields have become massive and only one gauge field remains massless. (This is particularly obvious if we choose the gauge $\phi^+ = \phi^- = \zeta = 0$.) Thus the original symmetry has spontaneously broken down to U(1). Thus we say

$$\mathrm{SU}_L(2) \times \mathrm{U}_Y(1) \longrightarrow \mathrm{U}_{\mathrm{em}}(1) \tag{6.28}$$

The field A_μ can be identified with the photon so that we have the familiar result that in this theory even though isospin and hypercharge quantum numbers may be violated in some processes, the electric charge will be conserved. We also note that spontaneous symmetry breaking gives a mass to the electron through the Yukawa coupling whereas the neutrino remains massless. Note that the mass of the W and the Z-bosons are given by

$$M_W = \frac{gv}{2}$$
$$M_Z = \frac{(g^2 + g'^2)^{1/2} v}{2} \tag{6.29}$$

so that

$$\frac{M_W}{M_Z} = \frac{g}{\sqrt{g^2 + g'^2}} = \cos\theta_W \tag{6.30}$$

Both these masses and the Weinberg angle are, of course, well measured experimentally.

Let us next look at the part of $\mathcal{L}_{\mathrm{TOT}}$ describing the interaction of the fermions with the gauge fields.

$$\mathcal{L}_{\mathrm{int}} = \frac{g}{\sqrt{2}} W_\mu^+ \bar{\nu}_{eL} \gamma^\mu e_L + \frac{g}{\sqrt{2}} W_\mu^- \bar{e}_L \gamma^\mu \nu_{eL}$$

$$+ \frac{g}{2} W_\mu^3 \left(\bar{\nu}_{eL} \gamma^\mu \nu_{eL} - \bar{e}_L \gamma^\mu e_L \right) \tag{6.31}$$

$$+ \frac{g'}{2} Y_\mu \left(\bar{\nu}_{eL} \gamma^\mu \nu_{eL} + \bar{e}_L \gamma^\mu e_L + 2\bar{e}_R \gamma^\mu e_R \right)$$

We can rewrite this in terms of the variables in Eq. (6.26) as

$$\mathcal{L}_{int} = \frac{g}{\sqrt{2}} \left(W_\mu^+ \bar{\nu}_{eL} \gamma^\mu e_L + W_\mu^- \bar{e}_L \gamma^\mu \nu_{eL} \right)$$

$$+ \frac{Z_\mu}{2\sqrt{g^2 + g'^2}} \left(g^2 (\bar{\nu}_{eL} \gamma^\mu \nu_{eL} - \bar{e}_L \gamma^\mu e_L) \right. \quad (6.32)$$

$$\left. + g'^2 (\bar{\nu}_{eL} \gamma^\mu \nu_{eL} + \bar{e}_L \gamma^\mu e_L + 2\bar{e}_R \gamma^\mu e_R) \right)$$

$$+ \frac{gg'}{\sqrt{g^2 + g'^2}} A_\mu (\bar{e}_L \gamma^\mu e_L + \bar{e}_R \gamma^\mu e_R)$$

The first and the second terms in Eq. (6.32) exprerss the charged and neutral current structures of the weak interactions. The last term, on the other hand, has precisely the form of the coupling of electrons to photons if we identify the electromagnetic coupling to be

$$e = \frac{gg'}{\sqrt{g^2 + g'^2}} = g \sin \theta_W = g' \cos \theta_W \quad (6.33)$$

We have, of course, considered the simplest model with one family of leptons. One can add more families of leptons as well as quarks. We will then have a gauge theory of weak and electromagnetic interactions involving quarks and leptons. This is known as the standard model and seems to work well experimentally.

Exercises

1. Given $Tr\, \gamma^\mu = 0$
   Determine:
   $$Tr\, \gamma^\mu \gamma^\nu,\ Tr\, \gamma^\mu \gamma^\nu \gamma^\lambda,$$
   $$Tr\, \gamma^\mu \gamma^\nu \gamma^\lambda \gamma^\rho,\ Tr \gamma_5,$$
   $$Tr\, \gamma_5 \gamma^\mu$$

2. Using a cutoff, calculate the following scattering amplitude in the ϕ^4 theory:

3. Given
$$\int \frac{d^n k}{(2\pi)^n} \frac{1}{(k^2 + 2k \cdot p - m^2)^\alpha} = (-1)^\alpha \frac{i\pi^{n/2}}{(2\pi)^n} \frac{\Gamma(\alpha - n/2)}{\Gamma(\alpha)} \frac{1}{(p^2 + m^2)^{\alpha - n/2}}$$

evaluate

i) $\int \dfrac{d^n k}{(2\pi)^n} \dfrac{k_\mu}{(k^2 + 2k \cdot p - m^2)^k}$

ii) $\int \dfrac{d^n k}{(2\pi)^n} \dfrac{k_\mu k_\nu}{(k^2 + 2k \cdot p - m^2)^\alpha}$

4. Use dimensional regularization to calculate

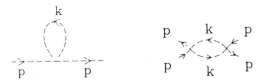

for a <u>2-dimensional</u> ϕ^4 theory.

5. Using the canonical commutation relations, check Eq. (3.17), namely, show that
$$\delta_\epsilon \phi(x) = -i[\phi(x), \epsilon Q]$$
$$\delta_\epsilon \phi^\dagger(x) = -i[\phi^\dagger(x), \epsilon Q].$$

6. From all the principles learnt so far, construct the simplest interacting theory of complex scalar fields and massless Dirac fermions which is renormalizable. Is there any internal symmetry in this case? What is the conserved charge? What additional consequences do you have if there is spontaneous symmetry breaking in such a case?

7. Consider
$$\mathcal{L}_{eff} = \mathcal{L}_{QED} + \partial^\mu F A_\mu + \frac{\alpha}{2} F^2 + \partial^\mu \bar{c} \partial_\mu c$$
(This differs from Eq.(4.17) by a total derivative.) This Lagrangian is invariant under the BRST symmetry. Construct the conserved charge, Q_{BRST}, associated with this from Noether's theorem.

8. For QED in <u>2-dimensions</u>, define

$$\Gamma_{\mu\nu}(p) = \underset{\gamma_5\,\gamma_\mu \quad k+p}{\underbrace{\qquad\qquad}}\underset{\gamma_\nu \qquad p}{\qquad\qquad}$$

Using dimensional regularization, calculate $p^\mu \Gamma_{\mu\nu}(p)$. (Use 2 dimensional γ − matrix identities. What you are calculating is the axial anomaly.)

INTRODUCTORY NOTES ON PARTICLE PHYSICS AND COSMOLOGY

Frank Wilczek

Institute for Advanced Study
Princeton, NJ 08540

1 INTRODUCTION

In the following pages you will find notes on two and a half of the four lectures I gave at the St. Croix summer school of 1990, taken by Michel Lefebvre and Harrison Prosper. The purpose of these lectures was to introduce students in experimental physics to some of the central concepts in cosmology, and perhaps inspire them to consider taking up some of the experimental challenges it poses. I think it is very important for young experimentalists, even (perhaps *especially*) those who consider their primary interest to be particle physics, to know something about cosmology. This is for two reasons.

First, a cultural one. With the development of QCD, which gives the microscopic theory of the nuclear force, it seems likely that the final chapter in the fundamental description of matter under "ordinary" conditions – even allowing a very liberal interpretation of the word "ordinary" – has been written. If the goal of further pursuit of particle physics was merely to determine what happens when you build extremely expensive big accelerators and subject matter to indignities that never occur in the natural world, the field would be rather hollow. Fortunately, this is not the case. Early in its history the Universe was a much hotter and denser place. Thus the proper arena for modern particle physics (beyond the standard model) is no less than the Universe as a whole, early in its history.

Second, a practical one. There are several important questions in cosmology which are at the limit of, or just beyond, the reach of current experimental technique. These include exploration of inhomogeneities in the microwave background (and possibly in the radio background; see below), mapping the distribution of galaxies, quasars, etc. in three dimensions, searches for dark matter candidates including axions and photinos, for a stochastic background of gravity waves, for the unusual "failed" galaxies suggested by the cold dark matter scenario, and others. It is important that the best and brightest young experimentalists think hard about these challenges and opportunities.

Due to pressures of time and space, the notes are appearing in a rather raw, terse form. I hope to produce something more digestible some time soon (and have hoped so for a long time.) Fortunately, there are two excellent texts in the field: Weinberg's *Gravitation and Cosmology*, whose last three chapters give an excellent account of classic cosmology and ideas about the very early universe just *before* the standard model of particle physics emerged; and Kolb and Turner's *The Early Universe*, which emphasizes recent developments.

Another terrific source should be appearing very soon. It is the proceedings of the 79th Nobel Symposium, which occurred just before the summer school. It will be published as a special issue of *Physica Scripta*. My original lectures included some material on the dark matter problem and on inhomogeneous cosmology. However, I would rather refer you to my lecture (and other lectures) at the Symposium for those topics.

2 EXPANDING UNIVERSES

2.1 Dynamical Equations

The dynamical equations of expanding universes can be motivated by quasi-Newtonian considerations. The result is the same as for General Relativity.

Consider the motion of a test particle moving with the cosmic fluid. From the conservation of energy for the test particle we derive:

$$\frac{1}{2}\dot{a}^2 - \frac{4\pi}{3}G\rho a^2 = \text{constant}$$

where a is the cosmic scale factor and ρ is the energy density. From the conservation of energy in the fluid we derive:

$$d\left(\frac{4\pi}{3}\rho a^3\right) = -P d\left(\frac{4\pi}{3}a^3\right).$$

Cleaning these up a bit, we obtain

$$\dot{a}^2 - \frac{8\pi}{3}G\rho a^2 = k \qquad (1)$$

and

$$\dot{\rho} = -3(P+\rho)\frac{\dot{a}}{a} \qquad (2)$$

where P is the pressure. The number $k \neq 0$, which here appears merely as an integration constant, actually is a measure of spatial curvature in the general relativistic treatment.

The formula for acceleration:

$$\ddot{a} = -\frac{4\pi}{3}G(\rho + 3P)a, \qquad (3)$$

which is a simple algebraic consequence of (1) and (2), is worthy of separate note. It is actually the form that arises most directly in the general relativistic treatment. The combination $\rho + 3P$ (rather than simply ρ) is the effective gravitating mass. This conclusion holds true for other situations. It is the main general relativistic correction in the theory of stellar structure, for example.

2.2 Solution for Popular Forms of Matter (mostly $k = 0$)

Equations (1) and (2) provide us with two first-order differential equations for the three variables a, ρ, ,p. Given an equation of state relating ρ and p, they will be complete. The core of classic cosmology is the study of model Universes, which are solutions of (1) and (2) with various equations of state.

2.2.1 Non-relativistic matter

For nonrelativistic matter, one has
$$P \ll \rho$$
Indeed, $P \sim nm\langle v^2 \rangle$ while the energy density is $\rho c^2 \sim nmc^2$, where n is the number density (and we have momentarily departed from our usual procedure of setting the speed of light $c = 1$.) From Equation 2, one then obtains
$$d(\rho a^3) = 0$$
or
$$\rho a^3 = \text{constant} \equiv \epsilon$$
This is easy to interpret; it is simply conservation of mass. If we now assume $k = 0$, then we obtain from Equation 1
$$a = (6\pi G \epsilon)^{1/3} t^{2/3}$$
that is,
$$a \propto t^{2/3}$$
This is a very important result, since it is probably true that the energy density of the Universe has been dominated by non-relativistic matter, and that k has been negligible, over most of its history – including the present.

2.2.2 Relativistic ideal gas

Another simple case, of great interest as an approximation to the Universe earlier in its history, is when the energy density is dominated by relativistic particles. In this "radiation dominated" regime we have
$$P = \frac{1}{3}\rho,$$
as may be familiar from the theory of black body radiation (the theory of the photon gas, which is always relativistic.) Using equation (2) we then find
$$d(\rho a^4) = 0$$
or
$$\rho a^4 = \text{constant} \equiv \lambda$$

This equation expresses the conservation of photons. It differs from what we found in the non-relativistic case, $\rho a^3 = $ constant. One way of thinking about the difference is that as the Universe expands, the wavelength of a photon increases along with it. Since the energy of a photon is inversely proportional to its wavelength, this accounts for the extra power of a. Another way of thinking about it, in line with (2), is that in expanding the photon gas does work against pressure – the pressure of the surrounding photon gas!

If we now assume $k = 0$, then using (1) we obtain

$$a = \left(\frac{32\pi}{3}G\lambda\right)^{1/4} t^{1/2}$$

that is, $a \propto t^{1/2}$. The expansion is slower than for the nonrelativistic case. This is because it fights against pressure, or alternatively (in line with $\rho_{\text{effective}} = \rho + 3P$) because the pressure exerts a gravitational pull.

2.2.3 Empty space

The case of empty space is very important in the inflationary universe model. If we assume that empty space is still four dimensional and Lorentz invariant locally, then the energy-momentum tensor is restricted to be of the form

$$\langle T_{\mu\nu}\rangle = \Lambda g_{\mu\nu}$$

with Λ a constant independent of time and space. This is equivalent to the rather bizarre equation of state

$$P = -\rho = -\Lambda$$

Here Λ is essentially the "cosmological constant". (Strictly speaking, the cosmological constant is actually $8\pi G\Lambda$.) An equation of state of this form arises when some scalar field has a constant, uniform expectation value – since such an expectation value does not violate Lorentz invariance. We shall assume $\Lambda \geq 0$.

The constancy of Λ is consistent with the dynamical equations – from Equation 2, we have $\dot{\rho} = 0$.

From Equation 1, and setting $k = 0$, we obtain

$$a \propto e^{Ht}$$

where

$$H = \sqrt{\frac{8\pi}{3}G\Lambda}$$

The Universe expands exponentially, yet the density is constant! This is possible, because work is done on the expanding fluid by the *negative* pressure from outside.

Note that from Equation 3, the acceleration is positive – gravity acts as a *repulsive* force – for positive Λ!

2.2.4 Curvature matter

It is sometimes a helpful heuristic to consider the effect of non-zero k – spatial curvature – as being due to a special fictitious sort of "curvature matter". The behavior of the curvature term can be mimicked by matter with the equation of state

$$P = -\frac{1}{3}\rho$$

which, from Equation 3, implies no acceleration.

We conclude this section concerning solutions of the dynamical equations for popular forms of matter by noting the " Hierarchy of Persistence "

$$\left(\frac{\dot{a}}{a}\right)^2 = \frac{8\pi}{3}G\left(\frac{\Omega_{\text{rad}}}{a^4} + \frac{\Omega_{\text{n.r.}}}{a^3} + \frac{\Omega_{\text{curv}}}{a^2} + \Omega_{\text{vac}}\right),$$

in the effects of different types of matter subjected only to the direct effects of expansion. Here the different Ωs are constants, equal to the values of the respective types of energy density – radiation, non-relativistic matter, curvature, cosmological term – at $a = 1$.

Two important implication of this hierarchy is that radiation will tend to be most important to the dynamics in the small early Universe, whereas non-relativistic matter, and if they are present curvature and vacuum energy density, take over later, as the Universe gets big.

(I should emphasize that the preceding equation, which assumes the density in different types of matter changes only due to expansion, is oversimplified – for two reasons. First, when the Universe is hot species that are nonrelativistic at low temperatures become relativistic, and species that are barely present at all – e.g. positrons, muons – become common. Second, the non-gravitational interactions of matter cannot always be neglected, particularly if there are phase transitions.)

2.3 Traditional "Dust" Models for Today's Universe

Much of classic cosmology is concerned with the analysis of models with $\Omega_{\text{rad}} = \Omega_{\text{vac}} = 0$. From a modern point of view, $\Omega_{\text{curv}} \approx 0$ and $\Omega_{\text{rad}} \approx 0$ are easiest to motivate for the present universe. Ω_{rad} is likely to be small for the reasons mentioned above, while Ω_{curv} is observed to be $\lesssim 1$ and is predicted to be very small in inflationary universe models. Similarly Ω_{vac} is observed to be $\lesssim 1$, but there is no good understanding of why it should be small. Indeed our standard model of particle physics, and its extension to unified models, features scalar Higgs fields that acquire enormous vacuum expectation values. On the basis of dimensional analysis, one might have expected $\Omega_{\text{vac}} \geq 10^{26}$ for electroweak symmetry breaking – and even much bigger for grand unified symmetry breaking. This is by far excluded by experiment; it would lead to a very short-lived Universe. That is the notorious cosmological constant problem, one of the main challenges facing present-day theoretical physics.

Anyway it is instructive to review the classical models, and we now turn to this.

2.3.1 $k = 0$: Einstein-de Sitter model

This is the leading candidate to describe the present Universe. It has the elegance of simplicity: no radiation, no curvature, and no cosmological term. In a field where the database is sparse and shaky (though improving) there is every reason to concentrate on the simplest model that is not obviously wrong.

As we shall see shortly, the Hubble parameter

$$H \equiv \frac{\dot{a}}{a} = \sqrt{\frac{8\pi}{3}\rho G} \qquad (4)$$

can be measured directly. From H, we can define a critical density which leads to spatial flatness ($k = 0$):

$$\rho_{\text{cr}} = \frac{3H^2}{8\pi G}$$

If in addition we suppose that the Universe has been dominated by nonrelativistic matter through most of its history, we deduce from $a \propto t^{2/3}$ a relationship between the Hubble parameter and the age of the Universe, to wit:

$$t_o = \frac{2}{3H}$$

This should be compared with the naive result, $t_o = 1/H$, that one would obtain by extrapolating $H = \dot{a}/a$ as a constant to $a = 0$. The difference reflects the fact that gravitational attraction has slowed down the expansions, so it must have been faster in the past.

For later use let us record the solution of the dynamical equation for expansion:

$$a = (6\pi G \rho_{\text{cr}} a^3)^{1/3} t_o^{2/3} \qquad (5)$$

Note that the term in parentheses is constant.

The deceleration parameter

$$q_o \equiv -\frac{\ddot{a}a}{\dot{a}^2} \qquad (6)$$

is also observable in principle. It is given by

$$q_o = \frac{\rho + 3P}{2\rho_{\text{cr}}}$$

(using Equation 3). It is equal to $1/2$ in the present model, where $P \ll \rho = \rho_{\text{cr}}$.

2.3.2 $k > 0$: open universe

For the case $k > 0$, the dynamical equations can be solved in the parametric form

$$t = \frac{GM}{k^{3/2}} \left(\sinh \eta \sqrt{k} - \eta \sqrt{k} \right)$$

$$a = \frac{GM}{k} \left(\cosh \eta \sqrt{k} - 1 \right)$$

where we have defined

$$M = \frac{4\pi}{3} \rho a^3$$

and

$$d\eta \equiv \frac{dt}{a}$$

Note that, in the limit $k \to 0$, these equations for t and a revert to Equation 5.

The Hubble parameter for the open universe becomes

$$H \equiv \frac{\dot{a}}{a} = \frac{k^{3/2}}{GM} \frac{\sinh \eta \sqrt{k}}{\left(\cosh \eta \sqrt{k} - 1 \right)^2}$$

Again, from H one obtains the age of the universe t_o using the relation between t and η. One finds
$$t_o > \frac{2}{3H}$$
and $H \to 1/t$ as $t \to \infty$.

The deceleration parameter can also be deduced easily:
$$q_o \equiv -\frac{\ddot{a}a}{\dot{a}^2} = \frac{1}{2}\left(\cosh\frac{\eta\sqrt{k}}{2}\right)^{-2} < \frac{1}{2}$$

2.3.3 $k < 0$: closed universe

For a closed universe, we obtain
$$t = \frac{GM}{|k|^{3/2}}\left(\eta\sqrt{|k|} - \sin\eta\sqrt{|k|}\right)$$
$$a = \frac{GM}{|k|}\left(1 - \cos\eta\sqrt{|k|}\right)$$

Note again that in the limit $k \to 0$, these equations for t and a revert to Equation 5. This illustrates the general principle that curvature becomes negligible early on, consistent with the hierarchy of persistence. Also note that these equations can be obtained from the equations for the open universe by substituting $k \to -k$.

For the case $k < 0$, the solution to the dynamical equations becomes singular near
$$\eta_{\text{end}} = \frac{2\pi}{\sqrt{|k|}}$$
because here we have $a = 0$ (Big Crunch).

For the closed universe, the Hubble and the deceleration parameters become
$$H \equiv \frac{\dot{a}}{a} = \frac{k^{3/2}}{GM}\frac{\sin\eta\sqrt{|k|}}{\left(1 - \cos\eta\sqrt{|k|}\right)^2}$$
$$q_o \equiv -\frac{\ddot{a}a}{\dot{a}^2} = \frac{1}{2}\left(\cos\frac{\eta\sqrt{|k|}}{2}\right)^{-2} > \frac{1}{2}$$

In this case, the relation between the Hubble parameter and the present age of the Universe is
$$t_o < \frac{2}{3H}$$
while the total lifetime of the Universe is given by
$$t_{\text{end}} = \frac{2\pi}{H}\frac{q_o}{(2q_o - 1)^{3/2}} = \frac{2\pi GM}{|k|^{3/2}}$$

2.4 Geometric (global) Content of the Models

Without derivations, I would now like to describe for you the global structure of space-time in the models we have been discussing. We have constructed solutions appropriate to the line elements

$$ds^2 = dt^2 - a(t)^2 d\mathcal{S}^2$$

where the spatial geometry is of one of three types:

2.4.1 $k = 0$: flat space

Then the line element is:

$$d\mathcal{S}^2 = dr^2 + r^2 \left(d\theta^2 + \sin^2\theta d\phi^2\right) = dx_1^2 + dx_2^2 + dx_3^2$$

with

$$x_1 = r \sin\theta \cos\phi$$
$$x_2 = r \sin\theta \sin\phi$$
$$x_3 = r \cos\theta$$

which corresponds to the geometry of flat three-dimensional Euclidean space.

2.4.2 $k > 0$: hyperbolic space

Here we have

$$d\mathcal{S}^2 = d\rho^2 + \frac{1}{k}\sinh^2\left(\rho\sqrt{k}\right)\left(d\theta^2 + \sin^2\theta d\phi^2\right) = -dx_4^2 + dx_1^2 + dx_2^2 + dx_3^2$$

with

$$x_1 = \frac{1}{\sqrt{k}} \sinh\left(\rho\sqrt{k}\right) \cos\theta$$
$$x_2 = \frac{1}{\sqrt{k}} \sinh\left(\rho\sqrt{k}\right) \sin\theta \cos\phi$$
$$x_3 = \frac{1}{\sqrt{k}} \sinh\left(\rho\sqrt{k}\right) \sin\theta \sin\phi$$
$$x_4 = \frac{1}{\sqrt{k}} \cosh\left(\rho\sqrt{k}\right)$$

which refers to a line element in a flat Minkowski space. Of course, the universe at a given value of t is described by three variables; it has the geometry of the surface

$$x_1^2 + x_2^2 + x_3^2 - x_4^2 = -\frac{1}{k}$$

in four-dimensional Minkowski space.

The space is clearly isotropic (invariant to rotations). It is also homogeneous (every point equivalent to every other point) because every point on the surface can be reached by a *boost* from any given point, and the boost is a symmetry (replacing translations)!

2.4.3 $k < 0$: spherical space

Here we have:

$$dS^2 = d\rho^2 + \frac{1}{|k|}\sin^2\left(\rho\sqrt{|k|}\right)\left(d\theta^2 + \sin^2\theta d\phi^2\right) = dx_4^2 + dx_1^2 + dx_2^2 + dx_3^2$$

with

$$x_1 = \frac{1}{\sqrt{|k|}}\sin\left(\rho\sqrt{|k|}\right)\cos\theta$$

$$x_2 = \frac{1}{\sqrt{|k|}}\sin\left(\rho\sqrt{|k|}\right)\sin\theta\cos\phi$$

$$x_3 = \frac{1}{\sqrt{|k|}}\sin\left(\rho\sqrt{|k|}\right)\sin\theta\sin\phi$$

$$x_4 = \frac{1}{\sqrt{|k|}}\cos\left(\rho\sqrt{|k|}\right)$$

which refers to a line element in a flat four-dimensional Euclidean space. The space is homogeneous and isotropic and invariant to rotations in the four dimensions. The space can be visualized as the surface

$$x_1^2 + x_2^2 + x_3^2 + x_4^2 = \frac{1}{|k|}$$

in four-dimensional Euclidean space.

2.5 Red-Shift

To analyze the propagation of light in our model universes, it is useful to introduce the variable $d\eta = dt/a$. In terms of this variable, the line element takes the form

$$ds^2 = dt^2 - a(t)^2 dS^2 = a(\eta)^2\{d\eta^2 - dS^2\}.$$

This line element is therefore said to be *conformal* to the simple line element we have in flat Minkowski space – it is a numerical function times that metric. This feature is extremely useful in problems involving light or causality, because Maxwell's equations, light cones, and causal relationships are insensitive to conformal transformations. We can therefore perform calculations in standard flat space, and then use the results in the model universes.

Let me take a moment to justify this approach. Maxwell's equations can be obtained from the action

$$S = \frac{1}{16\pi}\int F_{ik}F_{jl}g^{ij}g^{kl}\sqrt{-g}\, d^4x$$

where

$$F_{ik} = A_{k,i} - A_{i,k} = \frac{\partial A_k}{\partial x^i} - \frac{\partial A_i}{\partial x^k}$$

is the electromagnetic field strength and g is the determinant of the metric. This action, and therefore Maxwell's equations, are unchanged if the metric g_{ij} is multiplied by a function a^2: the determinant contains four factors of a^2, and its square root two, just cancelling the two factors of a^2 coming from the g^{ij}.

For flat space, plane waves are valid solutions:

$$A_\mu \propto \varepsilon_\mu \exp\{i(\omega_o \eta - k_o \chi)\} \equiv \varepsilon_\mu e^{i\phi}$$

where $d\chi = dx/a$. We can express the frequency ω_o in terms of the local or source frequencies as follows:

$$\omega_o = a_{local}\omega_{local} = a_{sent}\omega_{sent}$$

with the locally observed frequency given by:

$$\omega_{local} = \frac{d\phi}{dt} = \omega_o \frac{d\eta}{dt} = \frac{\omega_o}{a_{local}}$$

This can be rewritten in terms of a red-shift parameter z, and the source frequency, as follows:

$$\omega_{local} = \frac{a_{sent}}{a_{local}}\omega_{sent} \equiv \omega_{sent}\frac{1}{1+z}$$

where

$$\frac{a_{local}}{a_{sent}} = 1 + z.$$

We can also write

$$\frac{k_{local}}{k_{sent}} = \frac{1}{1+z}.$$

This relationship expresses formally the fact we mentioned before, that the wavelength of the photon stretches along with the expansion of the Universe. Of course, the frequency must adjust accordingly. This is an intuitively appealing way to understand the redshift. (By the way for non-relativistic matter it is still true that the deBroglie wavelength stretches – the momentum goes down – as the Universe expands, but the implications of this for the energy are a little different.)

The red-shift can be measured spectroscopically, of course, if a spectral line of known intrinsic frequency can be identified.

For a past time t that is not too early, we can usefully expand $a(t)$:

$$\begin{aligned}a(t) &= a_{local} + \dot{a}_{local}(t - t_{local}) + \frac{1}{2}\ddot{a}_{local}(t - t_{local})^2 + \cdots \\ &= a_{local}\{1 + H(t - t_{local}) - \frac{1}{2}q_o H^2(t - t_{local})^2 + \cdots\}\end{aligned}$$

where we have used Equations 4 and 6 to get the second equation. To first order we have:

$$\frac{a_{local}}{a(t)} = 1 + z \simeq 1 + H\Delta t = 1 + HR$$

where $R = \Delta t = t_{local} - t$ is the distance between source and local observer. We consequently obtain

$$H \simeq \frac{z}{R}$$

This shows shows how H can be measured, if one can determine the redshift for an object whose distance is known. This allows the overall framework to be tested, too, since one must get consistent values of H for different objects.

2.6 Classical Cosmology

Almost seventy years since Hubble's original determination (now though to be off by an order of magnitude), there are still significant uncertainties regarding the value of H. The great difficulty is in determining R reliably.

The Hubble parameter is observed to be

$$H \equiv 100h \text{ km/sec/Mpc}$$

where h is a fudge factor, with most probably

$$\frac{1}{2} \lesssim h \lesssim 1.$$

Here the astronomical unit of distance 1 Mpc (Megaparsec) is 3×10^{24} cm.

Using this value, the Einstein-de Sitter model yields:

$$\rho_{cr} = h^2 \, 1.88 \times 10^{-29} \text{ g/cm}^3$$

and

$$t_o = \frac{2}{3H} = \frac{2}{3h} \times 10^{10} \text{ years.}$$

These values are certainly of the right order of magnitude, which in itself is already amazing – the big bang picture really works! The best estimates of the age of the universe are obtained from measurements of natural radioactivity, stellar evolution (globular clusters) and white dwarf cooling (age of the galactic disk), and are found to be in the range

$$1.2 \times 10^{10} \text{ years} \lesssim t_o \lesssim 1.7 \times 10^{10} \text{ years.}$$

There are also looser (but not inconsistent) bounds based on observations of luminosity vs. red-shifts of stars and number vs. red-shifts. All these estimates of the age of the universe are consistent with the Einstein-de Sitter model, but suggest that h must be small.

The density of the universe is a complex and somewhat controversial subject. It is pretty firmly established that the fraction due to baryons is low:

$$0.02 \lesssim \frac{\rho_{\text{baryons}} h^2}{\rho_{cr}} \lesssim 0.03$$

and that there must be more matter in some other form (dark matter).

2.6.1 Exercise: mass of the visible universe

Now as an amusing and instructive exercise, I shall compute for you the mass of the visible universe (assuming the Einstein-deSitter model).

Because the position of a propagating wave is constrained by the relationship

$$\frac{r}{a} \pm \eta = \text{constant,}$$

looking back from the present η_o to a past η, the maximum distance r corresponds simply to

$$\frac{r}{a} = \eta_0 - \eta$$

Thus the increase in volume in any interval $d\eta$ is

$$dV = 4\pi r^2 \, dr = 4\pi a^3 (\eta_0 - \eta)^2 \, d\eta.$$

For non-relativistic matter, the density is proportional to a^{-3}. Consequently,

$$\rho = \frac{a_0^3}{a^3} \rho_0$$

The mass of the visible universe is now obtained as

$$\begin{aligned} M_{\text{vis}} &= \int \rho \, dV \\ &= 4\pi a_0^3 \rho_0 \int_0^{\eta_0} (\eta_0 - \eta)^2 \, d\eta \\ &= \frac{4\pi}{3} a_0^3 \rho_0 \eta_0^3 \end{aligned}$$

For the Einstein-de Sitter Model we know that $a \propto t^{2/3}$, and because $d\eta = dt/a$, we deduce that

$$\eta = 3 \frac{t_0^{2/3}}{a_0} t^{1/3}$$

and therefore

$$a_0 \eta_0 = 3 t_0$$

Using the value of the density and of the Hubble parameter for the Einstein-de Sitter model, we get

$$\begin{aligned} M_{\text{vis}} &= \frac{4\pi}{3} (3t_0)^3 \left(\frac{3H^2}{8\pi G} \right) \\ &= \frac{6t_0}{G}. \end{aligned}$$

Note the *linear* dependence on t_0. Using the values quoted above, we find that the mass of the visible universe is

$$M_{\text{vis}} \approx 10^{57} \text{ g} \approx 6 \times 10^{23} M_\odot$$

There are roughly an Avogadro's number of stars in the visible universe.

2.6.2 Exercise: Olbers' paradox

Why is the sky black at night? Assume an eternal and static universe of constant density, filled with a density N of standardized candles per unit *physical* volume. Then the total luminosity due to sources out to a distance R would be given by

$$L = \int dL \propto \int \frac{N}{R^2} 4\pi R^2 \, dR \to \infty$$

Let us now proceed using the Einstein-de Sitter Model. The luminosity of a single object is proportional to

$$\frac{1}{(1+z)}\frac{1}{(1+z)}\frac{1}{(\eta_o - \eta)^2}$$

The first factor comes from the red-shift, the second from the transmission rate and the third from the spread of surface areas. Then we have

$$dL \propto \frac{N}{a^3}\frac{1}{(1+z)^2}\frac{1}{(\eta_o - \eta)^2} a^3 (\eta - \eta_o)^2 \, d\eta$$

$$\propto \frac{N}{(1+z)^2} d\eta.$$

From the last example, and remembering that $a \propto t^{2/3}$ in the Einstein-de Sitter model, we get

$$\eta = 3\frac{t_o^{3/2}}{a_o} t^{1/3} = \frac{3t_o}{a_o}\left(\frac{a}{a_o}\right)^{1/2}$$

From Section 2.5, we know that $a_o/a = 1 + z$, therefore

$$d\eta \propto \frac{t_o}{a_o}(1+z)^{-3/2} dz$$

giving finally

$$dL \propto N\frac{t_o}{a_o}\frac{dz}{(1+z)^{7/2}}$$

The luminosity does grow linearly with z at first, as Olbers found, but then it cuts off *very* rapidly. Note the divergence for $t_o \to \infty$.

3 MATTER IN THE UNIVERSE

The overall philosophy in tackling the problem of matter in the Universe is to extrapolate back to times as early as 10^{-43} s using known physical laws and the simplest possible model universes – that is, the homogeneous and isotropic models we discussed above, based on Equations (1) and (2). We have more-or-less direct evidence that this extrapolation works well back to 10^{-2} s, based on the success of cosmological calculations of the synthesis of light nuclei. The justification for continuing this approach to much earlier times rests on a variety of arguments. The most important are: First, that inhomogeneities in the Universe tend to *grow* with time, due to gravitational instability. Thus the approximate isotropy and homogeneity of the galaxy distribution, and even more the observed high degree of isotropy of the microwave background (which tracks physical conditions about a hundred thousand years after the big bang, when the radiation gas ceased to interact significantly with matter) indicate that early in its history the Universe must have been homogeneous and isotropic to even higher accuracy. Second, that the standard model of particle physics becomes weakly coupled and thus *easier* to understand at ultra-high energies and densities (asymptotic freedom), so that it is possible to predict with some degree of credibility the behavior of matter in the early moments of the big bang.

3.1 Basic Concepts : Equilibrium and Freeze-out

Let there be several species of particles, labelled by an index i. Our previous treatment gives us the dynamical equations

$$\rho = \sum_i \rho_i$$

$$\dot\rho_i = -3(P_i + \rho_i)\frac{\dot a}{a}$$

and

$$\dot a^2 - \frac{8\pi}{3} G\rho a^2 = k$$

Now we will take a closer, more realistic look at the nameless, faceless, "particles".

An appropriate description of matter can often be given in terms of approximately ideal (weakly interacting) gases in equilibrium at temperature T. We have then the particle density

$$dn_i(p) = g_i \left(e^{\frac{1}{T}(E_p - \mu_i)} \mp 1\right)^{-1} \frac{d^3p}{(2\pi)^3}$$

where $n_i(p)$ is the density of particles of momentum p, with $E_p = (p^2 + m_i^2)^{1/2}$. The constant μ_i is the chemical potential and g_i is the number of degrees of freedom (dof). The negative sign is for Bose-Einstein statistics, while the positive sign is for Fermi-Dirac statistics. Using the relationship

$$\rho_i = \int E_p\, dn_i$$

for $T \gg m$ and $\mu_i = 0$, we find

$$\rho_i = g_i \frac{\pi^2}{30} T^4 \times \begin{cases} 1 & \text{for bosons} \\ \frac{7}{8} & \text{for fermions} \end{cases}$$

We also have:

$$P_i = \frac{1}{3}\rho_i$$

$$s_i^{\text{Bose}} = \frac{2\pi^2}{45} T^3 = \frac{8}{7} s_i^{\text{Fermi}} \qquad \text{per dof}$$

where s_i is the entropy density. When T falls below m_i, the equilibrium density of that species of particles falls precipitously. Unless there is an asymmetry ($\mu_i \neq 0$) or a lack of equilibrium (inert relic), the species ceases to be relevant for the dynamical evolution.

Thus we arrive at

$$\frac{\dot a}{a} = \sqrt{\frac{8\pi}{3} G\rho} = \sqrt{\frac{4\pi^3}{45} G\, N(T)^{1/2} T^2}$$

where

$$N(T) = \sum_{\text{Bose } m_i < T} g_i^{\text{Bose}} + \frac{7}{8} \sum_{\text{Fermi } m_i < T} g_i^{\text{Fermi}}$$

And since $d(\rho a^4) = 0$, then $\rho \propto a^{-4}$. But we just showed that $\rho \propto T^4$. Using $s \propto T^3$, we find that $s \propto a^{-3}$ or

$$d(sa^3) = 0$$

This is really the fundamental equation. It expresses that the expansion is adiabatic. This will be true to a very good approximation as long as either the expansion rate is much slower than the rate of interactions (then equilibrium is maintained) or the interactions may be neglected altogether. Since the rate of expansion of the Universe is very slow (for the precise sense of this, see immediately below) equilibrium is normally maintained – but its failures are especially interesting!

One simple way equilibrium may fail, is that some species may "freeze out". When the relevant rate of interactions

$$\Gamma = \sum_j \langle \sigma n_j(p) v \rangle$$

is smaller than the rate of expansion H, then the equilibrium generally fails to materialize. Thus for example if baryon number is violated, the equilibrium number of protons in the Universe is absurdly small – there is a Boltzmann factor of $\exp -(1\text{Gev}/3° K) \sim \exp -10^{12}$ – yet there is no immediate cause for alarm.

3.1.1 Examples for N(T)

First consider $m_e < T < m_\mu$. Then we have

$$N(T) = 2 + \frac{7}{8} \cdot 4 + \frac{7}{8} \cdot 3 \cdot 2 = \frac{43}{4}$$

since we must count 2 degrees of freedom for the photon, 4 for the electron and 2 for each of the 3 species of neutrino. If we include *all* standard model particles we have $N(T) = 427/4$, and for the SU(5) grand unified model we would have $N \geq 643/4$.

3.1.2 Example for equilibrium: renormalizable interaction at high temperature

A fundamental consideration concerns how the kinds of interactions we know how to handle in quantum field theory, that is renormalizable interactions, behave in the early universe. A renormalizable interaction is one governed by a dimensionless coupling (in particle physics units, with $\hbar = c = 1$) – call it g. At high temperatures, where the particles' mass may be neglected, their number density behaves as $n \propto T^3$. Also we have seen that

$$H \propto T^2 \sqrt{G} = T^2/M_{\text{Planck}}$$

Here M_{Planck} is the Planck mass, equal to about 10^{19}Gev. The cross section behaves as

$$\sigma \propto \frac{g^4}{T^2},$$

as one can see from a simple Feynman graph calculation. (The $1/T^2$ can be deduced simply by dimensional analysis. In the Feynman simplest graph one has two particles in, and two out, connected by a propagator – there are two vertices, giving g^2 in amplitude and g^4 in cross-section.) The competition between the rate of interaction and the rate of expansion is governed by the comparison of $n\sigma$ and H; the condition for equilibrium is

$$n\sigma \propto \frac{g^4}{T} \gtrsim \frac{t^2}{M_{\text{Planck}}}$$

or
$$g^4 \gtrsim \frac{T}{M_{\text{Planck}}}$$

This condition is satisfied even by rather weak interactions at low temperature, but becomes increasingly difficult as the temperature approaches the Planck mass.

Note that chemical equilibrium (common μ) is more demanding than kinetic equilibrium (common T). For only those interactions that change one species into another are effective in enforcing chemical equilibrium, while reactions of all kinds help maintain kinetic equilibrium.

Finally, equilibrium may fail spectacularly at a first order phase transition. Recall that a first order phase transition is one which requires a discontinuous change in state, and that under such conditions supercooling can occur. In the context of particle physics, this means that energy is locked up in the form of Higgs fields which have the "wrong" expectation value – trapped at a local minimum of the free energy, which is not the global minimum. If this condition persists for a long time, while more ordinary forms of energy are diluted by expansion of the universe, one can be fall into a vacuum-dominated phase. This leads to exponential expansion, as we have seen above.

3.2 Nucleosynthesis

3.2.1 General scenario

stage I : At high enough T ($T \gtrsim$ few MeV), neutrons and protons maintain equilibrium via the weak process
$$n + \nu \leftrightarrow p + e$$

(and also $n + e^+ \leftrightarrow p + \bar{\nu}$). During this stage, n/p $\equiv \exp(-\Delta m/T)$, where n/p is the neutron to proton density ratio and $\Delta m = M_n - M_p$.

stage II : The interactions drop out of equilibrium for $T \lesssim 1$ MeV, with n/p frozen out at $\sim 1/6$, and subject to neutron decay.

stage III : Nucleosynthesis commences when deuterium is stabilized. This requires that the back-reaction $D + \gamma \to p + n$ be slower than the synthesizing reaction. This occurs at $T \simeq 0.1$ MeV .

stage IV : Subsequently, reactions such as:
$$n + p \to D + \gamma$$
$$D + n \to T + \gamma$$
$$D + D \to T + p$$
$$D + D \to {}^4\text{He} + \gamma$$
$$D + D \to {}^3\text{He} + n$$
$$D + p \to {}^3\text{He} + \gamma$$
$$D + T \to {}^4\text{He} + n$$

$$D + {}^3He \rightarrow {}^4He + p$$

$$^3He + {}^3He \rightarrow {}^4He + 2p$$

$$\vdots$$

can occur rapidly.

To a first approximation, these reactions just funnel all available neutrons into ^{4}He. Because of gaps in nuclear stability at $A = 5$ and 8, cosmic nucleosynthesis proceeds no further — otherwise, we would all be men of iron!

The helium density is conventionally parametrized by the quantity (mass fraction)

$$Y \equiv \frac{4\,^4\text{He}}{p + 4\,^4\text{He}} = \frac{2\frac{n}{p}}{1 + \frac{n}{p}}$$

Its value is observed to be about 1/4, corresponding to $n/p = 1/7$.

In addition to ^{4}He, there is also some left-over D and ^{3}He ($\sim 10^{-5}$), as well as a very small but significant amount of ^{7}Li ($\sim 10^{-10}$) from the processes:

$$^4He + T \rightarrow {}^7Li + \gamma$$

$$^4He + {}^3He \rightarrow {}^7Be + \gamma$$

$$^7Be + n \rightarrow {}^7Li + p$$

3.2.2 Some estimates of key numbers

The value of the n/p ratio at freeze-out is obtained by requiring the interaction rate of $\nu + n \rightarrow e + p$ to be comparable with the expansion rate H. Using

$$\sigma \propto G_F^2 M_n T$$

and remembering that $H \propto T^2/M_{\text{Planck}}$, and $n \propto T^3$, where n is the density of neutrinos, we obtain the condition

$$T_*^2 \simeq \left(G_F^2 M_n M_{\text{Planck}}\right)^{-1} \simeq 10^{-9} \text{ GeV}^2$$

Therefore $T_* \simeq .1$ MeV. Since this is reasonably close to the n-p mass difference (and gets closer when various numerical factors are treated more carefully), we learn that the predicted n/p ratio is sensitive to the details of cosmology. Since this freeze-out occurred at $t \simeq M_{\text{Planck}}/T_*^2 \simeq 10$ s, we have a limited but clear view into the condition of the Universe at such early times.

Furthermore, with $n_B/n_\gamma \simeq 10^{-10}$ (at present, as well in past times), the freeze-out temperature for the reaction p+n→D+γ leads to the condition that $10^{10} \exp(2.2 \text{ MeV}/T) \simeq 1$, yielding $T \simeq 0.1$ MeV. This freeze-out corresponds to a time $t \simeq 10^3$ s. The synthesis of light nuclei commenced at that time.

3.2.3 Comparison with observations

The predicted abundances of the light nuclei ^{1}H, ^{2}H, ^{3}He, ^{4}He, ^{7}Li are sensitive to the baryon number density. This is because some of them (all but ^{1}H and ^{4}He) are produced in rare side-chains or are the results of incomplete burning – and higher density affects these (it means more complete burning, and also more side chains populated).

There is an excellent fit to the data for

$$2.6 \times 10^{-10} \leq \frac{n_B}{n_\gamma} \leq 4.6 \times 10^{-10}$$

corresponding to

$$0.02 h^{-2} \leq \Omega_B \equiv \frac{\rho_B}{\rho_{cr}} \leq 0.03 h^{-2}$$

Especially remarkable is the experimental result for ^{7}Li, the abundance of which is nearly at the minimum value possible for any value of n_B/n_γ.

3.2.4 Sensitivity to perturbations

The cosmic nucleosynthesis predictions are sensitive to various conceivable perturbations of the model parameters. For example, if the number of neutrino increases, or if G is bigger, or if there is a gravity-wave background, the expansion will occur faster. This would mean a poorer approximation to equilibrium at the crucial temperature, that is more n relative to p (it is harder for the ratio to n/p to decrease as fast as Boltzmann wants!). This would result in an increase in the density of ^{4}He. So would a longer lifetime of the neutron. Inhomogeneities would favor larger ^{7}Li abundance, since it is minimal for the assumed density.

For a long time cosmology provided the most stringent constraint on the number of neutrino species. Some cosmologists said it couldn't be more than 4 – others said 6, or 3, or even 2! All agreed, though, it was some small number. Direct measurements are now, of course, available – and confirm that the number is small: 3.

3.3 Generation of Microwave Background

The reaction

$$p + e \leftrightarrow H + \gamma$$

stops fairly abruptly when $E_\gamma \lesssim 13$ eV, corresponding to $z \sim 10^3$ (because $\bar{E}_\gamma$ is now $\simeq 3$ K $\simeq 3 \times 10^{-4}$ eV).

After almost 30 years of measurements, the microwave background still looks to be perfectly isotropic – a featureless haze. The measurements are now at the 10^{-5} level over a wide range of angular scales. If the large scale structure of the Universe resulted from the growth of small fluctuations due to gravitational instability – certainly the most straightforward and attractive conjecture – then these seeds must leave their imprint on the microwave background as inhomogeneities at the few $\times 10^{-6}$ level. There is a new sky – the sky of the early Universe – waiting to be seen.

3.4 Neutrino Background

Neutrinos freeze-out at $\gtrsim 2$ MeV. At that point, they have the same temperature as photons. However, photons are enriched by e^+e^- annihilation at $T \lesssim 1$ MeV. Because $s_i \propto g_i T^3$, using entropy conservation we get

$$\frac{T_\nu}{T_\gamma} = \left(\frac{g_\gamma}{g_{e+\gamma}}\right)^{1/3} = \left(\frac{2}{\frac{7}{8} \cdot 4 + 2}\right)^{1/3} = \left(\frac{4}{11}\right)^{1/3}.$$

Using $T_\gamma = 2.7$ K, we then obtain a prediction for the present neutrino background temperature of $T_\nu = 1.9$ K.

There is no very good idea for how to detect this radiation. Fame and glory await the lucky boy or girl who invents one.

3.5 Another Interesting Idea

Strictly speaking, photons do not decouple completely and abruptly when protons and electrons combine into electrically neutral atoms. Hogan and Rees have suggested that the 21 cm. absorption lines of neutral hydrogen gas clouds at extremely high red shift ($z \sim 10$) may detectable. They are back-lighted by the long-wavelength tail of the microwave background radiation. The microwave background radiation provides only a small component of the total radiation in the sky at these wavelengths, but the motion of the clouds would lead to angle-dependent spectral features that could be isolated. If this could be accomplished, it would give important information on a very interesting period in the evolution of the Universe that is difficult to study otherwise. This is the "dark ages", when matter was entirely neutral (before radiation from quasars, stars, and so forth re-ionized it.)

This idea deserves more attention, in my opinion.

3.6 Baryogenesis

Now I'd like to shift gears, and instead of suggesting experiments for the future talk about a big one that's already done.

The asymmetry between matter and antimatter in the Universe is, at least superficially, quite striking. Although the laws of physics are very nearly the symmetrical between matter and antimatter (by CP, or better yet CPT), the actual Universe is not. There are baryons, but very few antibaryons, in the Universe. This asymmetry begs for physical explanation.

Can a universe which is symmetric between matter and antimatter early on evolve into one that is not? There are three necessary conditions for baryogenesis:

1. B-nonconserving interactions,

2. violation of both C and CP,

3. significant departure from thermal equilibrium while the B-violating interactions were effective.

The first is clear. The second is also pretty obvious, since either C or CP, if they were strictly valid, would enable us to relate whatever processes which produce matter to equally likely ones which produce antimatter. The third is a bit more subtle. One way of seeing it is to realize that CPT (which should be rigorously valid) guarantees the equality of particle and antiparticle masses, and thus of their abundance in thermal equilibrium according to Boltzmann. Thus to deviate from this equality requires that one deviate from thermal equilibrium.

3.6.1 first realization

The observed asymmetry is really quite small! The proper quantitative measure of it is $n_B/n_\gamma \sim 10^{-10}$. For this is a quantity which is approximately conserved during the expansion of the universe, after baryon-number violating interactions have frozen out. And the physical meaning of its numerical value is that at temperatures $T \geq 1$Gev, when there were plenty of anti-quarks as well as quarks in equilibrium – comparable to photons – the number of quarks was greater than the number of antiquarks by a part in ten billion or so. The "striking" asymmetry we see today results from this tiny intrinsic asymmetry, as the vast bulk of quarks and antiquarks annihilate in pairs, leaving a tiny relic remainder of unpaired quarks.

3.6.2 second realization

The number of baryons B is almost certainly *not* conserved.

The strongest argument for this conclusion is based on the compelling idea that the gauge theories of the strong and electroweak interactions, which beg to be unified, are in fact unified. The most concrete and detailed idea along these lines is the Georgi-Glashow $SU(5)$ model. In it the three 'colors' of QCD's $SU(3)$ gauge group are united with the two of weak $SU(2)$ and the color-sensitive but not color-changing electromagnetic interaction, into the five-color theory of $SU(5)$. This is not the place to elaborate on the details of this model, but I would like to emphasize that it has two highly non-trivial and remarkable virtues. First, it rationalizes the peculiar haphazard hypercharge assignments that one has to make in the ordinary $SU(2)$: 1/6 for left-handed quarks, 2/3 and -1/3 for the right-handed quarks, 1/2 for the left-handed leptons, -1 for the right-handed leptons ... . In $SU(5)$ this scattered mess comes out in a simple and compelling way. Second, it predicts a numerical relationship between the strong, weak, and electromagnetic couplings that is very close to being satisfied in Nature. The logic of this is that by demanding that the three effective couplings – running, according to asymptotic freedom, as a function of energy scale – eventually converge on one value, one fits three quantities (the couplings) with two parameters (the value of the unified coupling, and the energy scale at which it is achieved.) The unification is achieved, in the simplest versions of $SU(5)$, at energy scales of approximately 10^{15}Gev, a fact we shall use below.

Within $SU(5)$, or most modifications of it that preserve its virtues, there are baryon-number violating interactions. These arise from the additional color-changing gauge bosons that are in $SU(5)$ but not in $SU(3)$ or $SU(2)$, and change strong into weak colors. These bosons are predicted to be extremely heavy (10^{15}Gev), and the baryon-number violating interactions they induce, including nucleon decay, are correspondingly tiny at present. In the very early universe, however, when plenty of energy was available to produce these bosons, their effects would be bigger and more obvious.

Unfortunately the greatest potential virtue of $SU(5)$: that it would predict the rate and branching fractions for baryon-number violating nucleon decays, has not come to fruition. Heroic experiments to search for the predicted decay have only succeeded in placing limits, at the level $\tau_{\text{nucleon}} \gtrsim 10^{31}$yrs. This pretty firmly rules out the simplest model implementations of $SU(5)$. Nevertheless I think we must take the hint that baryon number violation occurs, and is significant in the early universe, most seriously. There are modifications of $SU(5)$ which preserve its remarkable successes while allowing the proton to survive a bit longer. (The running of the couplings can be affected by the existence of very massive particles, of which we would have no hint at low energies. This turns out to change the energy scale of unification, and thus the mass of the heavy vector bosons responsible for nucleon decay and thus the lifetime of the proton, dramatically; but changes the rationalization of hypercharge not at all and the unification of couplings only slightly.) By the way, although I cannot do justice to the subject here I would like to call your attention to the fact that precision measurements of the strong, electromagnetic, and weak couplings at low energies can provide most important clues to discriminate among potential unification schemes.

3.6.3 third realization

C and CP are violated.

This story should be familiar to you. Of course CP is only violated a little bit, but according to the first realization that's really just what we want.

3.6.4 fourth realization

Thermal equilibrium is not accurately maintained during a period when baryon number is significantly violated.

Indeed, recall that we found a condition $g^4/T \geq M_{\text{Planck}}$ for thermal equilibrium. At the temperatures of interest for grand unification, $T \geq 10^{15}$Gev, this condition is barely satisfied if at all.

Putting together all these realizations, one realizes at last that there is no significant conceptual barrier to understanding how a universe asymmetric between matter and antimatter might have arisen from a symmetric starting point. Unfortunately, at this time we do not have all the ingredients necessary to formulate a compelling quantitative calculation. (We would need a detailed theory of CP violation at ultra-high energies, a sure list of the available particles, ...)

[Between the time these lectures were given and the preparation of these notes, some new ideas have come forth, which suggest that significant baryon number violation occurs at temperatures as low as 1 Tev, even within the standard model, and that in some special circumstances these might generate the observed baryon asymmetry. I believe the first of these conclusions is firm; but the second needs further work. In any case, the basic ingredients necessary for baryogenesis, as discussed above, remain the same, and interactions at the unification scale still fulfill them naturally.]

EXPERIMENTAL ASPECTS OF HADRON COLLIDER PHYSICS

Allan G. Clark

Département de physique nucléaire et corpusculaire
Université de Genève, Geneva, Switzerland
UGVA-DPNC 1991/142

ABSTRACT

These notes summarise a series of lectures given at the 1990 NATO Advanced Institute on Techniques and Concepts in High Energy Physics. Data on hard parton-parton collisions at hadron colliders are presented, and are compared with expectations from the Standard Model of strong and electroweak interactions. In making these comparisons, an attempt is made to identify existing theoretical and experimental limitations or uncertainties, and to give a perspective of future relevant measurements at existing hadron colliders. Also emphasized is the complementarity of measurements testing the validity of the Standard Model using (e^+e^-) colliders.

1. INTRODUCTION

Hadron colliders are and will continue to be an important tool to probe the validity of the Standard Model at large centre-of-mass energy ($\sqrt{s}$), that is at small interaction distances. In this role two parton constituents of the nucleons collide to produce high-transverse momentum (p_T) outgoing particles. Typical strong, electromagnetic, or electroweak processes, (fig. 1.1a) can be interpreted in terms of couplings between quarks and leptons with gauge bosons, or between gauge bosons themselves. Typical examples are shown in figure 1.1b.

The successful operation of the CERN $\bar{p}p$ Collider at $\sqrt{s}$ = 630 GeV, a factor 10 higher in energy than the CERN ISR, enabled the identification by the UA1 and UA2 experiments of the W and Z bosons, and the identification of hadron jets (see fig. 1.1) resulting from parton fragmentation. Using data collected between 1981 and 1985, a quantitative comparison with expectations of the Standard Model was possible. Within rather substantial experimental and theoretical uncertainties, the agreement was excellent. However, with improved theoretical evaluations and a commitment to test the Standard Model with improved precision, both detector and machine upgrades were necessary.

Both machine and detector developments have changed the emphasis of physics studies at existing hadron colliders since 1987 (see section 2).

i) The CERN $\bar{p}p$ Collider has been upgraded to provide peak luminosities

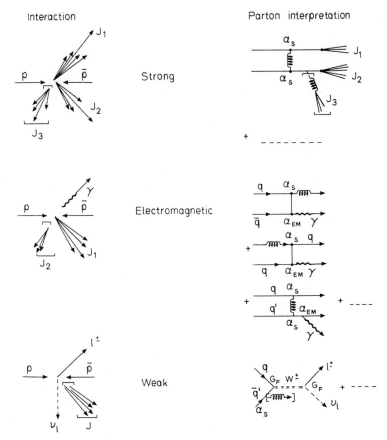

Figure 1.1. Typical strong, weak and electromagnetic processes in hard ($\bar{p}p$) collisions, and their interpretation in terms of typical parton–parton scattering diagrams. Not only valence quarks but also sea quarks and gluons in the nucleon may contribute to the hard scattering processes.

of $\mathcal{L} \simeq 3.10^{30}$ cm^{-2}s^{-1} (a factor 10 higher than previously). In 1990 a further factor 2 in peak luminosity is expected, during a final run of the UA2 detector. In the light of existing data, and to benefit from the improved luminosity, the UA2 detector has also been upgraded.

ii) The Fermilab Collider with the CDF detector has also been operated at similar peak luminosities since 1987, but with $\sqrt{s}$ = 1.8 TeV, extending the accessible distance scale proportionally. In 1991, a second major detector (called DØ) will be commissioned.

iii) The UA1 and UA2 detectors at CERN, and CDF at Fermilab, have each collected data corresponding to an integrated luminosity L = ∫ $\mathcal{L}$dt $\gtrsim$ 5 pb^{-1} (see table 1). An ambitious Fermilab program in 1991 and beyond aims to further augment the collected data by 2 orders of magnitude.

At the same time, however, the LEP and SLC e$^+$e$^-$ machines have started operating, also changing the focus of hadron collider activities. For processes accessible to (e$^+$e$^-$) machines (for example rare Z^0-decays, and the fragmentation of quarks from Z-decay) the advantages are obvious. Studies of hard collisions at hadron colliders suffer from :

i) the lack of a defined parton-parton centre-of-mass, and an unknown parton-parton centre-of-mass energy $\sqrt{\hat{s}}$ that is less than that of the total ($\bar{p}p$) energy ($\sqrt{s}$),

ii) the existence of soft collision products that are not associated with the basic scattering process and which may obscure or limit the study of hard collisions, and

iii) the need to identify relatively rare hard collisions of interest from a much larger rate of soft collisions (for example W or Z production occurs in approximately 1 in 10^7 $\bar{p}p$ interactions).

Though item (i) above complicates the interpretation of data, it does permit a range of effective $\sqrt{\hat{s}}$ values making colliders valuable exploratory machines, especially since high $\sqrt{s}$ values are technically more easily obtained at hadron colliders than at (e$^+$e$^-$) machines. Important validity checks of the Standard Model presently accessible only to $\bar{p}p$ colliders include (to give only a few examples)

i) a comparison of data with QCD predictions at very high jet transverse momenta. As discussed in section 3 the CDF jet-jet mass spectrum now extends to 1 TeV, that is one order of magnitude beyond the LEP mass scale,

ii) the identification of a heavy top quark and a measurement of the top quark mass,

iii) the identification of the triple electroweak gauge coupling (in particular WWγ coupling), and

iv) detailed W mass measurements prior to LEP2 operation.

Exploratory searches for new particle production (for example additional high mass gauge bosons or supersymmetric particles beyond the mass range of Z-decays where LEP data is more accurate) are also best made at hadron colliders. Finally a "grey" area exists where measurements from (e$^+$e$^-$) machines and hadron colliders are competitive; an example of this which takes advantage of the spectrometric capabilities of the UA1 and CDF detectors is the identification of B-mesons and the subsequent study of both their production and decay properties. The high rate of B-production at hadron colliders compensates for the complete event structure available in clean e$^+$e$^-$ events.

In section 2, a very brief summary is given of the CERN and Fermilab $\bar{p}p$ Colliders, with a subsequent discussion of criteria motivating the design of general purpose detectors (UA1, UA2, CDF, DØ) at these machines. Two such detectors (UA2, CDF) are described in detail, as examples.

Sections 3 and 4 describe the production characteristics of jets in parton-parton collisions, and compare these data with QCD predictions. Within quoted experimental and theoretical uncertainties, agreement is good. More sensitive QCD tests can be made using multijet final states, and by using photon or W/Z probes with associated high-p_T jet production. Again the agreement with QCD expectations is excellent.

Section 5 introduces the Standard Glashow-Weinberg-Salam (GWS) [1] model of electroweak interactions, and summarises the role of existing hadron colliders in testing the model. Following a technical discussion on the identification and momentum (energy) measurement accuracy of leptons ($e^\pm$, $\mu^\pm$, $\tau^\pm$ and ν) recent measurements (updated to include data available since these lectures) of the W and Z mass and width are compared with Standard Model expectations. The derived Standard Model parameters are compared with independent determinations (including LEP), and with an evaluation of $\sin^2\theta_W$ from the forward-backward Z-decay asymmetry at CDF. Lepton universality measurements from UA1 are presented and the identification of (W/Z $\rightarrow$ $\bar{q}q$) decay modes by the UA2 Collaboration is discussed.

Sections 6 and 7 are devoted to the identification and study of heavy-quark production and decay at hadron colliders. Searches for the top quark have proved elusive, and both direct and indirect top mass limits are described. Perspectives for future top quark identification at the Fermilab Collider are also considered. Existing measurements of b-quark production, however, are in excellent agreement with Standard Model expectations, and perspectives for b-quark studies at hadron colliders are very encouraging.

Sections 3 to 7 emphasise "standard" tests of the gauge theories of strong and electroweak interactions, and within experimental and theoretical uncertainties, no deviations are measured. However, many aspects of the Standard electroweak model remain untested, for example triple gauge coupling, the identification of the Higgs boson etc. Further, for a consistent gauge theory of strong and electroweak interactions, extensions to the existing models must exist at some mass scale. Section 8 is devoted to a summary of the status of (selected) searches for deviations or extensions to the Standard Model, and for so-far inaccessible tests of the Standard Model.

2. MACHINE AND DETECTOR CONSIDERATIONS

2.1 Machine considerations

Proton-antiproton collisions were originally proposed by Rubbia et al. [2] in 1976 as a realistic method to produce and subsequently detect the W and Z bosons. Using QCD evolved structure functions that are measured at lower energies, and assuming the validity of the Standard Model, the cross-section for W (Z) production is estimated [3] to be $\sigma \sim 6.5$ (2.0) nb at $\sqrt{s}$ = 630 GeV. For a mean luminosity $\langle \mathcal{L} \rangle \sim 5.10^{28} \text{cm}^{-2} \text{s}^{-1}$, as available in the first years of running, this implies a production rate

$$\bar{p}p \rightarrow W^\pm + X \; ; \; W^\pm \rightarrow e^\pm \nu_e \; : \; \sim 3.0 \text{ events/day}$$
$$\bar{p}p \rightarrow Z + X \; ; \; Z \rightarrow e^+ e^- \; : \; \sim 0.3 \text{ events/day}.$$

Such luminosities are only obtainable if high-intensity, monoenergetic and highly collimated bunches of protons and antiprotons are able to circulate and collide. High-intensity $\bar{p}$ bunches can be obtained by "electron cooling" [4], or by "stochastic cooling" [5]. The latter method was chosen for technical reasons. The technique measures fluctuations about the mean of a disordered beam over a fixed time interval and generates from this corrections to the beam trajectory. After repeated applications the momentum spread of the $\bar{p}$ bunches is reduced.

The CERN Collider is described in detail by Evans et al. and the Fermilab Collider with upgrade plans in ref. [6]. In these notes, we briefly describe the Fermilab Collider. At Fermilab, collisions are obtained (fig. 2.1) as follows:

i) Protons of 8 GeV are injected in the main ring and after acceleration to 150 GeV are focussed into a beryllium target to produce produce $\bar{p}$'s of approximately 8 GeV/c peak momentum. The $\bar{p}$'s are then focussed and collected into a "collector ring" at a rate of approximately 2.10^{10} $\bar{p}$'s hour^{-1} and cooled to produce a typical monoenergetic $\bar{p}$ stack of approximately 2.10^{11} particles.

ii) In successive main ring cycles, six proton and six antiproton bunches of typically 6.10^{10} (3.10^{10}) particles each are transferred to the main ring, accelerated to 150 GeV and then transferred to the Tevatron ring. Finally the bunches are simultaneously accelerated to 0.9 TeV in the Tevatron.

iii) In the vicinity of $\bar{p}p$ interaction regions, special low-β quadrupoles are used to reduce the bunch size and thus to increase the luminosity.

Table 1a) summarises the enormous improvement of machine performance since the initial CERN Collider operation in 1981, and table 1b) indicates future improvements of nearly two orders of magnitude planned for the Fermilab Collider.

Table 1a. Existing machine performance

Year	Machine	$\int \mathcal{L}dt$ pb^{-1}	$\int \mathcal{L}dt$ (UA2,CDF) pb^{-1}	$\mathcal{L}_{MAX}$ cm^{-2}s^{-1}	$\sqrt{s}$ (GeV)
1981 – 83		~ 0.18		~ $.10^{29}$	546
1984 – 85	CERN	~ 1.05		~ 4.10^{29}	630
1987 – 89	CERN	~ 9.7	7.8	~ 3.10^{30}	630
1987	Fermilab	~ 0.1	0.027	~ 2.10^{29}	1800
1988 – 89	Fermilab	~ 9.05	4.7	~ 2.10^{30}	1800

Table 1b. Expected future machine performance

Year	Machine	$\int \mathcal{L}dt$ pb^{-1}	$\mathcal{L}_{MAX}$ cm^{-2}s^{-1}	$\sqrt{s}$ (GeV)
1990	CERN	~ 7	~ 5.10^{30}	630
1991	Fermilab	> 25	~ 8.10^{30}	1800
1993	Fermilab	~ 100	> 10^{31}	2000
1995	Fermilab	~ 500	> 5.10^{31}	2000

2.2 Detector design considerations

A total of eight experiments have collected data at the CERN $\bar{p}p$ Collider; in two experiments (UA1 [7] and UA2 [8]) general-purpose detectors were used to study hard collision processes. At the Fermilab Collider, five experiments have been approved; of these CDF [9] and from 1991 DØ [10] are sophisticated general-purpose detectors.

The total cross-section for $\bar{p}p$ interactions ($\sigma_{tot} = 61.9 \pm 1.5$ mb at $\sqrt{s} = 630$ GeV and 72.1 ± 3.3 mb at $\sqrt{s} = 1.8$ TeV [11, 12]) implies an interaction rate for $\mathcal{L} = 5.10^{30} \text{cm}^{-2}\text{s}^{-1}$ of ~ 350 KHz. The cross-section seen by these experiments is dominantly the soft inelastic non-diffractive component (σ_{inel} ~ 42 mb at $\sqrt{s} = 1.8$ TeV). These "minimum bias" events are selected by requiring at least 1 particle in each of forward and backward small-angle scintillators (see figure 2.9 for the CDF experiment). These soft collisions have a particle density which falls exponentially with increasing p_T, with $<p_T>$ ~ 0.4 - 0.5 GeV as expected for a proton of radius ~ 1 fm. Since the p_T is limited, the longitudinal particle density is distributed according to phase space :

$$\frac{d^3p}{E} = d\phi \cdot \frac{dp_T^2}{2} \cdot \frac{dp_L}{E} = d\phi \cdot dp_T^2 \cdot dy \qquad (2.1)$$

$$y = \tfrac{1}{2} \ln \left[(E + p_L)/(E + p_L)\right]$$

implying an uniform particle density in rapidity (y). With increasing $\sqrt{s}$, both the magnitude and width in y of this distribution varies as ($\ell n s$) implying a $\ell^2 n s$ dependence on multiplicity. Figures 2.2 and 2.3 show the p_T and rapidity distributions for soft collisions at CDF, and figure 2.4 shows the approximate ($\ell n s$) dependence of the particle multiplicity at $\eta = 0$. Because of a logarithmic extension of the rapidity plateau, this implies a $(\ell n s)^2$ behaviour of the cross-section. The spectator distribution from hard collisions can be assumed to be the same as for minimum bias events at energy ($\sqrt{s} - \sqrt{\hat{s}}$). In practice some differences can be expected from differing colour recombinations of the spectator quarks, and the particle multiplicity underlying 2-jet events from parton-parton scattering (or W/Z events) is measured to be approximately twice (1.5 times for W/Z events) that of minimum bias events.

The soft particle distributions are of considerable theoretical interest and a review of CERN data by Ward [14] is suggested. However, discussion in following sections is restricted to experimental difficulties encountered as a result of soft collisions. Any general purpose detector should be able to select hard parton production processes in the presence of this background.

i) Figure 2.5 shows the lowest-order graphs for parton-parton scattering. Outgoing (coloured) parton fragmentation into (colourless) hadrons is a soft process, with limited p_T relative to the jet axis and with a uniform particle density in rapidity with respect to the jet axis. Consequently, above some E_T threshold (E_T ~ 10 GeV) determined by the soft collision background, partons can be reliably associated with jets of collimated particles, and can be detected above some E_T threshold using a granular calorimeter of reasonable spatial and energy resolution and reliable cell-to-cell calibration (to maintain a sharp threshold).

ii) The production cross-section of a number of typical hard scattering processes as a function of the p_T of the hard scattering are shown in figure 2.6 [15 - 17]. The dominant backgrounds to W and Z production (with sub-sequent leptonic or hadronic decay) are the QCD processes noted above. Therefore, as noted in sections 3 through 5, measurements of W and Z production using hadronic decay modes are

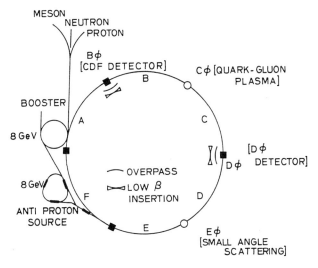

Figure 2.1. Overall site layout of machines used at Fermilab (main ring and Tevatron, antiproton cooling ring and booster ring) to supply protons and antiprotons to the Tevatron.

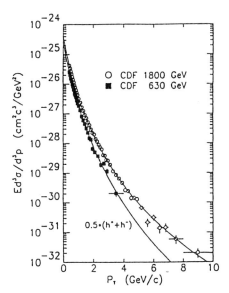

Figure 2.2. The p_T dependence of the inclusive cross-section for charged particle production at the CDF experiment. For $p_T \lesssim 1$ GeV the variation with p_T is exponential. For $p_T > 1$ GeV, there is a power law behaviour
$E\, d^3\sigma/dp^3 = A\, p_0^n/(p_T + p_0)^n$; $n \sim 8.26 \pm 0.08$
$p_0 \simeq 1.3$ GeV/c

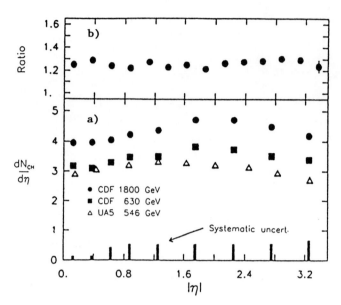

Figure 2.3. a) $dN_{ch}/d\eta$ measured by CDF at 1800 and 630 GeV, and by UA5 [13] at 546 GeV. b) The ratio of $dN_{ch}/d\eta$ at 1800 GeV to that at 630 GeV.

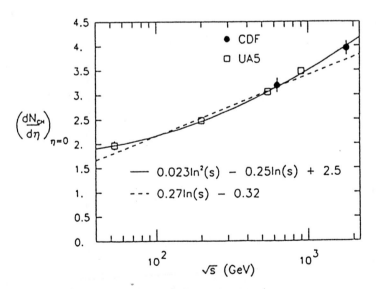

Figure 2.4. $dN_{ch}/d\eta$ at $\eta = 0$ as a function of $\sqrt{s}$ measured by UA5 [13,14] and CDF. The curves are the result of fits. An approximate but not exact ($\ell n s$) behaviour is measured.

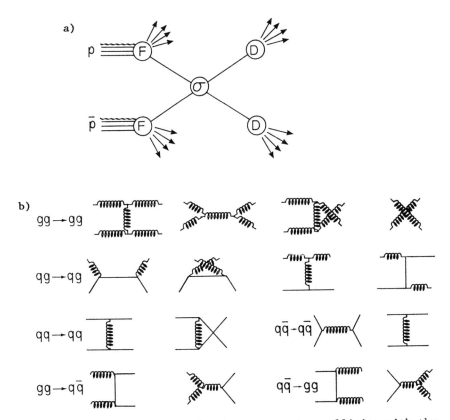

Figure 2.5. a) Schematic of a hard parton-parton collision with the subsequent fragmentation of quarks into jets. F represents the parton distribution in the nucleon, $\hat{\sigma}$ the elementary parton-parton scattering cross-sections, and D the fragmentation function of the outgoing partons. b) Lowest order parton-parton scattering diagrams contributing to the jet cross-section in $\bar{p}p$ collisions.

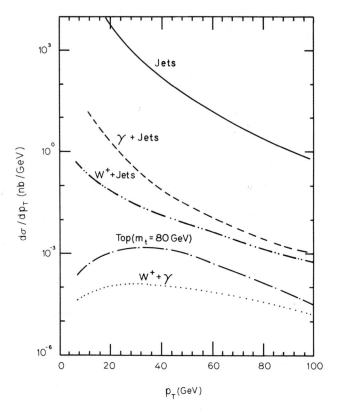

Figure 2.6. The production cross-section for a variety of $\bar{p}p$ physics processes as a function of p_t of the hard scattering system. As expected jet production dominates by several orders of magnitude. From Shapiro [15] using the PYTHIA [16] or Papageno [17] generators.

difficult and of limited accuracy. To identify clean W or Z samples, it is necessary to use their leptonic decay modes. To identify a W or Z signal using an inclusive single particle lepton signature ($e^{\pm}$, $\mu^{\pm}$, $\tau^{\pm}$) an (inclusive) rejection against fake electrons from QCD jets of $\gtrsim 10^4$ is required. For searches of the t-quark via its production and decay (for example) :

$$\bar{p}p \to \bar{t}t + x \; ; \quad t \to W^+b \quad ; \; \bar{t} \to W^-\bar{b}$$
$$\phantom{\bar{p}p \to \bar{t}t + x \; ; \quad t \to W^+b} \hookrightarrow \ell^+\nu_\ell \hookrightarrow \bar{q}q' ,$$

a rejection $\sim 10^6$ is required unless some additional characteristic event topology is used (see section 6).

iii) Neutrinos are not detected in the apparatus, and their existence can consequently only be interpreted in terms of identified missing energy or momentum. In a detector measuring every particle α produced,

$$\sum_\alpha \vec{p}_T^{\,\alpha} = 0 \quad \text{and} \quad \sum_\alpha \vec{p}_L^{\,\alpha} = 0. \tag{2.2}$$

In practice neither relation is valid. Because of small angle particle production from spectators which do not leave the vacuum pipe of the machine, $\Sigma \vec{p}_L^{\,\alpha}$ cannot be measured accurately. However, in the transverse plane, spectator partons are of limited p_T, and so the missing transverse momentum (or energy) can ideally be defined by

$$\vec{\not{p}}_T = - \Sigma \vec{p}_T^{\,i}$$

$$\vec{\not{E}}_T = - \Sigma \vec{E}_T^{\,i} \tag{2.3}$$

where the sum over i represents all detected particles. In the case of W-decay, $|\vec{\not{E}}_T|$ can be associated with E_T^ν. Of course in the case of several undetected particles, the interpretation is more complex (for example the existence of noninteracting photinos from postulated supersymmetric particle production, or the decay sequence $W \to \tau \nu_\tau; \tau \to e \nu_e \nu_\tau$). The evaluation of $\not{E}_T$ is affected by detector acceptance, and by detector measurement errors such as the calorimeter granularity and energy resolution, the low energy calorimeter response, false energy depositions such as noise and beam halo, the existence of undetected muons, etc. The existence of a central magnetic field further affects the measurement of $\not{E}_T$, since particles of p_T less than some threshold p_T^0 are not reconstructed. Using well-contained and well measured 2-jet events, figure 2.7 shows the gaussian $\not{E}_T$ - resolution for the upgraded UA2 detector [18]

$$\frac{dN}{d\not{E}_T^2} = \frac{1}{2\sigma^2} \cdot \exp(-\not{E}_T^2/2\sigma^2); \; \sigma = 0.8 \, (\Sigma E_T^i)^{0.4} \tag{2.4}$$

where the sum is over all calorimeter cells, i. Data from the UA1 and CDF experiments are of similar quality.

Figure 2.8 indicates the detection principle for different particle types in the case of the CDF detector [9] (detailed discussions will be given as relevant during these notes).

i) Electron identification requires an isolated electromagnetic cluster, with small hadronic leakage. A track with measured momentum satisfying E/P $\sim$ 1 should face this cluster with a good spatial match to the centroid of both the calorimeter and the signal from a

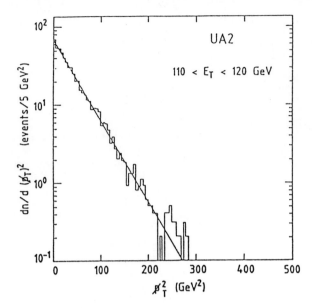

Figure 2.7. The distribution dN/dp_T^2 of well measured 2-jet events satisfying $110 < E_T < 120$ GeV, in the UA2 calorimeter. The distribution is fit using a parametrisation

$$dN/dp_T^2 = \frac{1}{2\sigma^2} \cdot \exp(-p_T^2/2\sigma^2) \; ; \; \sigma = 0.8 \, (\Sigma_i E_T^i)^{0.4}$$

where the sum is over all the calorimeter cells, i.

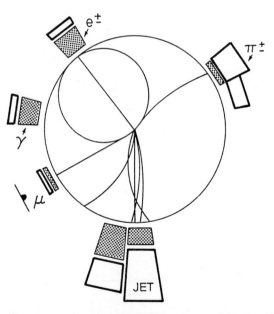

Figure 2.8. A schematic of the identification criteria used at CDF for the detection of electrons, muons, photons, isolated charged hadrons and jets. Shown schematically are the energy depositions for each particle type in the electromagnetic and hadron calorimeters, and the signal from muon chambers. Not shown is the signal from wire chambers embedded at a depth $\sim 5X_0$ inside the electromagnetic calorimeter.

cathode-strip chamber placed at a depth of ~ 5 radiation lengths (X_0) in the electromagnetic calorimeter. In the case of the UA2 detector, which has no central magnetic field, the momentum requirement cannot be applied, and the signal at $5X_0$ depth is replaced by a preshower signal following a 1.5 X_0 lead absorber preceding the calorimeter.

ii) Muon identification requires the reconstruction of a central track which when extrapolated through the calorimeter and a (possibly magnetised) iron absorber, matches well with a track stub in the muon chambers. The momentum determination may result from the measurement of the track momentum using a central magnetic field (CDF, UA1) and/or from a magnetised iron absorber behind the calorimeter (UA1, DØ). The dominant backgrounds are punch-through and decays. The former background is minimised by adequate absorber thickness and preferably sufficient calorimeter granularity to enable the requirement of a minimum-ionising signal in the cell traversed by the muon. The latter is minimised by maintaining a small decay distance, and by having adequate pattern recognition in the central tracking region to reconstruct decay kinks with good efficiency.

iii) Single hadrons are identified as for electrons, but with significant hadronic leakage and with a preshower (or cathode strip) signal as expected for a hadron.

iv) Jets are characterised by an extended cluster of hadronic and electromagnetic energy deposition, corresponding to the superposition of single neutral and charged particles with the appropriate momentum spectrum. Significant fluctuations of the shape, energy deposition and associated charged track multiplicity, exist.

v) "Neutrino" identification is as discussed above.

vi) No experiment is at present well-suited to photon-detection, because of poor rejection against a copious high-p_T π^0-background which fakes direct photon production. For both high-p_T photons and π^0's, an electromagnetic cluster with no associated charged track is expected. Photon identification is discussed in detail in section 4.1.

vii) The only published data from hadron colliders using identified τ's is from the UA1 experiment [19]. They identify the decay $W \to \tau\nu_\tau$ see section 5.1.3) using a requirement of large $\not{E}_T$ together with characteristic high-p_T 1-prong or 3-prong decays

$$\tau^\pm \to \pi^\pm + n\pi^0 + \nu_\tau \quad (\sim 50\ \%) \text{ or}$$

$$\tau^\pm \to 3\pi^\pm + n\pi^0 + \nu_\tau \quad (\sim 14\ \%) \qquad (2.5)$$

Similar identification studies in the CDF detector are in progress.

2.3 The CDF and UA2 detectors [9, 18]

Existing general-purpose collider detectors are of two types : those with or without a central magnetic field. In this section the CDF and UA2 detectors are described, these being "typical" second generation detectors of each type that have collected data since 1987. Descriptions of the DØ detector and the earlier UA1 and UA2 detectors are available elsewhere [10, 7, 8].

Figure 2.9 shows a view of the CDF experiment. Extending outward from the collision region are successively, a vertex TPC (time projection chamber) track detector, a central track detector in an ~ 1.4 tesla solenoidal magnetic field, and finally segmented electromagnetic and hadronic calorimeters. In the central region the calorimeters are of lead/scintillator and iron/scintillator construction, and at less than

30^0 to the beam direction the scintillator is replaced by proportional tubes. Outside the calorimeter are muon detectors in the rapidity region $|\eta| < 0.6$, and muon toroids exist in the forward direction. The central track detector has 84 wire layers arranged in 9 superlayers (of which 4 superlayers have small angle stereo). The r-ϕ resolution is approximately 150-200 μm per hit, enabling, with the additional constraint of beam position, a momentum measurement accuracy of $\delta p_T/p_t \simeq 0.0011 p_T$ (p_T in the GeV/c) in the central region. The momentum scale is known to better than 0.1 % (see section 5.1.4). Figure 2.10 shows the angular segmentation of a quadrant of the CDF calorimeter, and table 2 summarises construction details of these calorimeters. The mean energy scale of the central electromagnetic calorimeter has been maintained to better than ± 0.24 %, with cell-to-cell variations about this calibration scale of an r.m.s. variation ± 1.7 % (see section 5). The energy scale in the forward direction is unfortunately less well determined.

Figure 2.11 shows a schematic of future CDF detector upgrades. In particular, the installation of a silicon microstrip vertex detector is foreseen in 1991 to provide a tag for b-quark decays (section 7). Also, the muon coverage is being extended and in the central rapidity region an additional iron absorber is being added to improve the rejection against punch through (important to maintain low-p_T muon trigger thresholds at high machine luminosity). Finally, a preshower counter is being added immediately following the solenoidal coil (~ 0.9 X_0) to improve photon identification. This latter improvement is important for QCD studies using a photon probe (section 4.1), and for the study of γ production in association with W/Z for identification of the (WWγ) vertex (section 8).

Table 2. CDF Calorimeter Systems

Electromagnetic calorimeters	CEM	PEM	FEM	
Absorber thickness (rℓ)	18.0X_0	19.4X_0	25.7X_0	
Resolution at 50 GeV	2 %	4 %	4 %	
$|\eta|$ coverage	0-1.1	1.1-2.4	2.2-4.2	

Hadronic calorimeters	CHA	WHA	PHA	FHA
Absorber thickness (abs.ℓ)	5.3λ_0	6.0λ_0	6.3λ_0	8.1λ_0
Resolution at 50 GeV	11 %	14 %	20 %	20 %
$|\eta|$ coverage	0-.9	.7-1.3	1.3-2.4	2.3-4.2

```
CEM - central electromagnetic
PEM - plug electromagnetic
FEM - forward electromagnetic
CHA - central hadronic
WHA - wall hadronic
PHA - plug hadronic
FHA - forward hadronic
```

A plan view of the upgraded UA2 detector is shown in figure 2.12. Extending out from the interaction region, the field-free central tracking detector consists of two arrays of silicon pads separated by a cylindrical drift chamber. Outside these arrays two transition radiation detectors are installed, followed by a scintillating fibre detector con-

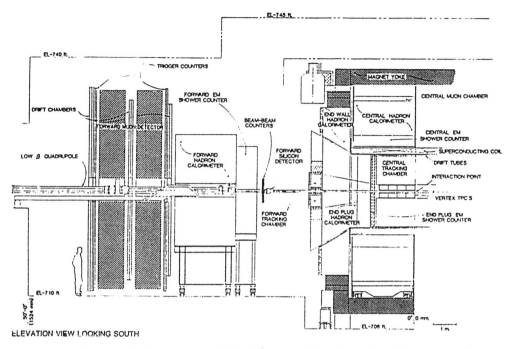

Figure 2.9. View of the CDF experiment at Fermilab.

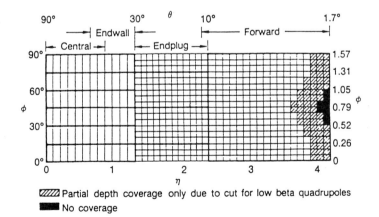

Figure 2.10. One quadrant of the CDF calorimeters, showing their segmentation in η–ϕ and regions of coverage.

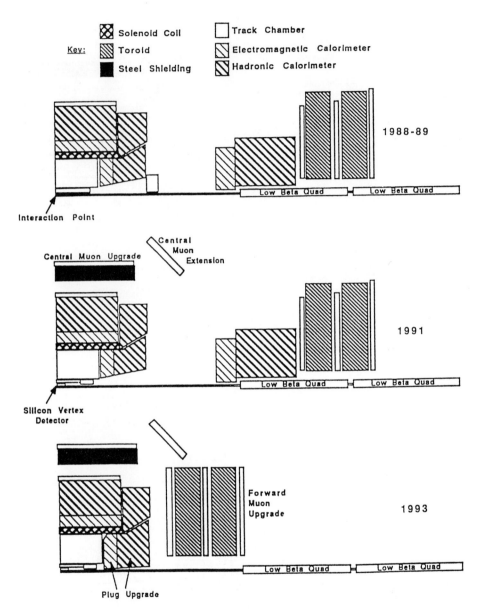

Figure 2.11. Schematic of planned upgrades for the CDF detector at Fermilab.

sisting of 18 tracking layers, a 1.5 X_0 lead converter and a further 6 layers of fibre to provide spatial and pulse-height information on electromagnetic showers produced in the converter. A forward track detector covers the range $1.1 < |\eta| < 1.6$. The hermetic calorimetry extends to $|\eta| = 2.5$ (electromagnetic) and $|\eta| = 3$ (hadronic), with a granularity of approximately $(\phi \times \Delta\eta) = 15^0 \times 0.2$. The construction is lead/scintillator (~ 17 radiation lengths in the central region) for the electromagnetic calorimeter, and iron/scintillator for the hadronic section. The energy scale of the electromagnetic calorimeter is known to better than ± 1 %. The UA2 detector has no muon detection capability. The experiment is expected to complete data taking at the end of 1990.

As noted in section 2.1, each experiment selects inelastic non-diffractive p̄p interactions using a beam-beam counter hodoscope at small angle to the beam line. These signals are usually (some exceptions to this exist) put into coincidence with trigger signals from the calorimetry and/or tracking to provide hard-scattering events of interest.

3. JET PRODUCTION AT HADRON COLLIDERS AND A COMPARISON WITH QCD

This section compares data on the longitudinal and transverse production of high-p_T jets with QCD expectations for hard parton-parton scattering. In general, only the most recent data from UA2 and CDF are presented and for a complete discussion of data collected prior to 1987 reference [20] is suggested.

In addition to the data of this section, QCD comparisons with collider data involving W/Z or direct γ production, and heavy quark production, are discussed (sections 4, 6 and 7). In this section an experimentally motivated discussion of hard parton scattering is followed by the presentation of data for inclusive jet production. Experimental corrections and uncertainties associated with these data are discussed in detail. Also shown are measurements of the 2-jet invariant mass distributions, and characteristic production variables of multi-jet events (n ≥ 3). From measurements of longitudinal jet production, a measurement of the effective parton distribution function can be compared with extrapolations from lower-energy data assuming the validity of QCD.

A discussion of parton fragmentation properties has been excluded from these notes, despite their inclusion in the original lectures. Fragmentation processes are soft, with limited p_T with respect to the jet axis. Consequently the strong coupling (α_s) is large and non-leading order QCD effects are important. The data from UA1 and UA2 [20] and CDF [21] are in good agreement with extrapolations from lower energies. However, the data are less accurate than data from (e^+e^-) experiments because of fluctuations of the spectator quark distribution. Accurate fragmentation studies at LEP remain an important priority.

3.1 The parton model for hard scattering, and QCD non-scaling effects

Figures 1.1 and 2.5 showed a representation of parton-parton scattering in p̄p collisions. The initial partons are distributed in the proton (F), $f_i(x)$ representing the probability that parton type i has fractional momentum between x and (x + dx). Partons may at this stage be sea or valence quarks and antiquarks, or gluons. The hard collision is described by the cross-section, $\hat{\sigma}$, and the outgoing partons fragment into jets (D). The inclusive jet cross-section can be evaluated by summing over all initial and final-state partons

$$E\frac{d^3\sigma}{dp^3} = \sum_{ij}\int dx_1 dx_2 f_i(x_1,Q^2) \cdot f_j(x_2,Q^2) \cdot \delta(\hat{s}+\hat{t}+\hat{u}) \cdot \frac{\hat{s}}{\pi} \sum_k \frac{d\hat{\sigma}}{d\hat{t}} (ij \to k)$$

with $\quad \dfrac{d\hat{\sigma}}{d\hat{t}} = \pi\,\alpha_s^2(Q^2) \cdot \dfrac{|M|^2}{\hat{s}^2}$,

and $\quad \alpha_s(Q^2) = 12\pi/[(33-2n_F)\,\ell n\,(Q^2/\Lambda^2)]$ (3.1)

In equation 3.1, n_F is the number of quark flavours and Λ^2 is a mass scale. The Mandelstam variables are described by $\hat{s}$, $\hat{t}$ and $\hat{u}$, and Q is the momentum transfer of the hard collision. The matrix element $|M|^2$ can be evaluated according to the rules of QCD. It is essential to the evaluation of the cross-section that the hard scattering is factorisable from the parton distributions, and the fragmentation functions (D).

i) In the parton model the nucleon distributions $f_i(x,Q^2)$ are independent of the momentum transfer, Q, of the hard scattering. The quark and antiquark functions are measured at low values of Q^2 in deep inelastic lepton scattering experiments and the gluon contribution is also inferred from the Q-dependence of these data in a model dependent way. Evaluations of the structure functions should in principle be possible using QCD, but the mathematical tools are not available since the evaluation is non-perturbative. In QCD the structure functions must be evolved to the Q^2 of the process considered (using the Altarelli-Parisi equations), by soft gluon emission. For hadron-hadron interactions calculated at leading order, the momentum transfer Q is not uniquely defined and therefore there exists a normalisation uncertainty in the term $(Q/\Lambda)^2$. The evolution is quantitatively sensitive to the scale (Q^2/Λ^2) used in the structure function parametrisations, on whether non-leading order QCD corrections have been made, and on experimental inaccuracies of the (inferred) gluon distribution (especially in the case of $\bar{p}p$ scattering). A detailed review is given in ref. [22].

ii) Figure 2.5 showed the $2 \to 2$ parton sub-processes and table 3 shows the contributions from each sub-process. At Collider energies, gluon-gluon and gluon-quark scattering predominate, and as already stressed the gluon contribution is at present poorly measured.

iii) The sub-processes of figure 2.5 are leading-order (LO) graphs. Several hundred next-to-leading order (NLO) graphs also contribute. As noted above, in LO calculations the value (Q^2/Λ^2) is ambiguous (excepting that the value of $\alpha_s(Q^2)$ should allow perturbative calculations, and Q^2 should be of the order of the parton squared momentum transfer). If all orders are calculated the result should be scale-independent; it follows that if NLO terms are evaluated there should be less sensitivity to the scale choice. It also follows that a choice of scale which minimises the contribution of NLO terms is desirable. However, these non-leading corrections include bremsstrahlung terms that experimentally result in merged jets that must be treated in the same way for theoretical and experimental evaluations.

Experimental uncertainties will be considered in the next sub-section. However, using perturbative LO calculations with existing structure function measurements, the inclusive jet cross-section can be evaluated to within a scale uncertainty of approximately ± 50 %. Figure 3.1 compares a common LO evaluation with published collider data over a factor 30 in $\sqrt{s}$, (~ 7 orders of magnitude in cross-section) and for E_T up to 400 GeV.

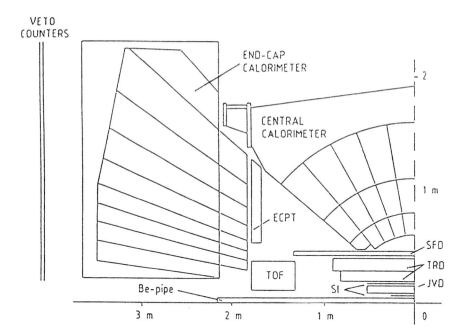

Figure 2.12. Plan view of one quadrant of the upgraded UA2 detector at the CERN Collider.

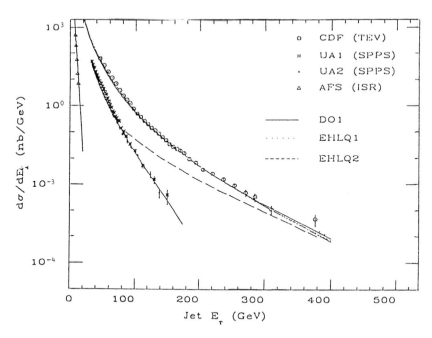

Figure 3.1. The distribution $d\sigma/dE_T$ for published UA1 and AFS data, and (since superceded) data from UA2 and CDF. Superimposed is (absolute scale) a LO QCD calculation using the DO1 or EHLQ structure function paramertisations.

3.2 Measurement of the inclusive jet cross-section at CDF

Any measurement of the inclusive jet cross-section requires an efficient identification of high p_T jets with a unique correspondence to both the parton direction from hard scattering and to the parton energy.

Table 3. 2-to-2 parton sub-processes. $|M|^2$ is the squared invariant matrix element. The colour and spin indices are averaged (summed) over initial (final) states. All partons are assumed massless. The scattering angle in the center-of-mass frame is denoted by θ^*. For $\theta^* = \pi/2$, $t = u = -s/2$.

Process	$\Sigma\|M\|^2$	$\theta^* = \pi/2$
$qq' \to qq'$	$\dfrac{4}{9}\dfrac{s^2+u^2}{t^2}$	2.22
$qq \to qq$	$\dfrac{4}{9}\left[\dfrac{s^2+u^2}{t^2} + \dfrac{s^2+t^2}{u^2}\right] - \dfrac{8}{27}\dfrac{s^2}{ut}$	3.26
$q\bar{q} \to q'\bar{q}'$	$\dfrac{4}{9}\dfrac{t^2+u^2}{s^2}$	0.22
$q\bar{q} \to q\bar{q}$	$\dfrac{4}{9}\left[\dfrac{s^2+u^2}{t^2} + \dfrac{t^2+u^2}{s^2}\right] - \dfrac{8}{27}\dfrac{u^2}{st}$	2.59
$q\bar{q} \to gg$	$\dfrac{32}{27}\dfrac{t^2+u^2}{tu} - \dfrac{8}{3}\dfrac{t^2+u^2}{s^2}$	1.04
$gg \to q\bar{q}$	$\dfrac{1}{6}\dfrac{t^2+u^2}{tu} - \dfrac{3}{8}\dfrac{t^2+u^2}{s^2}$	0.15
$gq \to gq$	$-\dfrac{4}{9}\dfrac{s^2+u^2}{su} + \dfrac{u^2+s^2}{t^2}$	6.11
$gg \to gg$	$\dfrac{9}{2}\left[3 - \dfrac{tu}{s^2} - \dfrac{su}{t^2} - \dfrac{st}{u^2}\right]$	30.4

Parton fragmentation into jets is a soft process of limited p_T ($<p_T> \sim 0.5$ GeV) with respect to the jet axis, and particles of fixed p_T will form a circle in $\eta-\phi$ space of $|\vec{P}_n| \simeq E_T [(\Delta\eta)^2 + (\Delta\phi)^2]^{\frac{1}{2}} = E_T R_c$, with $R_c \sim 1$. With this in mind jet identification algorithms proceed as follows :

i) After making a precluster of contiguous calorimeter towers satisfying $E_T^i > 1$ GeV an initial cluster centroid and direction is evaluated.
ii) Within radius $R_c = 0.7$ about this axis all cells satisfying $E_T > 0.1$ GeV are included and the centroid reevaluated. The procedure is iterated until the solution is stable (R_c can be varied).
iii) If two clusters have ≥ 75 % of common cells, the clusters are merged.

Below $E_T \sim 10$ GeV, the effect of fluctuations on both the jet fragmentation and the underlying event make the jet reconstruction and parton association inefficient. At $E_T = 15$ GeV, $\epsilon > 90$ % with less than 5 % "fake" jet clusters.

Of considerable importance is an accurate jet energy measurement, in particular because of the steep variation in E_T of the inclusive jet cross-section. A number of different energy corrections must be made.

i) Low energy response. Because of the soft fragmentation function for jets, they are composed of a large number of soft particles. Taking advantage of the excellent momentum resolution at CDF, isolated minimum-bias tracks are used in conjunction with test-beam data to map the response between 0.5 and > 230 GeV (fig. 3.2a). To relate this to jet energy corrections, a measurement of the fragmentation function (plus a Monte Carlo simulation thereof) is required. This has been done.

ii) Crack response. Figure 3.2b shows the variation of calorimeter response to azimuthal cracks in the calorimeter (ϕ-cracks) for electrons and pions. Again using a tuned jet fragmentation function, it is possible to correct for these cracks. Similar but less prominent response variations also exist for η-cracks.

iii) Electromagnetic response. Because the CDF calorimeter is not compensating, and jets include both charged hadrons and electrons/photons, the calorimeter response must be separately evaluated for each particle type. Using the measured jet fragmentation, an average correction can be evaluated.

iv) Clustering and the underlying event. Figure 3.2c shows the azimuthal response for typical 2-jet events. Energy should be added to account for the cell threshold used (~ 0.35 GeV) and (unless similarly treated in higher-order QCD evaluations) the cone $R_c = 0.7$ used to define the jet (+ 1.9 GeV). Energy should however be subtracted because of the underlying event ($- 1.52 \pm 0.45$ GeV) contribution within that cone.

Finally a correction relating the parton energy (assumed to be of zero mass) to the deposited calorimeter energy of an ideal detector is required.

The full energy correction, and the associated energy uncertainty from the different known effects, is shown in figure 3.3. In addition to these correction uncertainties, smearing uncertainties, a calibration scale uncertainty ($\leq \pm 1$ %) and an overall luminosity uncertainty (assumed ± 15 % for this analysis for CDF) are added in quadrature. Including all effects, the preliminary CDF cross-section uncertainty is estimated to vary between ± 28 % at $E_T^J \sim 40$ GeV and ± 36 % at $E_T^J \sim 400$ GeV.

The dominant detector improvements required in future calorimetric measurements are homogeneity (no cracks), and good linearity of the low-energy response.

The preliminary inclusive jet cross-section from CDF is shown in figure 3.4, where it is compared with a NLO calculation of Ellis et al. [23], using a range of structure function parametrisations [22]. The agreement with data is excellent, but the data are not yet of sufficient quality to require the NLO contribution. The effect of varying the scale parameter by a factor 2 about E_T changes the NLO fit by only ~ ± 5 %, compared with ~ ± 50 % for LO evaluations. Of more concern is the dependence of the inclusive cross-section on the cone size (R_c), where at present the cross-section variation with cone size differs between data and the QCD evaluation. The UA2 collaboration have recently presented similar data at $\sqrt{s} = 630$ GeV (see later in this section).

As noted in section 8, deviations of these data from QCD fits can be used to put a limit on the mass scale for quark substructure if it exists.

3.3 The dijet mass spectrum

Preliminary data are also available from CDF and UA2 for the dijet

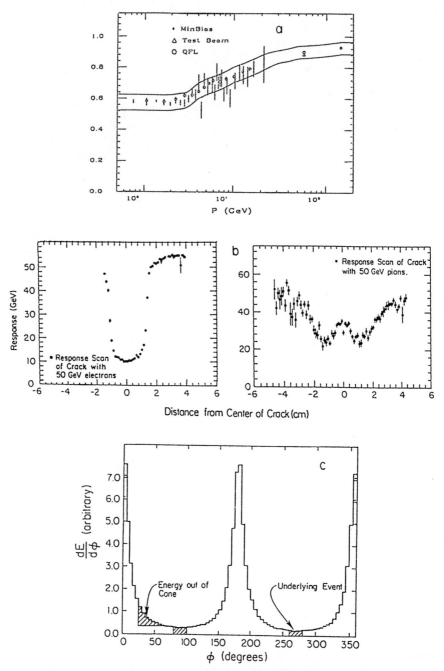

Figure 3.2. a) Calorimeter response at CDF as measured using pions from minimum bias events and from a test beam. Superimposed are response uncertainty limits, and the result of a response parametrisation in the CDF detector simulation. b) Response of the calorimeter to 50 GeV pions and electrons scanned across a ϕ-crack in the CDF central calorimeter.
c) Transverse energy deposited in the central calorimeter as a function of the azimuthal angle from the event thrust axis. The minimum at 90° is taken to be the energy from the underlying event. The cross hatched region at 40° would be roughly the amount of energy lost outside of the clustering cone.

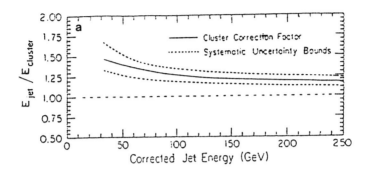

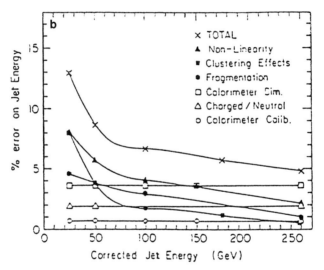

Figure 3.3. a) Jet energy correction expressed as a ratio of corrected to uncorrected jet energies. The curve is the result of a fit to the simulation results. The dashed lines represent the limit of the systematic uncertainty assigned to the correction. b) Contribution to the systematic uncertainty from contributing sources.

invariant mass spectrum in the central region. Preliminary CDF data are compared with LO calculations in figure 3.5 (NLO evaluations are in progress; the calculations are more complex than the inclusive case above because it is an exclusive 2-jet process). There is no evidence for high mass resonance structure, even at masses approaching 1 TeV. The observation of unbiased hadronic W/Z decays in CDF is not possible because of the existing calorimeter resolution and the lack of a suitable trigger. However, UA2 has observed a deviation from QCD behaviour in the dijet mass spectrum that can be ascribed to W/Z production. These data are described in section 5.3.

3.4 Longitudinal distribution of jet production

The angle distribution of jet production is determined by the parton-parton scattering amplitude, and by the longitudinal boost of the parton-parton centre-of-mass. In well balanced jet events for which all jets of the hard parton scattering are measured, the fractional momenta x_1 and x_2 of the incoming partons can be estimated. In the case of a 2-jet system

$$x_1 = [- P_L + (P_L^2 + m^2)^{1/2}]/\sqrt{s}$$

$$x_2 = [+ P_L + (P_L^2 + m^2)^{1/2}]/\sqrt{s} \qquad (3.2)$$

where P_L is the longitudinal momentum and m is the mass of the 2-jet system. Noting from table 3 that the dominant lowest-order scattering terms have an $\sim 1/E$ dependence from gluon exchange, the summation over quark types in eq. (3.1) may be replaced by a single average contribution, with some "average" structure function reflecting both the quark and gluon content of the nucleon and the differing coupling strength. Therefore we can write (following Ellis and Scott [20])

$$\frac{d^3\sigma}{dx_1 dx_2 d\cos\theta*} = \frac{F(x_1)}{x_1} \cdot \frac{F(x_2)}{x_2} \cdot \frac{d\hat{\sigma}}{d\cos\theta*}$$

with $$\frac{d\hat{\sigma}}{d\cos\theta*} = \frac{9}{8} \cdot \frac{\pi\alpha_s^2}{2\hat{s}} \frac{(3 + \cos\theta*)^2}{(1 - \cos\theta*)^2}$$

and $$F(x_i) = G(x_i) + \frac{4}{9}(Q(x_i) + Q(\bar{x}_i)) \qquad (3.3)$$

For 2-jet final states, CDF have recently presented preliminary data for the distribution of $\cos\theta*$, expressed in the form

$$\chi = \frac{1 + \cos\theta*}{1 - \cos\theta*} \sim e^{|\eta_1 - \eta_2|} \qquad (3.4)$$

Figure 3.6 shows for two different dijet mass bins the distribution $dN/d\chi$. Also shown are earlier data from UA1 [24]. The data are in good agreement with leading order QCD calculations and QCD non-scaling effects are clearly evident. UA2 have chosen to present their data on inclusive jet production in different η-bins [25] (fig. 3.7). Again agreement with both LO and NLO QCD is good within the available statistics, but the data are not yet of sufficient accuracy to require NLO QCD evaluations.

Assuming the validity of QCD (at some order), and the dominant gluon exchange terms as outlined above, it is possible to deduce an average structure function. Data are shown in figure 3.8 for UA1 data [20], and

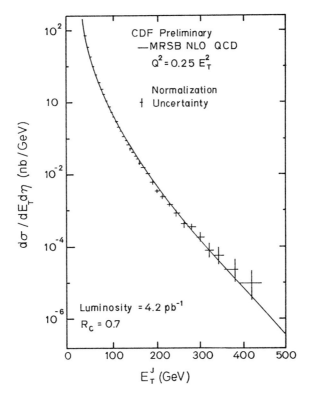

Figure 3.4. The measured inclusive jet cross-section plotted as a function of E_T^J. Superimposed is a fit using NLO QCD calculations of Ellis et al. with the MRSB structure functions. For both the data and QCD evaluation, the jet is defined within a cone $R_c = 0.7$.

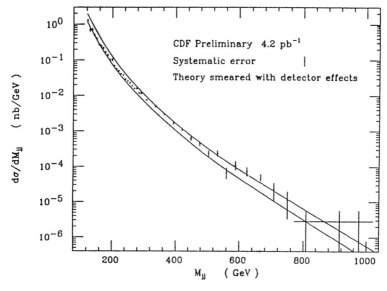

Figure 3.5. Preliminary CDF data for 2-jet cross-section as a function of the 2-jet invariant mass. Superimposed are LO QCD estimates, with the curves representing theoretical uncertainties from the choice of Q^2-scale and structure function parametrisation.

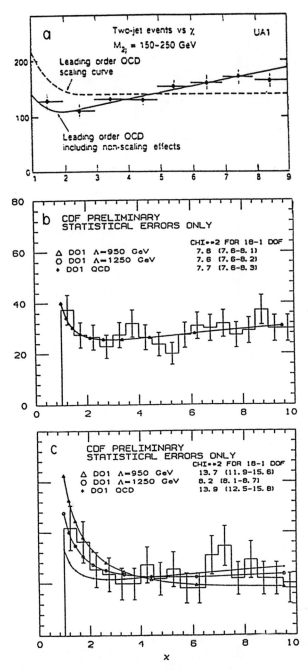

Figure 3.6. a) Distribution of the variable χ as measured by UA1, in the 2-jet mass range $150 \leq m_{JJ} \leq 250$ GeV. b) The distribution of χ from CDF data for $200 \leq m_{JJ} \leq 250$ GeV. c) As for (b), but $m_{JJ} > 550$ GeV.

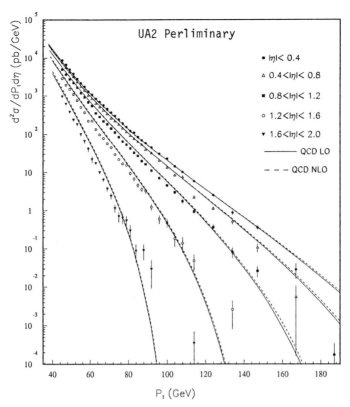

Figure 3.7. Preliminary inclusive jet distribution from UA2, plotted in several rapidity ranges. Superimposed are LO and NLO QCD evaluations.

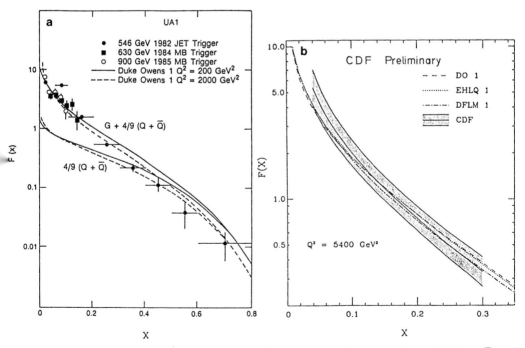

Figure 3.8. a) Points show structure function combination $[G + 4/9(Q + \bar{Q})]$ deduced from UA1 data. b) Preliminary CDF data with superimposed structure function parametrisations evolved to $Q^2 = 5400$ GeV^2.

for CDF (1987 data). Though of limited accuracy, this is an important pedagogic measurement :

i) The effect of structure function evolution at different Q^2-values is evident experimentally and is well verified theoretically.
ii) A substantial gluon contribution is essential.

It is important for CDF to pursue this analysis in different jet-jet mass bins using the most recent data.

If we now consider the same angle, $\cos\theta^*$, in the 3-jet final states (fig. 3.9a), the LO diagram should be characterized by bremsstrahlung off the initial state or final state partons and the $\cos\theta^*$ distribution should be similar for both 2-jet and 3-jet final states. As demonstrated by UA2 [26], this is indeed the case (fig. 3.9b).

3.5 Multi-jet Production

Using the nomenclature of incoming and outgoing partons of figure 3.9a, each of UA1, UA2 and CDF have compared the independent variables for 3-jet production with expectations of both phase space and tree-level QCD matrix elements for the main processes contributing to 3-jet production.

In addition to the fractional momenta of the final state particles (of which two are independent), $\cos\theta^*$ defines the centre-of-mass scattering angle, and Ψ defines the orientation of the 3-jet system with respect to the plane defined by the colliding beams. Writing E_i^{cm} for the outgoing energy of parton i in the centre-of-mass frame

$$x_i = 2E_i^{cm}/m^{3J}, \text{ and}$$

$$x_3 + x_4 + x_5 = 2. \tag{3.5}$$

The plot of x_3 vs. x_4 from CDF (fig. 3.10) shows the fractional sharing of final state parton energies. This figure uses events selected according to the following criteria :

i) $m^{3J} > 250$ GeV, with $E_T^i > 15$ GeV for each outgoing jet,
ii) $x_3 < 0.9$ (this ensures an adequate separation between the jets 2 and 3),
iii) $30^0 < \Psi < 150^0$ (this removes uninteresting initial state bremsstrahlung contributions), and
iv) $|\cos\theta^*| < 0.72$ (this requires that the leading jet is centrally produced).

Superimposed on the projections of this Dalitz plot are tree-level QCD calculations and (uncorrected) phase space expectations. These data, and similar recent data from UA2, are in excellent agreement with the QCD expectations, and the requirement of a non-constant matrix element.

The UA2 Collaboration [25] has measured the inclusive p_T spectrum for exclusive n-jet final states ($3 \leq n \leq 5$). These data, shown in figure 3.11, used the following event selections :

i) $E_T^i > 15$ GeV $i \leq n$ for an n-jet final state,
ii) $E_T^i < 10$ GeV $i > n$ (that is no additional cluster exceeding 10 GeV),
iii) $|\eta_i| < 2.0$ (that is each jet is centrally produced),
iv) all jets are well-measured with no-high E_T jet lost from the

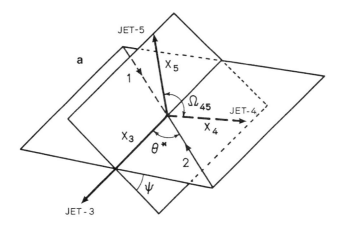

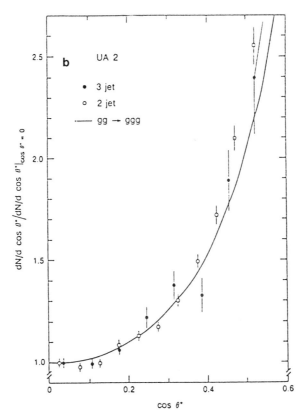

Figure 3.9. a) Momentum and angular variables defined in the process $1+2 \rightarrow 3+4+5$, where jets 3 through 5 are E_T-ordered. b) The distribution of $\cos\theta^*$, normalised at $\cos\theta^* = 0$ and corrected for detector and acceptance effects, for 3-jet and 2-jet events in the central region ($|\eta_i| < 0.8$), and requiring a jet E_T of $E_T^J > 10$ GeV, with $E_T^{TOT} > 70$ GeV.

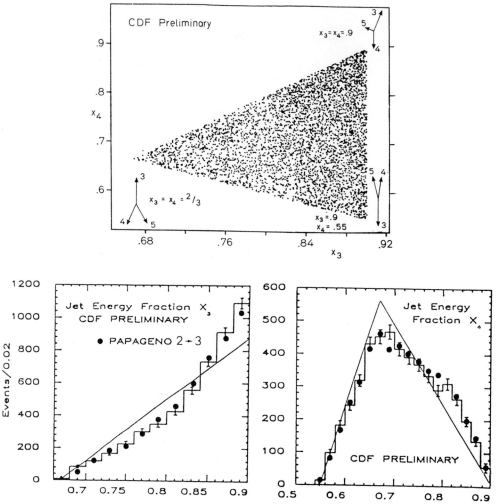

Figure 3.10. The distribution x_3 vs x_4 for 3-jet events from CDF satisfying selection requirements described in the text. Superimposed are (uncorrected) phase space and QCD expectations of the x_3 and x_4 projections.

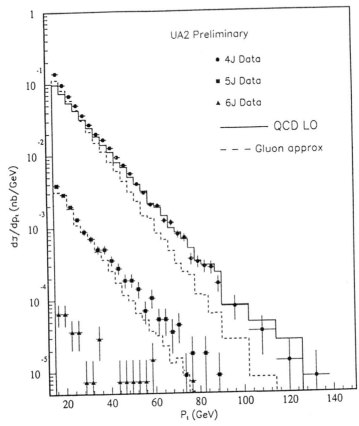

Figure 3.11. The inclusive p_T spectrum for four (circles) five (squares) and six (triangles) jet events. The solid histogram is the absolute LO prediction of QCD. The dashed histograms are the prediction of the purely gluonic approximation.

acceptance, and in particular $\not{E}_T < 3\sigma$ where $\sigma = 0.8 \, [\Sigma_i E_T^i]^{0.4}$ as described in section 2.2,

v)   the event energy is predominantly associated with hard parton scattering, with a quantitative selection

$$\sum_{i=1}^{n} E_T^i / E_T^{TOT} > 0.4, \qquad (3.6)$$

where i is summed over the jets and E_T^{TOT} is the total E_T of the event.

Predictions at tree-level have been made by Kunzst et al. [27] for n = 4, and a comparison with data is excellent after a full simulation of the UA2 detector. For n ≥ 4, Kunzst et al. and Parke et al. [28] have computed the contribution from gluon diagrams. A comparison is shown for n = 5 at the parton level, but after accounting for detector energy and angular resolution. The predictions are normalised to the exact n = 4 calculation; given that these calculations are incomplete, the comparisons are encouraging. Since these lectures, Kuijf and Berends [29] have extended the full calculation to 5-jet final states, with agreement to the data.

Historically, the UA1 and UA2 Collaborations paid considerable attention to 3-jet final states in order to measure α_s. They assumed a simple QCD n-jet cross-section model of the form

$$\sigma_n^{LO} = \frac{[\alpha_s]^n}{\hat{s}} \cdot \int \sum_{ij} \frac{F_i(x_i)}{x_i} \frac{F_j(x_j)}{x_j} Q_n^{ij} \, \Phi_n \, dx_i \, dx_j \quad , \qquad (3.7)$$

where $F_i(x_i)$ is the nucleon structure function discussed earlier and Φ_n is n-body phase space. Quite apart from the fact that only LO evaluations are made, the QCD matrix elements Q_n^{ij} diverge (the dominant terms being gluon bremsstrahlung for 3-jet final states), and kinematic cutoffs are required for their evaluation. As noted below similar experimental cutoffs are also needed. Then, for n-jet final states,

$$\sigma^{obs}(\bar{p}p \to n \text{ jets}) = K_n \sigma_n^{LO} \qquad (3.8)$$

with the K-factor K_n in principle calculable from QCD but in practice dependent on the assumptions of the calculation and the experimental cuts applied.

The value of α_s is determined from the experimental cross-section ratio

$$R = \frac{\sigma^{obs}(\bar{p}p \to 3 \text{ jets})}{\sigma^{obs}(\bar{p}p \to 2 \text{ jets})} \quad , \qquad (3.9)$$

after correction for contamination of the n-jet sample by

i)   true (n + 1)-jet final states in which either one jet is outside the acceptance or 2 of the jets merge (mainly from gluon bremsstrahlung) and

ii) true (n − 1)-jet final states resulting (mainly at low E_T) from jet-splitting, or from fluctuations of the underlying event which create a "fake" jet unassociated with the QCD scattering.

From this ratio, the UA1 and UA2 Collaborations have published [24, 26, 30] :

UA1 : K3/K2 α_s = 0.22 ± 0.02 ± 0.03 at $<Q^2>$ ~ 4000 GeV²

UA2 : K3/K2 α_s = 0.23 ± 0.01 ± 0.04 at $<Q^2>$ ~ 1700 GeV² (3.10)

With the recent NLO calculation of the 2-jet production cross-section described above, the systematic uncertainty would be reduced, since a major uncertainty of the analysis is the evaluation of (K3/K2). A new analysis, however, could not compete with measurements of α_s using (W + jet) events by the UA2 Collaboration [31], or accurate event-shape analyses from hadronic Z^0-decays at LEP.

The UA2 and CDF Collaborations have pursued event shape analyses of 4-jet events. The most recent data are from UA2 [25]. Figure 3.12 shows, for the 4-jet selection described above, the distributions Ω_{ij} (the angle between E_T-ordered outgoing jets i and j). The normalised LO QCD predictions that are superimposed are in good agreement with the data.

In addition to standard QCD processes, multiple (double) parton interactions (DPS) should exist, and naively

$$\sigma_{DPS} = \frac{1}{2} \cdot \frac{\sigma^2_{2-jet}}{\sigma_{eff}} \quad \text{with} \quad \sigma_{eff} \simeq 40 \text{ mb}. \quad (3.11)$$

The event topology, however, is quite different from standard multi-jet QCD processes, with balanced 2-jet pairs being expected. A sensitive variable to evaluate the possible DPS contribution is the variable S, defined by [25, 32]

$$S = \frac{1}{2} \min \left[\frac{|\vec{P}_T^i + \vec{P}_T^j|}{|P_T^i| + |P_T^j|} + \frac{|\vec{P}_T^k + \vec{P}_T^\ell|}{|P_T^k| + |P_T^\ell|} \right] \quad (3.12)$$

where S is minimised over all allowed jets i, j, k and ℓ. Figure 3.13 shows the distribution in log (S) with superimposed QCD and DPS expectations using data from the UA2 experiment. Two models have been used to simulate DPS scattering. In the "2-jet" model, data from uncorrelated 2-jet events have been superimposed. In the "Pythia" model, DPS events have been faked using the Pythia Monte Carlo program [16] by requiring a hard collision every pp collision. Ingoing and outgoing partons are then treated as for the standard "Pythia" program. From the $\ln(s)$ distribution, UA2 obtains preliminary 95 % upper limits for the DPS cross-section of :

"Pythia" model σ_{DPS} < 0.51 nb σ_{eff} > 20 mb

"2-jet" model σ_{DPS} < 0.26 nb σ_{eff} > 39 mb   (3.13)

3.6 Some remarks

The data on hard parton-parton scattering presented in this section are in spectacular qualitative agreement with leading-order QCD evaluations, and no evidence exists for deviations from QCD. This qualitative agreement extends beyond the inclusive distribution in E_T for jet production, to details of the event structure for exclusive multi-jet final states.

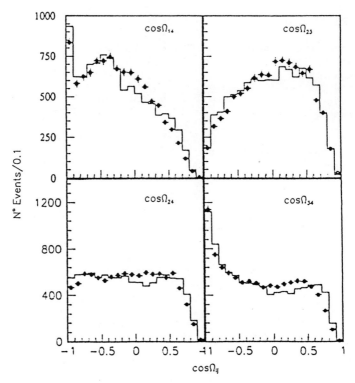

Figure 3.12. Space angles of jet pairs in the centre-of-mass system from the UA2 experiment. The solid histograms are the normalised LO prediction of QCD. Only statistical errors are shown.

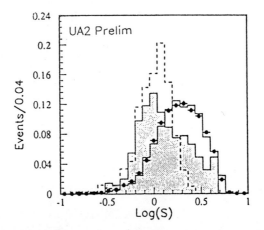

Figure 3.13. Distribution of log (S) from UA2. The LO QCD prediction is the solid histogram, and the dashed histogram is the DPS prediction using merged 2-jet events. Results of a modified PYTHIA is shown as a shaded histogram.

However, the precision of these comparisons remains inadequate both experimentally and theoretically. In particular, the data quality do not require NLO QCD contributions, even though comparisons with the data are improved, and theoretical scale uncertainties are much reduced.

For detailed QCD comparisons, the number of contributing QCD sub-processes, plus the dependence except at the highest p_T-values on the (poorly measured) gluon structure function, will continue to limit the accuracy of QCD comparisons with the data. Detailed comparisons are best made using processes for which few QCD processes contribute (see the next section). Aspects which should be pursued using these data include

i) improved comparisons with multi-jet final states, including limits on double parton scattering,
ii) improved common fits to data from the CERN and Tevatron Colliders,
iii) detailed comparisons with QCD at the highest E_T^j values to search for deviations of the data from QCD, or at large multi-jet (or 2-jet) masses to search for high-mass resonance structure. Studies of this type are noted in section 8.

Despite these rather negative comments, these collider data have played a crucial role in verifying the QCD strong interaction gauge theory. The data are complementary to QCD tests at (e^+e^-) machines, and to other collider QCD studies described in the following sections.

4. QCD TESTS INVOLVING (INVERSE) ELECTROMAGNETIC OR ELECTROWEAK PROBES

As already described, experimental uncertainties on the inclusive jet production cross-section at hadron colliders result mainly from limitations of the jet energy measurement. While the measurement accuracy can and should be improved in future experiments by using modern calorimetric techniques, fluctuations of the jet fragmentation and of the associated underlying event will always remain as significant limitations to the measurement accuracy. Furthermore the QCD comparisons described in section 3 theoretically involve a plethora of contributing sub-processes.

At the expense of some rate (see fig. 2.6), the inclusive production of direct photons or W/Z bosons in the reactions

$$\bar{p}p \rightarrow \gamma + X \quad \text{and} \quad \bar{p}p \rightarrow W/Z + X \tag{4.1}$$

can be described theoretically using fewer sub-processes, and experimentally can be more accurately measured in energy. Sensitive QCD tests involving the production of γ and W/Z bosons in association with jet activity are therefore possible.

4.1 The reaction $\bar{p}p \rightarrow \gamma + X$

Direct photons are produced in lowest order QCD by ($\bar{q}q$) annihilation, or by gluon Compton scattering (fig. 4.1a). Higher order contributions involving coupling terms $\alpha_{EM}\alpha_s^2$ or $\alpha_{EM}\alpha_s^3$ are shown in figure 4.1b. These terms complicate the cross-section evaluation for single γ production, but have been evaluated by Aurenche et al. and others [33]. Additional diagrams exist of the bremsstrahlung type (fig. 4.1c). These diagrams are important because they cancel divergences in the diagrams of figure 4.1b. However, the photons are characterised by a bremsstrahlung p_T-distribution with respect to the jet (quark) axis and therefore may not be localised from jet activity. Theoretically, these contributions are difficult to calculate using perturbative QCD; comparisons with the data should take this into account. Experimentally, a substantial fraction of bremsstrahlung photons will not be identified.

A significant background to isolated direct photons are hadron jets, since the high-p_T π^0 and η^0 production tail of jet fragmentation may not be experimentally resolvable from photons because of the limited calorimeter granularity. Since the jet production rate exceeds that of direct photons by $\sim 10^4$ the photons from π^0 and η^0 production are therefore an important background. Fortunately, the photons are not generally isolated, and therefore the requirement of isolation is a powerful identification tool. Isolated high-p_T π^0 and η^0 production will nevertheless remain as a background which must be subtracted event-by-event or statistically as appropriate to the detection technique (described below).

Each of the UA1, UA2 and CDF detectors have been used to identify photons, though none is optimised for this task. The construction of a large-acceptance detector optimised for direct photon studies would have very useful collider applications.

i) Conversion techniques. The photon identification signature has already been schematically shown in figure 2.8. An isolated electromagnetic calorimeter cluster, having a lateral and longitudinal profile characteristic of photons, is associated with a possible signal in a thin preshower detector preceding the calorimeter, but has no track association within a cone of typically the cluster radius. The UA2 experiment uses a $1.5X_0$ lead converter with associated scintillating-fibre track and preshower layers. The CDF experiment utilises the $0.18X_0$ outer wall of the central track chamber, which is followed by a drift-tube array. For the 1991 run, CDF is also installing pad chambers immediately following the magnet coil. If ε_γ (ε_π^0) is the probability of $\gamma(\pi^0)$ conversion in the preconverter, and if α is the fraction of photon candidates with a preshower signal,

$$\alpha = \frac{N \text{ (conversion)}}{N \text{ (conversion)} + N \text{ (non-conversion)}} \text{ , and} \quad (4.2)$$

$$b(p_T) = \frac{\alpha(p_T) - \varepsilon_\gamma}{\varepsilon_\pi^0 - \varepsilon_\gamma}$$

is the fractional π^0 background to the γ signal for large-p_T π^0's, after correction for large-angle π^0-decays. The identification efficiency and background rejection of this method improves with increasing E_T^γ.

ii) Longitudinal sampling. The UA1 Collaboration initially selected electromagnetic clusters of $E_T > 20$ GeV, with additional isolation requirements ΣE_T (calorimeter) < 2 GeV and Σp_T (tracks) < 2 GeV within a cone $R_c = 0.7$ about the cluster centroid. They then constructed a likelihood function taking account of the expected photon isolation in cones of $R_c = 0.7$ and $R_c = 1.0$, and the fractional energy deposition in the first of 4 longitudinal electromagnetic calorimeter samplings.

iii) Shower profile. At low E_T values ($E_T < 30$ GeV), the CDF Collaboration has used information from the strip chamber at a depth of approximately $5X_0$ within the electromagnetic calorimeter. Because of the excellent spatial resolution of this chamber, isolated electromagnetic clusters with no associated track and with a strip profile characteristic of single photons can be identified. The estimated detection efficiency is shown in figure 4.2 for electrons, γ's, and background events.

Preliminary data are shown for UA2 [25, 34], and CDF [35] in figure 4.3. The uncertainties shown for UA2 are statistical; systematic uncertainties are not quoted, but as a guide the cross-section uncertainty of earlier published data was $\sim \pm 20$ % including $\sim \pm 10$ % from each of the

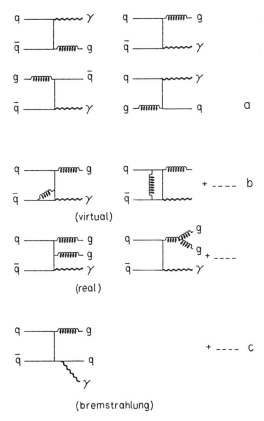

Figure 4.1. a) Lowest order (LO) diagrams for direct photon production via the annihilation or Compton processes. b) Real and virtual higher-order processes contributing to direct photon production. c) The production of direct photons via typical bremsstrhlung processes from incoming or outgoing quark lines.

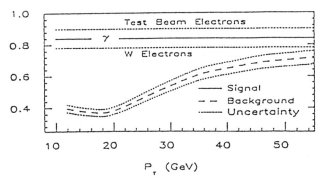

Figure 4.2. The efficiency for direct photon production at CDF and associated background as a function of E_T^γ, after a chisquare requirement $\chi^2 < 4$ on the longitudinal and lateral profile of the electromagnetic shower.

energy scale, background estimate, and conversion efficiency. Systematics at low E_T are shown for CDF data and result mainly from background substraction uncertainties. At high E_T^γ, the systematics arise from uncertainties of the E_T measurement. Superimposed QCD fits of Aurenche et al. [33] and Eichten et al. [36] using several different parton structure functions are shown in figure 4.3 and the agreement with data is excellent.

The UA2 experiment has measured, in the central rapidity region ($|\eta| < 0.7$), the ratio of photon to jet production at $\sqrt{s}$ = 630 GeV,

$$R = \frac{\sigma(\bar{p}p \to \gamma + X; \quad E_T^\gamma > 30 \text{ GeV})}{\sigma(\bar{p}p \to \text{JET} + X; \quad E_T^J > 30 \text{ GeV})} \tag{4.3}$$

Naively, one would expect $R \sim (\alpha_{EM}/\alpha_s)^2 \sim 3.10^{-3}$. However, because of the larger number of sub-processes contributing to jet production,

$$R \text{ (theor)} \approx 4 \cdot 10^{-4} \text{ and experimentally}$$

$$R \text{ (meas)} = [4.6 \pm 0.8 \text{ (stat)} \pm 2.1 \text{(sys)}] \, 10^{-4} \tag{4.4}$$

The UA2 Collaboration has also measured the angular dependence of the ratio R. To minimise systematic uncertainties, π^0-like jets are used to compare with direct photon events. For kinematic selections (γ applies to respectively the "direct γ" and "π^0" samples here)

$E_T^J > 10$ GeV; $m^{\gamma J} > 30$ GeV

$50° < \theta_J < 130°$

$\Delta\phi_{\gamma J} > 140°$ (the photon and jet back-to-back)

$E_T^J < 0.5 \, E_T^\gamma$ (both the γ and jet well measured),

figure 4.4 shows the normalised dependence of R on $\cos\theta^*$, the centre-of-mass production angle in the Collins-Soper [37] approximation. While jet-jet data result predominantly from gluon exchange diagrams, those of the dominant Compton diagrams for direct γ production involve spin-½ quark exchange. A lowest order QCD evaluation using the structure functions of Duke and Owens [22] is in good agreement with the data.

Two-photon production has also been observed and after selecting $E_T^\gamma \geq 10$ GeV, UA2 and CDF each have ~ 40 events under study. QCD expectations [38] using the processes of figure 4.5 are in reasonable agreement with earlier data from UA2.

4.2 The reaction $\bar{p}p \to W/Z + X$

The lowest order diagram for W and Z production, shown in figure 4.6a, is the weak analog of Drell-Yan (D-Y) production. The process is purely electroweak at the parton level. Of course, strong interaction evolutions of the nucleon quark distributions are indirectly very important.

For fractional quark momenta x_1 and x_2, the longitudinal momentum of the produced boson ($W^\pm$, Z, γ) is

$$P_L = (x_1 - x_2) \frac{\sqrt{s}}{2} = x \frac{\sqrt{s}}{2} \tag{4.5}$$

The outgoing fermion-antifermion mass is

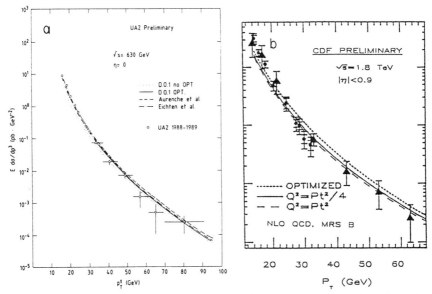

Figure 4.3. a) Preliminary corrected inclusive cross-section for photon production from UA2, with superimposed curves from Aurenche et al. [33] and Eichten et al. [36] using various structure function parametrisations. b) Preliminary corrected inclusive cross-section from CDF (using both conversion and profile analyses) with superimposed curves from Aurenche et al. [33] and various choices of P_T-scale using the MRSB structure functions.

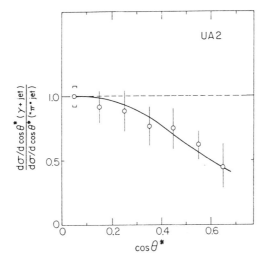

Figure 4.4. Data from the UA2 experiment comparing the centre-of-mass angular distributions of photons and jets.

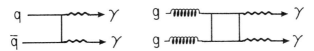

Figure 4.5. Major contributing processes to 2-photon production at collider energies.

109

$$m_{f_1}^2 \bar{f}_2 = x_1 x_2 s = \tau s. \tag{4.6}$$

In the case of electromagnetic D-Y production,

$$\frac{d^2\sigma}{dx_1 dx_2} = \frac{1}{3} \cdot \frac{4\pi\alpha_{EM}^2}{3sx_1 x_2} \sum_i e_i^2 R_i(x_1, x_2),$$

or

$$m^3 \frac{d^2\sigma}{dxd\tau} = \frac{1}{3} \frac{8\pi\alpha_{EM}^2 \tau}{3\sqrt{x^2 + \tau^2}} \sum_i e_i^2 R_i(x_1, x_2),$$

with $R_i(x_1, x_2) = q_i(x_1) \bar{q}_i(x_2) + (1 \leftrightarrow 2)$,

$q_i(x)$ = the probability of finding quark q_i with fractional momentum in the range $(x, x + dx)$, and

$e_i = 1/3$ or $2/3$ according as the quark charge. (4.7)

We stress that the D-Y process is electromagnetic; therefore the cross-section evaluation is averaged over colour. At the lowest order it probes $\bar{q}q$ fusion of a given flavour (predominantly $\bar{u}u$ with a small $\bar{d}d$ contribution at $\bar{p}p$ colliders). It is of course well known that higher order contributions to the D-Y cross-section are important. Following Altarelli et al. [39] for the simple lowest order D-Y production of a virtual photon γ^* of mass m_v, regardless of its decay, using (4.5) – (4.7) above,

$$\sigma_v = N_v \int \frac{dx_1}{x_1} \cdot \frac{dx_2}{x_2} \left\{ \sum_i e_i^2 R_i(x_1, x_2) \right\} \delta \left(1 - \frac{\tau}{x_1 x_2}\right)$$

with $N_v = 4\pi^2 \alpha_{EM}/3s$ (4.8)

Analogously for W and Z production it can be easily shown that

$$N_W = \pi^2 \alpha_{EM}/3s \sin^2\theta_W$$

$$N_Z = \pi^2 \alpha_{EM}/3s \sin^2 2\theta_W \tag{4.9}$$

and the sum over the quark content of the nucleon is modified to take account of the allowed weak coupling, that is:

W^+ : $(u_1 \bar{d}_2 + c_1 \bar{s}_2) \cos^2\theta_c + (u_1 \bar{s}_2 + c_1 \bar{d}_2) \sin^2\theta_c$

$\approx u_1 \bar{d}_2 \cos^2\theta_c$ with $\theta_c = \theta$(Cabibbo)

Z : $\sum_f [1 + (1 - 4|Q_f|\sin^2\theta_W)^2] q_f \bar{q}_f$ (4.10)

with Q_f being the quark charge

While at lowest order both the D-Y and Z production processes are most sensitive to $(\bar{u}u)$ fusion, the relative W^+ and W^- production rates are a sensitive probe to the u:d ratio in the nucleon at at given value of x.

Higher order processes of the type shown in figure 4.6b contribute significantly to the production cross-section. At $\bar{p}p$ collider energies, the small value of α_s allows a perturbative evaluation of these terms. The full $O(\alpha_s)$ calculation has now been made by Matsuura et al., and others [3], together with a partial $O(\alpha_s^2)$ evaluation that allows, as previously discussed for jet production and direct photon production, less dependence on the choice of the Q^2-scale. These higher-order contributions to W and Z production result in a large-p_T tail which increases with $\sqrt{s}$; the QCD evaluation of large-p_T W/Z production is therefore sensitive to both structure function parametrisations and to the inclusion of higher order QCD contributions.

Large p_T W production in association with jet activity can (see section 3.5) also be used to measure α_s at $Q^2 \sim m_W^2$. Such a measurement has been made by UA2 [31]. It is not described in these notes for reasons of brevity, and because recent LEP data will enable more model independent measurements of α_s.

As noted earlier and again in section 5, the selection of clean W and Z data samples requires an identification of their leptonic decays to reduce the background from standard QCD processes. So far the most accurate W and Z production measurements use data from the reactions

$$\bar{p}p \rightarrow W^\pm + X \quad ; \quad W^\pm \rightarrow e^\pm \nu_e$$

$$\bar{p}p \rightarrow Z + X \quad ; \quad Z \rightarrow e^+ e^- \qquad (4.11)$$

The identification criteria used for electron detection (and where relevant neutrino identification), and details of the energy measurement, are discussed in section 5 where the data are compared with electroweak expectations. With the 1988–89 runs by UA2 and CDF the W and Z data samples increased by more than one order of magnitude on what previously existed. Following a short discussion of these data samples, we compare σ_W^e and σ_Z^e with QCD calculations. A comparison is also made with $d\sigma/dp_T$ measurements of both W and Z production. Using published UA1 and preliminary CDF data we show the sensitivity of longitudinal W and Z production properties to the nucleon structure.

4.2.1. The UA2 and CDF data samples

The UA2 experiment identifies electrons from W-decay in 3 separate regions within the pseudorapidity range $|\eta| < 1.6$, namely

a) the central calorimeter region, $50^0 < \theta < 130^0$,
b) the overlap region between the central and forward calorimeters, and
c) the forward region.

In each region, the electron detection efficiency and the associated rejection against QCD background faking the electron signature differs and the cross-section is therefore evaluated separately in each region. Selecting $p_T^e > 20$ GeV, $p_T^\nu > 20$ GeV and $m_T^{e\nu} > 40$ GeV, a total of 2041 events remain in the final event sample (see table 4a). This includes an estimated background of 75.7 ± 1.8 events from leptonic τ-decay.

The CDF experiment identifies 2664 electrons in the rapidity region $|\eta| < 1.1$ (that is the central calorimeter) after requiring electron selections as in section 5, and kinematic selections $p_T^e > 20$ GeV and $p_T^\nu > 20$ GeV. The estimated background for the CDF sample is $(238 \pm ^{53}_{62})$ events. This background results in part from QCD background (following less stringent selection criteria and the lack of an $m_T^{e\nu}$ cut as used by UA2), and in part from the limited central acceptance with a consequent $Z \rightarrow e^+e^-$ background contribution.

In each experiment, the Z data samples involve the selection of a well identified electron of $p_T^e > 20$ GeV as above, with the identification of a second electromagnetic cluster of $p_T > 20$ GeV (10 GeV in the case of CDF) satisfying less stringent selection requirements. As summarised in

111

Table 4a. Evaluation of σ_W^e

	UA2 $\sqrt{s} = 630$ GeV	CDF (prelim) $\sqrt{s} = 1.8$ TeV
No events	2041	2664
Backgrounds : $W \to \tau\nu_\tau$	75.8 ± 1.7	90 ± 10
$Z \to ee$	–	40 ± 15
$Z \to \tau\tau$	–	8 ± 4
top	–	0 ± 3^1_0
QCD	–	100 ± 50
Total	75.8 ± 1.7	$238 \pm ^{62}_{53}$
Signal	1965 ± 45 ± 1.7	$2426 \pm 52 \pm ^{53}_{62}$
Acceptance (inc. vertex)	0.619 ± 0.019	0.352 ± 0.015
Efficiency	~ 0.7 (η-dependent)	0.81 ± 0.04
Luminosity (pb^{-1})	7.4 ± 0.4	4.03 ± 0.28
Cross-section (pb)	660 ± 15 (stat) ± 37 (sys)	2210 ± 40 (stat) ± 230 (sys)

table 4b), a total of 169 (243) leptonic Z-events are identified by UA2 (CDF), in the mass range 76 < m^{ee} < 110 GeV for the UA2 experiment, and 65 < m^{ee} < 115 GeV for the CDF experiment. In each experiment, the background is small.

4.2.2. Measurement of σ_W^e and σ_Z^e [40, 41]

The leptonic W cross-section is evaluated for each experiment according to the formula

$$\sigma_W^e = (N_W - N_{BKGRD})/\varepsilon.\eta.L, \qquad (4.12)$$

where ε is the electron identification efficiency, η is the acceptance for geometric and kinematic selections and L is the integrated luminosity. In each experiment, detailed Monte Carlo simulations are used to estimate the acceptance, and ε is determined from the data themselves. The UA2 integrated luminosity was measured to be L = 7.4 ± 0.4 pb^{-1}, where the error is dominated by a 2.3 % uncertainty on the luminosity telescope acceptance and a 4.7 % uncertainty on the measured inelastic cross-section. This small uncertainty is facilitated by measurements of the total cross-section by the UA4 Collaboration [11] in the same intersection region, and using the same luminosity telescope. Unfortunately a similar measurement does not yet exist at CDF and the preliminary luminosity uncertainty is ± 7 %. For CDF, L(prelim) = 4.03 ± 0.28 pb^{-1}*. Work to reduce the luminosity uncertainty of each experiment remains an important experimental priority if the s-dependence of QCD-tests is to be properly exploited. The UA2 and preliminary CDF cross-sections are shown in table 4a and figure 4.7, together with statistical and systematic uncertainties. Note that the UA2 cross-section has been evaluated separately for the different acceptance regions, and combined statistically to give the final result.

---

*Data in the written version of this talk are updated by the most recent analyses (August 1990).

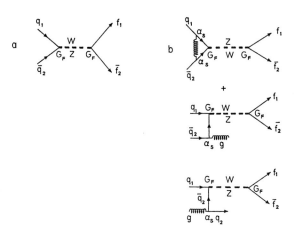

Figure 4.6. Lowest order (LO) and selected higher order (NLO) contributions of order α_s or α_s^2 to W and Z production at $p\bar{p}$ colliders.

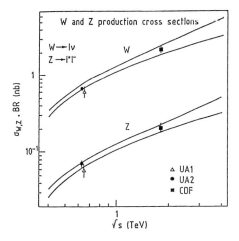

Figure 4.7. Measured cross-sections from UA2 and CDF for the production and subsequent electron decay of W and Z bosons. Superimposed estimates at NLO by Matsuura et al. are shown for the (extreme) structure function parametrisations. Also shown are data from the UA1 experiment using pre-1987 data [42].

Similar evaluations are made for the Z cross-section by each experiment, using the formula

$$\sigma_Z^e = (N_Z - N_{BKGRD}) (1 - f_\gamma^*)/\varepsilon.\eta.L \qquad (4.13)$$

where f_γ^* is the contribution from virtual photon exchange and γ^*z interference. Results are shown in table 4b.

The data for σ_W^e are compared with the QCD evaluations of Matsuura et al. in figure 4.7. In addition to using the expected W-decay widths assuming $m_t > m_W$ with $0(\alpha_s)$ corrections when appropriate, values $m_W = 80.5$ GeV and $M_Z = 91.15$ GeV are used and α_s is evaluated assuming a momentum transfer scale $Q^2 = M_W^2$ and Λ_{QCD} as used in recent NLO structure function parametrisations [22]. A major uncertainty of the QCD comparison is the discrepancy between different structure function sets, and a concern is the lack of quoted (and evolved) uncertainties associated with these structure function parametrisations. In figure 4.7, the range was determined from 14 different structure functions giving an uncertainty at $\sqrt{s} = 1.8$ TeV of $\sim \pm 12$ % [43]. The Born level evaluation of σ_W can be excluded, but the data and QCD fits are not yet of sufficient quality to require $0(\alpha_s^2)$ contributions. Similar comparisons are made for σ_Z in figure 4.7. This latter comparison can benefit from improved statistics.

Table 4b.   Evaluation of σ_Z^e

	UA2 $\sqrt{s} = 630$ GeV	CDF (prelim) $\sqrt{s} = 1.8$ TeV
Events	169	243
Backgrounds : $Z \to \tau\tau$	–	< 0.5
W + jet	–	1 ± 1
QCD	2.39 ± 0.3	5 ± 3
Signal	166.6 ± 13 + .3	238 ± 16 ± 3
Acceptance (inc. vertex)	.496 ± .01	.371 ± .007
Efficiency	η-dependent	η-dependent
Luminosity (pb^{-1})	7.4 ± 0.4	4.03 ± 0.28
Cross-section (pb)	70.4 ± 5.5 (stat) ± 4.0 (sys)	210 ± 13 (stat) ± 18 (sys)

4.2.3. Transverse motion of $W^\pm$ and Z production [44, 45]

As already noted, the p_T-dependence of W and Z production provides a sensitive QCD test because of the emission of initial state gluon radiation. At low p_T-values, multiple gluon emission dominates the spectral shape. Because it is a low-p_T sub-process, it cannot be reliably calculated using perturbative QCD, but resummation techniques have been used by Altarelli et al. [46] to successfully predict the p_T-spectrum. At higher transverse momenta, perturbative calculations are reliable and a complete $0(\alpha_s)$ evaluation has been made with (partial) $0(\alpha_s^2)$ terms. Arnold et al. [47] and others have successfully predicted the high-p_T tail for p_T^W and p_T^Z production, and the calculations have only limited sensitivity to the choice of Q^2-scale. While this gluon radiation process is the same process as that of multi-jet Z-decay at LEP it remains important to obtain accurate p_T^Z and p_T^W measurements because

a) the accessible transverse-momentum range exceeds that of LEP by one order of magnitude,
b) deviations of high-p_T W and Z production from QCD expectations may signal new physics processes beyond or in conflict with the Standard Model, and
c) technically the limited knowledge of the p_T^W spectrum due to systematic measurement uncertainties and p_T^Z due to the small statistical sample, remains a limiting uncertainty for measurements of m_W.

The evaluation of p_T^Z is straight-forward since both electrons from Z-decay are well measured. Using standard selection criteria, preliminary CDF data are shown in figure 4.8a, after correction for known biases, including acceptance and resolution or smearing effects. In addition to statistical uncertainties of between ± 20 % and ± 70 %, systematic uncertainties contribute as in table 5.

Table 5. Fractional systematic uncertainties in the $d\sigma_z/dp_T^e$ measurement at CDF

	Fractional Uncertainty
Acceptance evaluation (structure function dependence)	± 0.02
ε_z evaluation	± 0.02
Resolution uncertainties	± 0.03
Smearing corrrections	± 0.01
Luminosity	± 0.07

The data are in good agreement with the evaluations by Arnold et al. that are superimposed on figure 4.8. No significant deviation from QCD expectations is measured for $p_T^Z \leq 150$ GeV/c. Similar data are available from the UA1 and UA2 experiments. The superimposed low-p_T predictions of Altarelli at al. are also in good agreement with the data, after correction for resolution and acceptance effects.

The measurement of p_T^W is more difficult, since p_T^ν is inferred from the transverse momentum imbalance of the event. In practice,

$$\vec{p}_T^W = \vec{p}_T^e + \vec{p}_T^\nu = - \vec{p}_T^{UL}, \qquad (4.14)$$

where the transverse momentum contribution from the underlying event, $\vec{p}_T^{UL}$, is evaluated as in section 2 from the calorimetric energy deposition, and in the case of an associated track momentum measurement, from the track momenta. Thus, the uncertainty of p_T^W is determined by the uncertainty of the p_T^{UL} evaluation, and in particular the relation

$$p_T^{UL}(\text{true}) = f [p_T^{UL}(\text{meas})]. \qquad (4.15)$$

Systematic corrections and uncertainties affecting (4.15) include

a) the effect of magnetic field (CDF) resulting in a low-p_T detection threshold; the fractional correction is approximately inversely proportional to p_T^W, and

b) jet measurement uncertainties as described in section 3 that include crack effects, energy non-linearity etc. For $p_T^W \geq 20$ GeV the p_T^{UL}(meas) uncertainty is dominantly that of the uncertainty of the gluon jet energy measurement.

Additional systematic corrections include

a) acceptance losses especially at small p_T^W when gluon radiation is at small angle,
b) smearing effects on a fast-falling p_T^W spectrum, and
c) p_T-dependent QCD backgrounds.

Using an independent data sample without a p_T^ν selection (and therefore dominantly jets misidentified as electrons), the UA2 estimate of QCD background does not exceed 10 % in any bin. Similar studies using a sample of identified non-isolated electrons are made in the CDF experiment, with similar resulting background estimates.

Preliminary corrected data from CDF extend to $p_T^W \sim 150$ GeV and are compared with QCD calculations from Arnold et al. in figure 4.9. No significant deviation is measured. Uncorrected data from UA2 are compared in figure 4.10 with predictions from Altarelli et al. at low p_T and Arnold et al. at high p_T after acceptance and smearing corrections. Again, no significant discrepancy is measured. Detailed systematic studies for data from UA2 are summarised in figure 4.10 and indicate substantial uncertainties, especially at low p_T^W.

Because of increasing sea-valence contributions at $\sqrt{s} = 1.8$ TeV, $<p_T^W>$ and $<p_T^Z>$ are expected to increase with $\sqrt{s}$. The data from UA1, UA2 and CDF confirm this.

As discussed, high-p_T W or Z production is inevitably associated with jet production. Figure 4.11 shows preliminary CDF data for the number of associated jets [p_T^j (meas) > 10 GeV/c], identified according to the criteria of section 3. Recent calculations have been made by Berends et al. [48]; they compute at the parton tree-level the probability of observing n ($\leq$ 3) jets satisfying $p_T^j > p_T^{min}$. Also shown is a direct comparison with the data after a full CDF detector simulation using the matrix elements of Ellis et al. [49] with the Papageno Monte Carlo [17]. The data and calculations are in excellent agreement.

4.2.4. Longitudinal motion of W and Z production at $\sqrt{s} = 630$ GeV and $\sqrt{s} = 1.8$ TeV

Data from UA1 [42] are shown in figure 4.12 for the distribution of fractional longitudinal W and Z momentum, x_W and x_Z. The data are in agreement with lowest order QCD-evolved nucleon structure function expectations. In evaluating x_W, it is assumed that $m_W = m^{ev}$; two values of p_L^ν are therefore allowed, corresponding to different values of P_L^W. In practice at $\sqrt{s} = 630$ GeV, one solution is unphysical 70-80 % of the time, and when two solutions are allowed that of smallest p_L^W is most likely. Similar data have been published by the UA2 experiment.

Assuming that the $W^\pm$ is dominantly produced by valence quark annihilation as shown in figure 4.6a, it is possible from the sign of the lepton to evaluate x_q and x_q^- from relations 4.5 and 4.6. Figure 4.13 shows data from UA1 (assuming $dN/dx_q = dN/dx_q^-$), together with superimposed QCD-evolved lowest-order structure function parametrisations. Given the limited statistical accuracy, agreement is good.

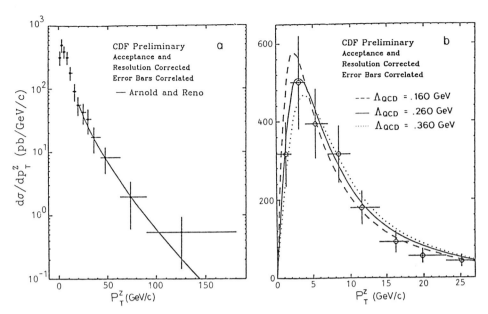

Figure 4.8. a) Differential cross-section for Z^0 production. Errors bars include statistical and systematic uncertainties and are correlated. Curve is $O(\alpha_s^2)$ QCD prediction. The theoretical uncertainty is $\sim \pm 15$ %. b) Differential cross-section for Z^0 production for low $p_T^{Z^0}$ region. Errors are statistical and systematic and are correlated. Curves are from Altarelli et al. [46] with three choices of Λ_{QCD}.

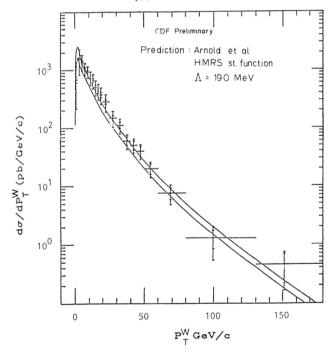

Figure 4.9. Differential cross-section for $W^{\pm}$ production from the CDF experiment. Superimposed predictions are from Arnold et al., using HMRS structure functions.

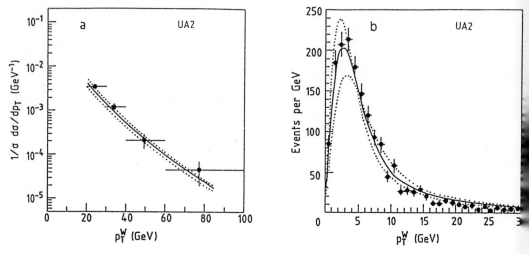

Figure 4.10. The differential cross-section for $W^{\pm}$ production from the UA2 experiment, in a) the high-p_T and b) low-p_T regions. Detector resolution effects have not been corrected. The superimposed curves show the effect of allowed detector response uncertainties, especially in the low-p_T region.

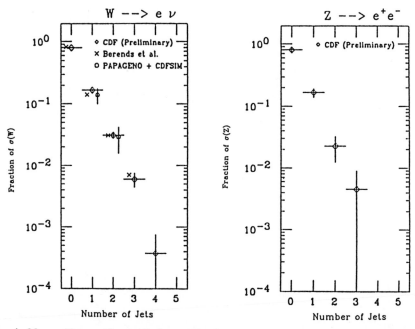

Figure 4.11. The number of jets of $p_T^j > 10$ GeV in association with W and Z production. Comparison is made with tree-level expectations of Berends et al., and a full detector simulation including the matrix elements of Ellis et al.

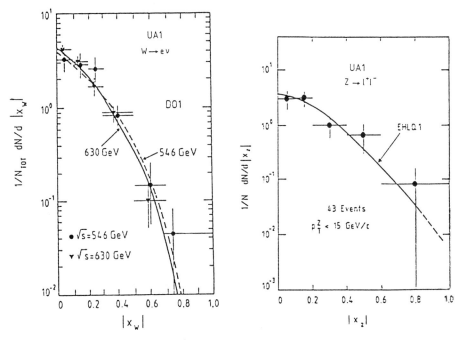

Figure 4.12. Data from UA1 showing the fractional longitudinal momentum for W and Z production. Superimposed are lowest-order predictions using the DO1 and EHLQ1 nucleon structure functions.

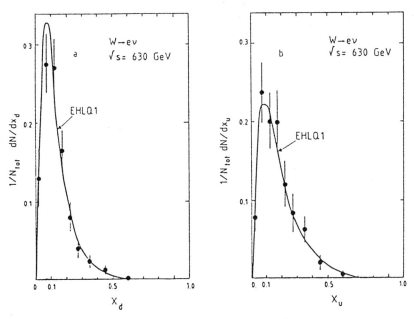

Figure 4.13. Measurements of u-quark and d-quark content from UA1, with superimposed lowest-order QCD-evolved EHLQ1 structure functions.

119

It is less interesting to pursue these studies at $\sqrt{s} = 1.8$ TeV (that is at CDF energies) because of the increased contribution of valence-sea interactions. The competing dependence of valence and sea contributions to $W^{\pm}$ production has been evaluated by Berger et al. [50], and is shown in figure 4.14. At $\sqrt{s} = 1.8$ TeV, $W^{\pm}$ production results dominantly from valence-sea quark and gluon contributions.

The rapidity distribution of W production is sensitive to the parton ratio $d(x)/u(x)$ in the nucleon. Since on average u quarks have higher momentum than d quarks, the $W^{+}(W^{-})$ will be boosted in the $p(\bar{p})$ direction. In a CDF analysis [51], both electron and muon W-signatures are used (see table 6) and a requirement is made that no jet of $E_T^J > 10$ GeV is detected in the event. The QCD background for these data samples is estimated to be less than 1 %. The resulting data samples are also shown in table 6.

Table 6. Kinematic selections and resultant data samples for forward-backward asymmetry measurement by CDF.

	p_T^W (GeV)	p_T^{ν} (GeV)	m_T^W (GeV)	$|\eta^{\ell}|$	n_w+	n_w-
e	> 20	> 20	> 50	< 1.0	923	994
μ	> 20 < 60	> 20	> 50	< 0.7	411	386

As noted above, the W-rapidity can be reconstructed with a two-fold ambiguity that is usually resolved at $\sqrt{s} = 630$ GeV; however at $\sqrt{s} = 1.8$ TeV, the kinematic ambiguity is difficult to resolve. Therefore the comparison of CDF data with QCD uses the measured lepton rapidity distribution. Furthermore, to minimise any systematic uncertainty resulting from variations of the reconstruction efficiency etc., an asymmetry is evaluated in different bins of $|\eta|$:

$$A(\eta) = \frac{\sigma_+(\eta) - \sigma_-(\eta)}{\sigma_+(\eta) + \sigma_-(\eta)} \qquad (4.16)$$

The measured asymmetry $A(\eta)$ is shown in figure 4.15. By requiring no associated jet activity, higher-order contributions to $A(\eta)$ are small. QCD predictions are superimposed on figure 4.15, using as input several structure function parametrisations. The parametrisations of Duke and Owens (DO2, DO1) can be excluded. With improved statistical accuracy these data will provide a powerful constraint on the $[d(x)/u(x)]$ ratio of structure function parametrisations.

4.3 Summary

With a recent improvement by one order of magnitude in the integrated luminosity recorded by the UA2 and CDF experiments, careful comparisons of W/Z production have been made with higher order QCD calculations. There is no significant evidence for deviations from Standard Model behaviour. The data do not yet require $O(\alpha_s^2)$ contributions in QCD comparisons but with improved data from the 1990 UA2 run and the 1991 CDF run, this comparison will be important.

Theoretically, these processes remain an ideal testing ground for comparison with higher-order QCD evaluations, in a kinematic regime far

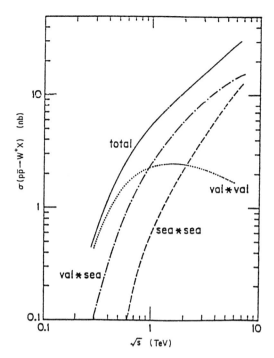

Figure 4.14. The $\sqrt{s}$-dependence of valence and sea contributions in $W^{\pm}$ production at $\bar{p}p$ colliders. From Berger et al. [50], using the MRS structure function parametrisations.

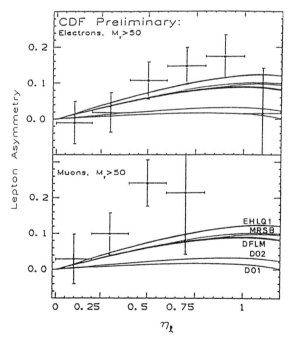

Figure 4.15. Measured asymmetry for charged leptons from $W^{\pm}$ decay, plotted as a function of the lepton rapidity. Superimposed expectations of several structure function parametrisations are shown; the data are sensitive to the relative d- and u-quark distributions.

removed from that of LEP. Further, with increasing statistics possible deviations from QCD can be probed at very high p_T^W or p_T^Z values. A major obstacle to theoretical predictions is the uncertainty of existing structure function parametrisations. With improved statistics collider data can provide a powerful constraint on the [d(x)/u(x)] ratio in the nucleon, and on the shape of the gluon structure function.

Experimentally, major uncertainties remain in the evaluation of p_T^W (and similarly m_w) arising from the quality of existing calorimetric methods for both the measurement of jets and for the measurement of low-p_T particles. Significantly improved calorimetry at the DØ experiment should enable an improved p_T^W measurement accuracy.

Although neither CDF nor UA1/UA2 are optimised for direct γ detection, the agreement of data with calculations from Aurenche et al. is encouraging. It should be noted, however, that theoretical uncertainties persist on how to treat the bremsstrahlung contribution to direct γ production. In the 1991 run, CDF will use its new preshower detector to extend accurate QCD comparisons to the highest E_T^γ values (in a search for deviations from QCD). They also intend to extend measurements to low E_T values where the rate of direct γ production is a sensitive probe to the gluon content of the nucleon.

5. TESTING THE STANDARD ELECTROWEAK MODEL AT HADRON COLLIDERS [52]

From our current knowledge of weak interactions, we can group the known fermions into three identical families of quarks and leptons containing left-handed (LH) doublets and right-handed (RH) singlets. We can assign a charge quantum number to each element of the group, and it is convenient to define a "weak-isospin" quantum number whose third component, I_3, is shown in table 7.

Of the three multiplets shown, only the t-quark and ν_τ have not been directly identified. Within the electroweak model no requirement is placed on the number of families; measurements of Z-decays from LEP, however, indicate only 3 kinematically accessible neutrino families. In table 7, both lepton and quark doublets are shown. The following comments apply to lepton families and are later extended to quark families.

i) It is assumed and well justified experimentally that weak charged currents act with identical coupling strength, g_1, on any lepton multiplet and that the interaction involves only pairs of a given multiplet (lepton conservation). Furthermore, it is assumed and well verified in the Q^2-range of β-decay and neutrino experiments that charged currents act on only left-handed doublets. The Glashow-Weinberg-Salam (GWS) model [1] of electroweak interactions relates the ratio of charged weak coupling to the coupling, e, of electromagnetic interactions, in the relation

$$\sin^2\theta_W = \frac{e^2}{g_1^2} = \frac{\pi\alpha_{EM}}{\sqrt{2}G_F} \cdot \frac{1}{m_W^2} \tag{5.1}$$

ii) The weak neutral current acts on both left-handed doublets and right-handed singlets. Introducing a triplet of fields W_μ coupling to fermions via the isospin operator with the familiar coupling g_1 and a field B_μ coupling with strength g_2 via the weak hypercharge operator Y, one can write the neutral current as

$$J = g_1 \, J_\mu^{(3)} \, W_\mu^{(3)} + g_2 \, j_\mu^{(Y)} \, B_\mu \tag{5.2}$$

where $J_\mu^{(3)}$ and $j_\mu^{(Y)}$ are the neutral isospin and hypercharge currents. The GWS model relates $W_\mu^{(3)}$ and B_μ to the physical photon field A_μ (which couples equally to ℓ_L and ℓ_R but to neither ν_L nor ν_R) and the physical Z field Z_μ, by the Weinberg angle $\sin\theta_W$:

$$\begin{pmatrix} B \\ W^{(3)} \end{pmatrix} = \begin{pmatrix} \cos\theta_W & -\sin\theta_W \\ \sin\theta_W & \cos\theta_W \end{pmatrix} \begin{pmatrix} A \\ Z \end{pmatrix} \qquad (5.3)$$

The neutral current can then be written

$$J = e\, j_\mu^{em} A_\mu + (g_1/\cos\theta_W)(J_\mu^3 - \sin^2\theta_W\, j_\mu^{em})Z_\mu,$$

with

$$g_1 \sin\theta_W = g_2 \cos\theta_W = e$$

and

$$j_\mu^{em} = J_\mu^3 + j_\mu^Y \qquad (5.4)$$

Defining

$$J^{NC} = J_\mu^{(3)} - \sin^2\theta_W\, j_\mu^{(em)}$$

it can be shown that the weak vector and axial couplings are (table 7)

$$g_V = I_3 - 2Q \sin^2\theta_W \qquad g_A = I_3 \qquad (5.5)$$

iii) The generation of W and Z masses via spontaneous symmetry breaking, while at the same time retaining a massless photon, can be shown to require in the simplest case

$$m_W = \tfrac{1}{2}\, v g_1$$
$$m_Z = \tfrac{1}{2}\, v'\, (g_1^2 + g_2^2)^{\tfrac{1}{2}} \text{ with}$$
$$g_1 \sin\theta_W = g_2 \cos\theta_W = e \text{ and } v,\ v' \text{ real.} \qquad (5.6)$$

This implies

$$\sin^2\theta_W = 1 - \frac{m_W^2}{\rho m_Z^2}$$

with $\rho \equiv 1$ in the GWS model (neglecting small higher order corrections) if $v = v'$. The parameter ρ, specifying the relative charged and neutral weak couplings, need not be equal to 1 in less economical symmetry breaking mechanisms. To maintain local gauge invariance at least one additional scalar field, or Higgs particle is required. The Higgs mass is not specified, but the couplings are well defined to both fermion and vector boson pairs (and to the Higgs itself) with a strength proportional to their masses.

The items (i) to (iii) relating $\sin^2\theta_W$ to both charged and neutral electroweak couplings, and to the value of the W and Z masses, are essential aspects of the Standard Model.

A consequence of the non-abelian character of the (SU(2)xU(1)) gauge theory describing the Standard electroweak model is the self-coupling in 3-point and 4-point vertices of gauge bosons (γ, W, Z).

Table 7. Quantum numbers for different fermion families

	Families				I_3	Q	g_{LR}^C	g_{LR}^N
$\begin{pmatrix}\nu\ell\\\ell\end{pmatrix}_L$	$\begin{pmatrix}\nu_e\\\ell\end{pmatrix}_L$	$\begin{pmatrix}\nu_\mu\\\mu\end{pmatrix}_L$	$\begin{pmatrix}\nu_\tau\\\tau\end{pmatrix}_L$		$\frac{1}{2}$ $-\frac{1}{2}$	0 -1	1 1	$\frac{1}{2}$ $-\frac{1}{2}+x$
$\begin{pmatrix}q_1\\q_2\end{pmatrix}_L$	$\begin{pmatrix}u\\d'\end{pmatrix}_L$	$\begin{pmatrix}c\\s'\end{pmatrix}_L$	$\begin{pmatrix}t\\b'\end{pmatrix}_L$		$\frac{1}{2}$ $-\frac{1}{2}$	$\frac{2}{3}$ $-\frac{1}{3}$	1 1	$\frac{1}{2}-\frac{2}{3}x$ $-\frac{1}{2}+\frac{1}{3}x$
ℓ_R	e_R	μ_R	τ_R		0	-1	0	x
q_{1R}	u_R	c_R	t_R		0	$\frac{2}{3}$	0	$-\frac{2}{3}x$
q_{2R}	d'_R	s'_R	b'_R		0	$-\frac{1}{3}$	0	$\frac{1}{3}x$

	g_A	g_V
ν	$\frac{1}{2}$	$\frac{1}{2}$
e, μ, τ	$-\frac{1}{2}$	$-\frac{1}{2}+2x$
u, c, t	$\frac{1}{2}$	$\frac{1}{2}-\frac{4}{3}x$
d', s', b'	$-\frac{1}{2}$	$-\frac{1}{2}+\frac{2}{3}x$

$x = \sin^2\theta_W$; $g_L^N = I_3 - Qx$; $g_R^N = -Qx$; $g_V = g_L + g_R$; $g_A = g_L - g_R$.

The above discussion has been limited to a single lepton pair (ν_ℓ, ℓ) whose elements differ in charge by one unit. Three lepton families are well established : (ν_e, e^-), (ν_μ, μ^-) and (ν_τ, τ^-), and these are included in the model by postulating the same weak quantum numbers for each family. All three lepton families are assumed to couple with the same strength (this is termed lepton universality) and lepton changing currents such as for example the coupling $(\bar\nu_e \gamma_\mu \tau^-)$ are assumed not to exist.

In the Standard Model, the quarks shown in table 7 are also grouped into pairs (q_1, q'_2), differing by one unit of electric charge, and are also assumed to occur as left-handed weak isospin doublets and right-handed weak isospin singlets. All known quarks have charge $Q = -\frac{1}{3}$ or $+\frac{2}{3}$ and the standard assignments are given in the bottom half of table 7. A more open question is how to choose (q_1, q'_2) in terms of the known mass eigenstates revealed by hadron spectroscopy. Three quarks with $Q = -\frac{1}{3}$ (namely d,s and b) are well established. Three $Q = +\frac{2}{3}$ states should also exist (u, c, and t) and only the t-quark remains unidentified.

It has long been established that the charge-changing weak interactions do not respect quark flavour. Therefore, the most general possible assumption is to relate the $Q = -\frac{1}{3}$ weak eigenstates to the $Q = -\frac{1}{3}$ mass eigenstates via a unitary transformation :

$$\begin{pmatrix}d'\\s'\\b'\end{pmatrix} = U_{KM} \begin{pmatrix}d\\s\\b\end{pmatrix} \tag{5.7}$$

The 3 x 3 matrix U_{KM}, called the CKM matrix and introduced by Kobayashi and Maskawa [53] is characterised by three real angles (one of them the Cabbibo angle) and a complex phase which can account for CP violation (see section 7). The introduction of a second transformation between the $Q = \frac{2}{3}$ weak and mass eigenstates is redundant, and can be absorbed into U_{KM}. More detailed discussions of the CKM matrix are outside the scope of these notes.

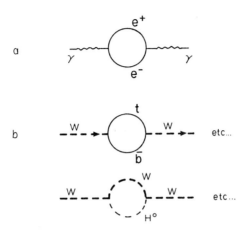

Figure 5.1   Typical electromagnetic (a) or weak (b) loop corrections.

The formulae given above must be modified for radiative corrections, either electromagnetic (for example fig. 5.1a) or weak (for example fig. 5.1b). Electromagnetic corrections are dominant and can be accurately evaluated (~ 6 % at m_z). Uncertainties of the weak correction mainly result from uncertainties of the t-quark and Higgs masses (m_t and m_H respectively). The qualitative dependence of ($m_z - m_W$) on m_t and m_H is shown later (see fig. 5.10); the difference is sensitive to heavy t-quark masses but rather insensitive to m_H.

The partial width for the decay $W \to f_1 \bar{f}_2$ can be shown to be

$$\Gamma(W \to f_1 \bar{f}_2) = \frac{G_F m_W^3}{6\pi\sqrt{2}} \cdot C, \qquad (5.8)$$

where C is a colour factor (= 1 for leptons, = 3 for fermions). For heavy fermion decays this is modified by a phase-space factor

$$H = (1-x)^2 (1 + x/2) \text{ where } x = (m_f/m_W)^2. \qquad (5.9)$$

The equivalent formula for Z-decay is

$$\Gamma(Z \to f\bar{f}) = \frac{G_F m_Z^3}{6\pi\sqrt{2}} \cdot C \cdot (g_V^2 + g_A^2) \qquad (5.10)$$

Formulae (5.8) and (5.10) are only slightly modified by higher-order electroweak and QCD corrections.

One would like to test the Standard Model as completely as possible. With the initial observation of the W and Z at approximately the expected masses at the CERN Collider, and the observation of neutral coupling of strength as expected for the Standard Model in several interactions over an extended kinematic range, in particular

i)   atomic parity violation,
ii) deep inelastic lepton and neutrino scattering,
iii) (ν_ee) and (ν_μe) scattering,
iv) the reactions $\bar{p}p \to Z + X$ and $\bar{p}p \to W^\pm + X$, and
v) e^+e^- interactions (in particular LEP),

the model is well verified. It is now important to determine if the charged and neutral couplings and the boson masses are exactly as predicted by the Standard Model, or if more exotic couplings are required.

In this section we first digress to describe the accurate identification and energy measurement of leptons in hadron collider experiments. We then describe the status of hadron collider measurements for

i)   the W and Z boson masses and widths,
ii) the W and Z decay asymmetry, and
iii) the independence of the charged and neutral couplings on the lepton or fermion family involved.

The accuracy of these measurements (present and future) are briefly compared with equivalent measurements from other machines, especially e^+e^- colliders. We defer discussion of the t-quark and its role in the Standard Model until section 6, and we discuss important future searches for weak gauge coupling and Higgs boson production in section 8.

5.1. A measurement of m_W and m_Z

In this subsection, the identification of leptons is described in detail, with emphasis on electron and muon detection at the CDF detector. Similar considerations are appropriate to UA1 and UA2 analyses. The identification of τ-leptons at the UA1 experiment is also described. A discussion of energy corrections to the electron and muon samples (again using CDF as an example) is then given, and finally a measurement of the W and Z mass and width is summarised, using the most recent results from the UA2 and CDF detectors.

5.1.1. Electron identification at CDF [9, 54]

Figures 2.8 and 5.2a show the detection principle for electron identification in existing hadron collider experiments. As already noted, the existence of an electromagnetic cluster with small lateral and longitudinal leakage is required, with an associated high-p_T track spatially matched to the cluster. Specifically for the W and Z data samples, the following selection criteria are used by CDF in the central rapidity region $|\eta| < 1.1$ (according to the physics analysis the acceptance and applied cuts may differ slightly) :

i)   a small hadronic leakage (E(had)/E(em) < 0.1) of the electromagnetic shower,
ii) a lateral shower profile consistent with electromagnetic clusters,
iii) a spatial match between an associated high-p_T track and a signal from the strip chamber embedded in the electromagnetic calorimeter of

$|\delta Z \text{ (trk-strip)}| < 30$ mm (parallel to beam axis)

$|\delta y \text{ (trk-strip)}| < 15$ mm (azimuthal direction),

iv) a charge profile of the strip chambers that is consistent with an electron shower, using a χ^2 test in each projection,
v) a good energy-momentum match (E/p < 1.4), and
vi) a track impact within the calibrated fiducial region of the calorimeter cells, in order to achieve good energy measurement.

The efficiency of these selections is ~ 80 %, depending on the exact cuts applied for a given analysis, and as well on the kinematic and acceptance regions involved. The most accurate evaluation of the electron identification efficiency is required for cross-section evaluations (see table 4a and section 4.2).

In some studies where well-isolated electrons are expected (for example (e-μ) pairs in $\bar{t}t$ production) an additional isolation requirement is made by requiring a limited energy deposition in the calorimeter in a cone in η-ϕ space about the electron. Typically, one requires

$$\frac{E \ (R_c < 0.4)}{E \ \text{(electron)}} < 0.1 \tag{5.11}$$

Additional kinematic criteria are required by the CDF experiment to obtain the Z and W event samples used for the evaluation of m_Z and m_W.

i) Two identified electrons of $E_T > 20$ GeV are required in the Z event sample. In the range $50 \leq m_{ee} \leq 150$ GeV, 73 electron pairs are measured of which 65 events have both tracks of adequate reconstruction quality to satisfy a beam-constrained fit. Of these 65 events, 58 events are in the mass range $75 \leq m_{ee} \leq 110$ GeV and are used for fits to the Z-mass.

ii) For the W event sample, it is required that $p_T^e > 25$ GeV and $p_T^\nu > 25$ GeV (as defined below) with no additional calorimeter energy deposition of $E_T > 7$ GeV in the event ($E_T > 5$ GeV within $\pm 30^0$ in azimuth from the electron direction).

5.1.2. Muon identification at CDF [9, 54]

Figures 2.8 and 5.2b show the principle of muon detection, in the central region ($|\eta| < 0.6$) of the CDF experiment. A final muon selection is made for the samples used for the W and Z mass evaluation by requiring the following selections.

i) A requirement $p_T^\mu > 11$ GeV ($p_T^\mu > 3$ GeV for $\mu\mu$ triggers) is made with an energy deposition in the calorimeter cell traversed by the muon of < 2 GeV (electromagnetic) and < 6 GeV (hadronic), and with a total energy deposition not exceeding 3.5 GeV.
ii) A good match of the selected muon central-detector track with the primary event vertex (< 5 mm transverse to the beam axis, and < 5 cm longitudinally, and with the muon track stub ($|\delta x| < 1.5$ cm azimuthally))
iii) For some analyses an isolation requirement is made to reduce punch-through by requiring no reconstructed jet axis ($E_T^J > 15$ GeV) within 10 degrees of the selected muon track, and in addition a cone requirement as for the electron samples.
iv) Cosmic rays provide a major background to identified high-p_T muons, and are rejected by requiring no track of $p_T > 10$ GeV/c within $\pm 3^0$ opposite the muon.

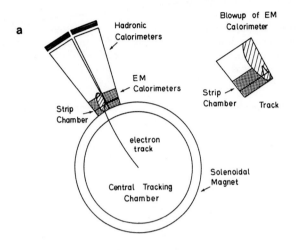
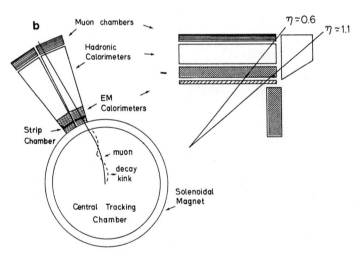

Figure 5.2. Detection of electrons (a) and muons (b) in the central region of the CDF detector.

Additional kinematic requirements for the W and Z samples are as for electrons. For the W and Z event samples, it is estimated that the muon identification efficiency is $\varepsilon = (0.98 \pm 0.015)$, before acceptance or additional isolation corrections. As for electrons, the selections described above differ slightly according to the analysis made.

Within the rapidity range $0.6 \leq |\eta| \leq 1.1$, no muon chambers presently exist (see fig. 2.11). However, if one lepton is already identified in the central region, for example an electron or muon in the dilepton decay of $t\bar{t}$ final states, the muon can be tagged if it is isolated from its associated minimum-ionising signal in the calorimeter. A rejection exceeding 100 against hadronic jet backgrounds is achieved using this selection.

5.1.3. Tau lepton identification at UA1 [19]

So far only the UA1 experiment has published data for the decay $W \to \tau \nu_\tau$. This decay is characterized for hadronic decays by a narrow low-multiplicity hadronic jet having a significant electromagnetic energy deposition, and with associated missing transverse momentum [55]:

$\tau^\pm \to 1\pi^\pm$ + neutrals + ν_τ : BR ~ .503 ± .006

$\tau^\pm \to 3\pi^\pm$ + neutrals + ν_τ : BR ~ 0.138 ± 0.003 (5.12)

Using an initial selection of 57 events obtained by requiring

i) a transverse missing momentum $\rlap{/}p_T > 15$ GeV and $\rlap{/}p_T > 4\ \sigma(\rlap{/}p_T)$,
ii) at least one jet of $E_T^J > 12$ GeV, and
iii) no additional identified electron or muon,

the UA1 Collaboration constructed from theoretical expectations the likelihood function

$L_\tau = \ell n \{(dN/dF) \cdot (dN/dr) \cdot (dN/dN_{CH})\}$, where (5.13)

$(dN/dK$ is the expected frequency distribution for variable K), and

i) F is the energy within radius $R_c = [\delta\phi^2 + \delta\eta^2]^{\frac{1}{2}}$ of the jet axis (a measure of jet collimation),
ii) r is the separation between the jet axis and the highest-energy track, and
iii) n_{CH} the charged particle multiplicity within the cone $R_c = 0.4$.

Figure 5.3 shows the distribution in L_τ with superimposed predictions for both τ-decays and hadronic QCD jets. Also shown is the distribution in $m_T^{\tau\nu}$ for events satisfying $L_\tau > 0$, with the superimposed prediction for $W \to \tau\nu_\tau$ decay.

5.1.4. The measurement of momentum and energy in the CDF experiment

Charged particle momenta are measured in the central detector of the CDF experiment with a momentum resolution (that includes a constraint on the transverse beam position) of

$\delta p_T / p_T = 0.0011 p_T$ (p_T in GeV), (5.14)

and a momentum scale uncertainty of less than < ± 0.1 %. This latter uncertainty is estimated from a comparison of reconstructed muon pairs with the expected J/Ψ and Y mass, and (more recently) the Z^0 mass from LEP experiments (see table 8).

Table 8. Comparison of measurement mass for J/Ψ, Y and Z^0 resonances, from reconstructed muon pairs, with PDG values [55].

	m(J/Ψ) GeV	m(Y) GeV	m(Z) GeV
m(CDF)	3.0963 ± 0.0005	9.457 ± 0.005	90.71 ± 0.45
m(PDG)	3.0969 ± 0.0001	9.460 ± 0.0002	91.161 ± 0.031

The electron energy scale and its associated uncertainty has been estimated in several successive steps.

i) Using test beam data, the response map of each cell type has been measured with an accuracy of < ± 1% over the cell area. In addition the mean calibration was monitored using radioactive sources and a laser flash system.

ii) Using an independent sample of 17000 inclusive electrons satisfying E_T > 15 GeV, all cells have been normalized to within an overall calibration to a statistical precision of ± 1.7 %. With increased integrated luminosity, this precision can be improved.

iii) The overall energy scale results from a comparison of the reconstructed energy E and momentum p for W electrons, with a prediction that includes radiative effects. The bremsstrahlung radiation of high energy electrons, and in addition the internal bremsstrahlung radiation from W and Z decay, means that E(cal) > p(track). From a gaussian fit to the E/p peak, the central peak value shifts by < 1 % from E=p, while the radiative tail shifts the mean value by ~ 3 %. Combining the statistical uncertainty of the measured E/p peak (± 0.16 %) with the momentum scale error and systematic uncertainties due to uncertainties of the material thickness and data selection criteria (± 0.15 %), the overall uncertainty of the energy scale is estimated to be ± 0.24 %.

In the upgraded UA2 experiment, the lack of a central magnetic field means that no comparison is available of the deposited electromagnetic energy and its associated track momentum. Therefore the scale uncertainty (< ± 1 %) results from frequent test-beam calibrations of parts of the calorimeter, and from the monitoring of calibrated calorimeter cells using radioactive sources and light flashers that are monitored by stable photomultipliers.

Because of the lack of (or less accurate) associated charged particle momenta, the CDF gas calorimeters are less well calibrated in energy.

5.1.5. Measurement of p_T^ν at CDF

We can write as in section 2 (equ. 2.3)

$$\vec{p}_T (\text{uncorrected}) = - \sum_i \vec{p}_T^i \qquad (5.15)$$

with $\vec{p}_T^i$ being the transverse momentum (or transverse energy E_T^i) associated with calorimeter cell i excluding that of the identified electron or muon. Then the corrected $\rlap{/}p_T$ is related functionally to $\rlap{/}p_T$ (uncorrected) as in equation (4.15) for the underlying event

$$\rlap{/}p_T^{corr} = f[\rlap{/}p_T \text{ (uncorrected)}] \tag{5.16}$$

The CDF experiment defines $p_T^\nu = 1.4\, \rlap{/}p_T$ (uncorrected), where the factor 1.4 accounts in an average way for the low-energy hadron response of the CDF calorimeter. Using minimum bias and well balanced 2-jet events, the CDF experiment measures the resolution of the uncorrected $\rlap{/}p_T$ to be

$$\sigma\,(\rlap{/}p_T \text{(uncorrected)}) = (0.47 \pm 0.03)\, \sum_i E_T^i, \tag{5.17}$$

with E_T^i being the uncorrected calorimeter cell energy for cell i. In the case of W production, the value of p_T^ν is at best only partially corrected with additional average or event-by-event corrections including for example the energy deposition nearby the electron or muon, the effect of cracks in the calorimeter, and the effect of magnetic field. At CDF, these effects on p_T^ν are corrected on average using a realistic model of $W^\pm$ production in the CDF detector.

Measurements of similar quality are available from the UA1 and UA2 experiments. It should be noted, however, that the absence of a magnetic field in the UA2 detector means that the correction required for low-p_T tracks that escape or modify calorimetric detection is avoided.

5.1.6. Measurement of the Z mass, m_Z

Using the previously discussed electron and muon samples (fig. 5.4), a fit is made to CDF data in the mass range $75 < m_{\ell\ell} < 105$ GeV, to a relativistic Breit Wigner shape convoluted with a gaussian resolution. The fitted mass is shown in table 9 for each of the ($\mu^+\mu^-$) and (e^+e^-) channels, with associated corrections and uncertainties.

Table 9. Fitted Z mass from the CDF experiment

Number of	$Z \to \mu^+\mu^-$ (tracking)		$Z^0 \to e^+e^-$ (calorimeter)	
events	123		65	
Fitted mass (GeV)	90.41 ± 0.40		91.18 ± 0.34	
Radiative corr.	+ 0.22	± 0.03	+ 0.11	± 0.03
Structure function	+ 0.08	± 0.03	+ 0.08	± 0.03
Mass scale		± 0.08		± 0.24
Corrected mass (GeV)	90.71 ± 0.4 ± 0.09		91.37 ± 0.34 ± 0.24	

Table 10 compares the CDF results [54] with recent data of similar quality from the UA2 experiment [56], and with the average of recent LEP data [55]. Hadron collider data is no longer competitive with measurement accuracies achievable at LEP, but they nevertheless remain important because of measurements of (m_W/m_Z) required to evaluate $\sin^2\theta_W$.

By allowing a 2-parameter fit, a direct measurement of the Z-width is obtained. The UA2 experiment measures

$$\Gamma_Z = 2.96 \pm {0.98 \atop 0.78}\, \text{GeV}, \tag{5.18}$$

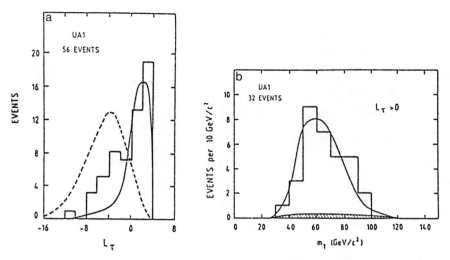

Figure 5.3. a) Distribution of the τ-likelihood for UA1 data. The superimposed solid curve is the Monte Carlo expectation for $W \to \tau\nu \to$ hadrons + $\bar{\nu}\nu$. The broken curve is that measured for hadronic jets in UA1, normalised to the total event sample. b) Transverse mass distribution for events satisfying $L_\tau > 0$. The solid curve is that expected for $W \to \tau\nu \to$ hadrons + $\bar{\nu}\nu$, normalised to the expected number of 28.7 events. The shaded area is the expected hadronic background of 2.7 events.

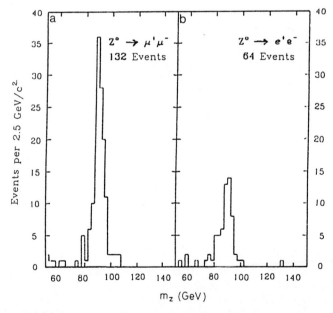

Figure 5.4. Published data from CDF showing the dilepton invariant mass distribution for (a) $Z \to \mu^+\mu^-$ and b) $Z \to e^+e^-$ candidates, evaluated using the track momenta.

consistent with the expected value Γ_Z = 2.5 GeV when $Z \to t\bar{t}$ is kinematically forbidden, but of limited accuracy when compared with comparable LEP measurements (Γ_Z = 2.55 ± .03 GeV) [57].

Table 10. Comparison of m_Z measurements at CDF, with data from the UA2 experiment and from LEP experiments.

	m_Z (GeV)
UA2 ($Z \to e^+e^-$)	91.49 ± 0.35 (stat) ± 0.12 (sys) ± 0.92 (scale)
CDF ($Z \to \mu^+\mu^-$)	90.71 ± 0.40 (stat) ± 0.09 (sys + scale)
CDF ($Z \to e^+e^-$)	91.37 ± 0.34 (stat) ± 0.24 (sys + scale)
LEP (PDG)	91.161 ± 0.031 (scale)

5.1.7. Measurement of the W mass m_W

The W mass is measured from a comparison of the distribution of data in one or more of the variables p_T^ℓ, p_T^ν or $m_T^{\ell\nu}$, with a full model of W production and decay that includes a realistic simulation of the underlying event, and major detector effects. Although p_T^ℓ is the most accurately measured quantity, an evaluation of m_W from the p_T distribution is sensitive to the shape of the production p_T distribution which has significant theoretical uncertainties and (in the absence of adequate Z-statistics for a measurement of $d\sigma/dp_T^Z$) is poorly measured (see section 4.2). The distribution of

$$m_T^{\ell\nu} = 2 p_T^\ell p_T^\nu [1 - \cos\phi_{\ell\nu}] \qquad (5.19)$$

where $\phi_{\ell\nu}$ is the difference in azimuth between the lepton and neutrino directions is less sensitive to p_T^W, but is sensitive to the resolution (gaussian and non-gaussian) of the p_T^ν measurement. Figure 5.5 shows the $m_T^{\mu\nu}$ and $m_T^{e\nu}$ distributions from the CDF experiment with a superimposed model prediction optimised in a maximum likelihood fit.

The major contributions to the W-mass uncertainty include :

i) the effect of background, in particular from QCD, Z-decay, the decay $W^\pm \to \tau^\pm \nu_\tau$; $\tau^\pm \to \ell^\pm \nu_e \nu_\tau$, and cosmics in the case of muons,
ii) the effect on m_W of using different structure function parametrisations,
iii) the effect of uncertainties of the p_T^e or p_T^μ measurements,
iv) the parametrisation used for p_T^W, and
v) uncertainties of both magnitude and direction of the $\vec{p}_T^\nu$ measurement.

Item (v) is sensitive to the deposited underlying event energy in the cell traversed by the W charged lepton, and to the leakage of the deposited lepton energy into neighbors of the traversed cell. A summary of the estimated fitting uncertainties are shown for the CDF experiment in table 11.

It is amusing to "guess" how the mass uncertainty might improve with a factor 5 improvement in statistics (as foreseen for the 1991 CDF run). Assuming a continued mass scale uncertainty of ± 80 MeV from tracking, and mainly statistical improvements to the p_T^W evaluation and estimates of $\vec{p}_T$ balance, a reasonable estimate for CDF is that of table 12. This

Table 11. Uncertainties in fitting the W mass using a fixed W width $\Gamma_W = 2.1$ GeV at the CDF experiment. All uncertainties are quoted in units of MeV. In parentheses are the statistical and overall mass uncertainties if Γ_W is determined as an additional parameter of the fit. The scale uncertainties are common with the scaled Z mass measurement. The contributing uncertainties which are the same for both samples are listed as common.

Uncertainty	Electrons	Muons	Common
STATISTICAL	350 (440)	530 (650)	
ENERGY SCALE	190	80	80
1. Tracking chamber	80	80	80
2. Calorimeter	175		
SYSTEMATICS	240	315	150
1. Proton structure functions	60	60	60
2. Uncertainties of p_T^e, p_T^ν and p_T^W	145	150	130
3. $\not{p}_T$ balance	170	240	
4. Background	50	110	
5. Fitting	50	50	50
OVERALL	465 (540)	620 (725)	

Table 12. Approximate uncertainty in MeV expected on W mass measurement at CDF, for an integrated luminosity of $L = 25$ pb^{-1}. Items 4 and 5 may have less uncertainty at the DØ experiment because of improved calorimeter measurements in the absence of a magnetic field.

	Electrons	Muons
Statistical uncertainty	140	215
Scale uncertainty	140	80
Systematics	163	163
1. Proton structure func.	50	50
2. Fitting	40	40
3. Background	50	50
4. Resolution effects	100	100
5. $\not{p}_T$ balance	100	100
Total	256	281

"guesstimate" may be optimistic if with the expected event pileup resulting from higher machine luminosity, p_T^ν measurements are degraded. This is not competitive with the accurate m_W measurements expected from LEP 2.

Table 13 shows the measured W-mass from CDF and equivalent measurements from the UA2 experiment. Because of a better p_T measurement and improved statistics the data from UA2 are slightly more accurate (excluding the scale uncertainty that is not important for measurements of the mass ratio).

Table 13. Measured W mass from the UA2 and CDF experiments.

	$m_W[W \to e\nu]$ (GeV)	$m_W[W \to \mu\nu]$ (GeV)	Combined (GeV)
UA2	80.79 ± 0.31 (stat) ± 0.21 (sys) ± 0.81 (scale)		
CDF	79.91 ± 0.35 (stat) ± 0.24 (sys) ± 0.19 (scale)	79.90 ± 0.53 (stat) ± 0.32 (sys) ± 0.08 (scale)	79.91 ± 0.39 (stat + sys + scale)

5.1.8. Measurement of Γ_W and Γ_Z

Values of Γ_W and Γ_Z can in principle be measured using the same maximum likelihood fits as those used for the determination of m_Z and m_W. However, since the experimental mass resolution is of the same order as the expected W and Z widths, these measurements are very sensitive to a proper knowledge of the measurement errors.

Direct measurements by UA2 (the systematic uncertainty is not quoted) and CDF of Γ_Z are

$$\text{UA2} : \Gamma_Z = 2.96 \, ^{+0.98}_{-0.78} \text{ (stat) GeV},$$

$$\text{CDF} : \Gamma_Z = 3.8 \pm 0.8 \text{ (stat)} \pm 1.0 \text{ GeV}. \tag{5.20}$$

These are to be compared with the weighted average of SLC and LEP results (which are compatible with Standard Model expectations)

$$\text{SLC + LEP} : \Gamma_Z = 2.546 \pm 0.032 \text{ GeV}. \tag{5.21}$$

Similarly a direct measurement of Γ_W for the UA2 experiment [56] is :

$$\Gamma_W = 1.89 \, ^{+0.47}_{-0.40} \text{ (stat) GeV}. \tag{5.22}$$

However, using result 5.21 above, and measurements of σ_W^ℓ and σ_Z^ℓ, it is possible within the context of the Standard Model to evaluate an accurate Γ_W value. From the values of σ_Z and σ_W described in section 4.2, $R = \sigma_W^e/\sigma_Z^e$ is measured for each of the UA1 [58], UA2 [40] and CDF [41, 59] experiments (see table 14).

Table 14. Measurement of $R = \sigma_W^e/\sigma_Z^e$

Exp.	R
UA1	$9.5 ^{+1.1}_{-1.0}$ (stat + sys)*
UA2	$9.38 ^{+.82}_{-.72}$ (stat) ± 0.25 (sys)
CDF	10.2 ± 0.8 (stat) ± 0.4 (sys)

* combined μ and e data

In the analysis of R in each experiment, stringent efforts have been made to minimise those uncertainties not common to both the $W^\pm$ and Z event samples. In practice, systematic uncertainties result mainly from differences in the W and Z acceptances, and differences in the selection efficiencies. The data are compared with expectations of the Standard Model following QCD corrections, in figure 5.6, as a function of the top quark mass [60]. The data of each experiment are compatible with 3 neutrino families as measured at LEP, and a heavy top quark mass. All experiments would benefit from improved statistical and systematic uncertainties. The theoretical uncertainties of figure 5.6 result in part from the range of available structure function parametrisations since, as noted previously, the ratio R is sensitive to the relative u and d quark contents of the nucleon. Now, R can be expressed [50, 60] as

$$R = \frac{\sigma_W^e}{\sigma_Z^e} = \frac{\sigma(\bar{p}p \to W + X)}{\sigma(\bar{p}p \to Z + X)} \cdot \frac{\Gamma(W \to e\nu)}{\Gamma(Z \to ee)} \cdot \frac{\Gamma(Z)}{\Gamma(W)} \qquad (5.23)$$

Since ratios of the total cross-section and partial widths are evaluated, theoretical uncertainties are now minimised, and using result (5.21) for Γ_Z together with measured R values,

$$\Gamma_W \text{ (UA1)} = 2.18 ^{+0.26}_{-0.24} \text{ (stat)} \pm .04 \text{ (theory) GeV}$$

$$\Gamma_W \text{ (UA2)} = 2.30 \pm 0.19 \text{ (stat)} \pm 0.06 \text{ (sys) GeV}$$

$$\Gamma_W \text{ (CDF)} = 2.17 \pm 0.20 \text{ (stat)} \pm 0.10 \text{ (sys) GeV} \qquad (5.24)$$

scaled from published data to take account of different Γ_Z values used by UA2 and CDF. This compares with a Standard Model expectation of Γ_W = 2.07 GeV assuming m_W = 80 GeV and no kinematically allowed t-quark decays. Future data will hopefully allow a more stringent comparison of Γ_W with Standard Model expectations.

5.2 Lepton Universality

It is an assumption of the Standard Model that (neglecting quark mixing) weak transitions occur within a fermion doublet independently of the doublet involved. To a good approximation, the squared lepton couplings are proportional to the associated gauge boson production cross-section :

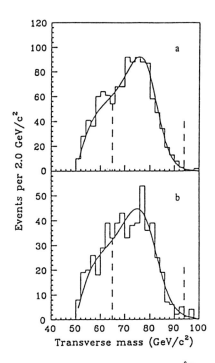

Figure 5.5. The transverse mass distribution $m_T^{\ell\nu}$ for (a) electron and (b) muon decays $W^\pm \to \ell^\pm \nu_\ell$ from CDF. The superimposed solid curve is a best fit to the data. The dashed lines represent the transverse mass range used for this fit.

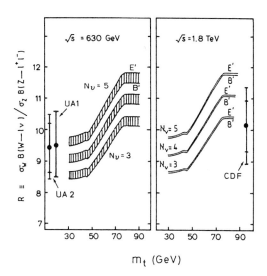

Figure 5.6. Dependence of $R = \sigma_W^e/\sigma_Z^e$ as a function of the top quark mass as calculated by Martin et al. [60] using the MRSE' and MRSB' structure functions. Also shown are measured values from CDF, UA1 and UA2.

137

$$g_1/g_2 = \left[\sigma_W^{\ell_1\nu}/\sigma_W^{\ell_2\nu}\right]^{\frac{1}{2}} \quad \ell^{\pm} = e^{\pm}, \mu^{\pm}, \tau^{\pm}$$

$$k_1/k_2 = \left[\sigma_Z^{\ell_1^+\ell_1^-}/\sigma_Z^{\ell_2^+\ell_2^-}\right]^{\frac{1}{2}} \tag{5.25}$$

Only the UA1 Collaboration [42] has so far published data on the cross-section ratios. They measure

$$g_\mu/g_e = 1.0 \pm 0.07 \pm 0.04$$

$$g_\tau/g_e = 1.01 \pm 0.10 \pm 0.06$$

$$k_\mu/k_e = 1.02 \pm 0.15 \pm 0.04 \tag{5.26}$$

A similar analysis presently in progress using CDF data should result in an improved statistical and systematic uncertainty on the (g_μ/g_e) and (k_μ/k_e) ratios. A better determination of g_τ/g_e is also of importance, and is currently being attempted using CDF data.

Both (g_1/g_2) and (k_1/k_2) are already well measured at LEP and other e^+e^- machines using the reactions ($e^+e^- \to \ell^+\ell^-$; $\ell = \mu, e, \tau$), and in lepton-nucleon interactions, with good agreement to the Standard Model. Deviations involving τ at the CERN and Fermilab $\bar{p}p$ Colliders might signal extensions or departures from the Standard Model, for example the existence of charged Higgs particles (see section 8).

5.3 Quark couplings to the W and Z gauge bosons

Tests of the quark universality of the neutral and charged couplings have been made at low Q^2 using e^+e^- and $\ell^{\pm}N$ interactions, and LEP has already provided accurate Z-decay measurements. At hadron colliders, the agreement between the measured and predicted cross-sections for W and Z production provides good evidence for the validity of Standard Model couplings. However, direct quark decays of the W and Z are more difficult to identify since although (excluding top decays)

$$\Gamma(Z \to \bar{q}q)/\Gamma(Z \to e^+e^-) \sim 21.1 \quad \text{and}$$
$$\Gamma(W \to q_1\bar{q}_2)/\Gamma(W \to e\nu) \sim 6.25 \tag{5.27}$$

the background from standard QCD processes exceeds 100:1. Any attempt to observe unbiased hadronic W and Z decays therefore requires

i) a good and unique association of quarks and their fragmentation products and a consequent optimisation of the (jet-jet) mass resolution and
ii) a good control of both the magnitude and shape of the QCD background and a minimisation of that background.

Only the UA2 Collaboration [61] has so far succeeded in identifying hadronic W and Z decays. In order to optimise the jet identification algorithm to minimise the energy resolution (see section 3.2), they use $W^{\pm}$ events from the Pythia event generator to define the cone size of the jet algorithm, taking into account realistic jet fragmentations and a realistic underlying event. For their calorimeter granularity, a cone size $R_c = 0.64$ resulted in an optimal (jet-jet) mass resolution.

At the trigger level, two-jet events were selected by requiring that 2 leading jets opposite in azimuth and defined by a rectangular window $(\delta\theta, \delta\phi) = (70°, 75°)$ should exceed 13 (10) GeV for the normal and pre-scaled triggers, respectively. Additional off-line selections were made to obtain a data sample having a well-controlled QCD background, using the jet reconstruction algorithms of section 3.2 :

i) $|Z(\text{vertex})| < 20$ cm,
ii) two centrally produced jets ($|\eta^J| < 0.7$) with no additional jets of $E_T^J > 20$ GeV elsewhere in the calorimeter,
iii) a requirement of between 20 and 80 % of the jet energy deposition in the electromagnetic calorimeter, to reject misidentified leptons and non-beam induced backgrounds such as cosmic rays.

Using these selections, the data of figure 5.7 have been fitted in the mass range $48 < m^{JJ} < 200$ GeV with a background of the form

$$\text{BKGRD} = A \cdot m^{-\alpha} \exp(-\beta m - \gamma m^2), \qquad (5.28)$$

and resonance line shapes about m_W and m_Z that assume

i) a gaussian resolution for each resonance, consistent with the experimental resolution.
ii) $m_Z/m_W = 1.\underline{14}$
iii) $\sigma_Z(q\bar{q})/\sigma_W(q\bar{q}) = 0.43$, and include
iv) QCD interference terms with the W and Z.

Excellent fits to the data are obtained and two typical fits are shown in table 15.

Table 15. Typical W-Fit measurements

BACKGROUND	Fixed resolution	Variable resolution
α	8.1 ± 0.02	7.8 ± .02
β (10^2)	− 3.65 ± 0.04	− 2.93 ± 0.04
γ (10^5)	9.93 ± 0.16	8.17 ± 0.18
SIGNAL		
N (ev)	6091 ± 1062	5618 ± 1334
σ_m (%)	10.0 (fixed)	9.9 ± 2.5
m_W (GeV)	79.9 ± 1.1	79.2 ± 1.7
χ^2/NDF	115/122	114/121

Expected uncertainties in mass reconstruction accuracy have been estimated using the Pythia event generator. The result is shown in table 16, and suggests a bias of up to 5 GeV resulting from the underlying (and rapidly falling) QCD background. Additional systematic uncertainties include the calorimeter response uncertainty, and energy scale uncertainties (see table 16).

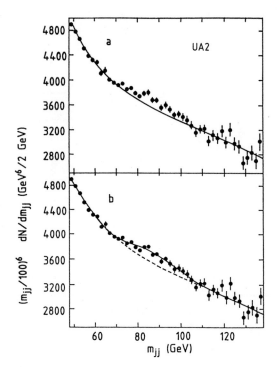

Figure 5.7. The 2-jet mass spectrum ($48 < m_{jj} < 138$ GeV) from the UA2 experiment. The vertical axis is the number of events in each bin, scaled by ($m_{jj}/100)^6$. a) Background fit, excluding the range $70 < m_{jj} < 100$ GeV. b) Combined fit to QCD background and W, Z signal described by 3 parameters explained in the text. The dashed line shows the background contribution.

Table 16. The reconstructed mass from the Pythia event generator for a generated mass $m_W = 81$ GeV, together with associated experimental uncertainties.

	Reconstructed W mass (GeV)
data	78.9 ± 1.5
PYTHIA ($m_W = 81$ GeV)	76.4 ± 1.0
underlying event uncertainty	± 2.0
calorimeter response uncertainty	± 2.5
single particle calibration uncertainty	± 1.2

An evaluation of the cross-section for W and Z production has been made following careful efficiency and background studies. They obtain

$\sigma \cdot B(W/Z \to 2\text{-jet}) = 9.6 \pm 2.3$ (stat) ± 1.1 (sys) nb
(mass resolution not constrained)

10.5 ± 1.3 (stat) ± 2.5 (sys) nb
(mass resolution constrained) (5.29)

This is a factor $K = 1.71 \pm 0.45$ larger than that expected from recent $O(\alpha_s^2)$ calculations of the W and Z cross-section as described in section 4, that is approximately a 1.5 Standard deviation effect.

This measurement demonstrates the difficulty of precision quark-coupling measurements at hadron colliders. More importantly, however, it demonstrates the feasibility of jet spectroscopy at hadron colliders. In particular, the DØ experiment, with its excellent calorimeter coverage and good calorimeter resolution, should be well prepared for exploratory high-mass jet spectroscopy studies.

5.4 The charge asymmetry of $W^\pm$ and Z production decay

As already noted (fig. 4.14) the $W^\pm$ cross-section at collider energies is dominated by quark fusion involving at least one valence quark. Therefore, if as expected the $W^\pm$ couples to left handed doublets the charged decay lepton should be aligned with respect to the incident proton in the $W^\pm$ rest fame with angle on θ^* given by

$$dN/d\cos\theta^* = (1 - q\cos\theta^*)^2 + 2\alpha\cos\theta^*, \qquad (5.30)$$

where $q = -1$ for electrons and $\alpha = 0$ for (V-A) coupling. The value of θ^* is not defined for $p_T^W \neq 0$ and the Collins-Soper [37] convention is used with a constraint $m^{e\nu} = m^W$ (see section 4.2). At CERN energies, the choice of the solution having minimum $|p_L^W|$ for ambiguous solutions gives the correct solution $\gtrsim 70$ % of the time; at the Tevatron, this is < 50 % because of the increasing sea contribution. As already noted in section 4.2, the expected asymmetry of (5.30) is consequently modified by structure function effects in localised η-ranges. Taking these effects into account, data from UA1 [42] shown in figure 5.8 are in good agreement with the Standard Model.

Because Z-coupling is not purely left-handed, the asymmetry is small. Using g_V and g_A as defined earlier, and defining θ^* to be the angle between the outgoing electron and incoming quark in the electron pair rest frame, it can be shown that at lowest order [62]

$$\frac{d\hat{\sigma}}{d\cos\theta^*} = N_c \int dx_a \int dx_b \sum_q \left[f_q^p(x_a) \, \bar{f}_q^{\bar{p}}(x_b) + f_q^p(x_a) \, \bar{f}_q^{\bar{p}}(x_b) \right]$$

$$\frac{\pi\alpha^2}{2\hat{s}} [A (1 + \cos^2\theta^*) + B \cos\theta^*]. \qquad (5.31)$$

In this formula N_c is a colour factor, $\hat{s}$ is the squared parton centre-of-mass energy, and $f(x)$ is the parton distribution function in the proton or antiproton. The distribution is integrated over the nucleon quark distributions and summed over available quark flavours. The coefficients A and B are

$$A = Q_f^2 + 2Q_f \, g_V^f \, g_A^\ell \, \mathrm{Re}(\chi) + (g_V^{\ell 2} + g_A^{\ell 2})(g_V^{f 2} + g_A^{f 2}) |\chi|^2$$

$$B = 4 \, g_A^\ell \, g_A^f \, \mathrm{Re}(\chi) + 8 \, g_V^\ell \, g_A^\ell \, g_V^f \, g_A^f \, |\chi|^2$$

with $\chi = \chi(M_Z, \hat{s})$ \hfill (5.32)

As a result, the forward-backward asymmetry can easily be shown to be

$$A_{FB} \propto \frac{g_V^\ell \, g_A^\ell \, g_V^f \, g_A^f}{\left[g_V^{f 2} + f_A^{f 2} \right] \left[g_V^{\ell 2} + g_A^{\ell 2} \right]} \cdot K(m_Z, \hat{s}) \qquad (5.33)$$

The most accurate Z-asymmetry measurements are from the CDF experiment [63]. An event sample of 232 dielectron events is selected satisfying

i) $75 < m^{ee} < 105$ GeV,
ii) $p_T^e > 15$ GeV, and
iii) $|\eta^e| < 1.0$ for at least one identified electron.

The $\cos\theta^*$ distribution for this sample is shown in figure 5.9. From this,

$$A_{FB} = 0.053 \pm 0.060 \text{ (stat)}, \qquad (5.34)$$

corresponding after correction for higher-order QED and QCD processes and using the Marciano-Sirlin [64] definition of $\sin^2\theta_W$ appropriate to $\bar{p}p$ colliders, to (eqn. 5.5)

$$\sin^2\theta_W = 0.231 \pm 0.016 \text{ (stat)} \pm 0.002 \text{ (sys)}. \qquad (5.35)$$

Since this measurement is insensitive to systematic effects (for example the parton distribution), the measurement will with improved statistics be competitive with that measured directly from the mass ratio, noted below. Similar asymmetry measurements at LEP have been made, giving when combined with determinations from the Z leptonic width [57]

$$\sin^2\theta_W = 0.2302 \pm 0.0021 \text{ (stat + sys)} \qquad (5.36)$$

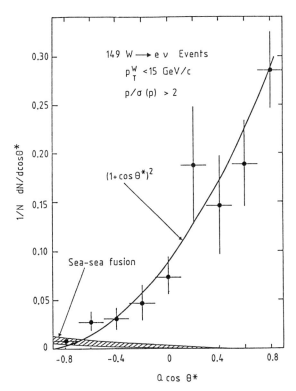

Figure 5.8. Decay angle distribution of $W \to e\nu_e$ decays in the UA1 experiment. The shaded band shows the expected contribution from annihilation processes involving wrong polarity sea-quarks only.

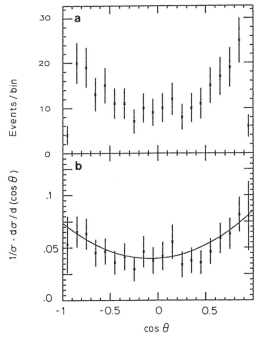

Figure 5.9. The angular distribution of electrons from the decay $Z \to e^+e^-$ before (a) and after (b) acceptance corrections. The solid line is the result of a likelihood fit. CDF preliminary.

The agreement of LEP and CDF data is additional evidence for the validity of the Standard Model couplings g_V^f and g_A^f. However, because of the large Z sample at LEP, existing hadron colliders can no longer provide a competitive $\sin^2\theta_W$ measurement at momentum transfer $Q^2 = m_Z^2$.

5.5 Measurement of $\sin^2\theta_W$

From section 5.4, $\sin^2\theta_W$ was measured in the Marciano-Sirlin framework at $Q^2 = m_Z^2$, from asymmetry measurements of Z^0 decay, to be

$$\sin^2\theta_W = 0.231 \pm 0.016 \text{ (stat)} \pm 0.002 \text{ (sys)} \text{ (CDF)}$$

From direct measurements of m_W and m_Z, in the CDF experiment (after combining the the electron and muon decay channels), equation (5.7) can be used for a $\sin^2\theta_W$ measurement. Combining the statistical and systematic uncertainties, and using the world average Z mass (table 8)

$$\sin^2\theta_W \text{ (CDF)} = 0.2317 \pm 0.0075 \tag{5.37}$$

The equivalent UA2 result is [56] :

$$\sin^2\theta_W \text{ (UA2)} = 0.2202 \pm 0.0095. \tag{5.38}$$

Both results may be compared with the world average [65] derived from non-LEP neutral current experiments but corrected for the renormalisation of Marciano and Sirlin appropriate to the definition $\sin^2\theta_W = (1 - m_W^2/m_Z^2)$

$$\sin^2\theta_W \text{ (world)} = 0.2309 \pm 0.0029 \text{ (stat)} \pm 0.0049 \text{ (sys)}. \tag{5.39}$$

The corresponding result at LEP, derived from the Z leptonic width and corrected for the definition above, is [57, 43] as in (5.36) above

$$\sin^2\theta_W = 0.2302 \pm 0.0021 \text{ (stat + sys)}.$$

The data of UA2 and CDF are compared in figure 5.10 after scaling to the LEP mass measurement for the Z. The variation of m_W with m_{top} is also shown in figure 5.10. From CDF data alone, a limit $m_t < 230$ GeV (95 % CL) can be interpreted if $m_H < 1$ TeV. That limit can be significantly more stringent if the world average is used, and this has been studied by Langacker and others [57, 66], who suggest a preferred top mass $m_t \sim 140$ GeV. Combining the data of the LEP experiments, the data of UA1 and UA2, and the neutrino data of CDHS [67] and CHARM [68], m_t is predicted in the context of the Standard Model to be $m_t = 137 \pm 40$ GeV. It is unlikely from collider measurements to obtain an m_W accuracy $\delta(m_W) \lesssim 150$–200 MeV. MeV. If however the W mass can be measured to ± 50 MeV at LEP 2, and if the top mass can be identified measured to within ± 5 GeV, realistic constraints result for the mass of the Higgs (fig. 5.10b), within the context of the Standard Model.

6. HEAVY QUARK PHYSICS : SEARCHING FOR THE TOP QUARK

The top quark is an essential ingredient of the Standard Model. As already noted, the GWS model assumes that quarks and leptons are arranged in families consisting of a left-handed doublet and right-handed singlets. It is assumed that the coupling of different families by the vector bosons (γ, $W^\pm$ and Z) are identical. Of the elements shown in table 7 of section 5.1, only the ν_τ and t-quark have not been directly identified.

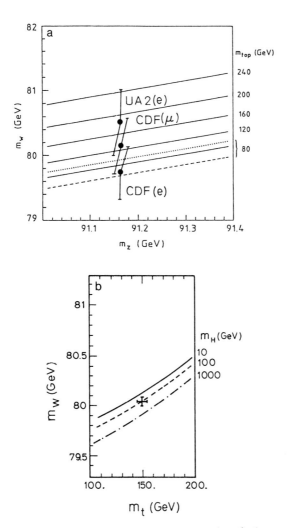

Figure 5.10. a) UA2 and CDF mass measurements (scaled to the LEP mass average), with predictions of the Standard Model in a range of m_t and values (m_H = 100 GeV) and for m_H = 10 (1000) GeV with m_t = 80 GeV. b) Expected sensitivity of LEP2 measurements, assuming a W mass measurement to within ± 50 MeV, and a top quark mass measurement of within ± 5 GeV, to estimates of the Higgs mass for the Minimal Standard Model (m_Z = 91 GeV).

The t-quark has been the object of intensive searches. Using the reaction

$$e^+e^- \to Z^0 \to t\bar{t} \tag{6.1}$$

at LEP and SLC, a limit $m_t > 45.8$ GeV has been obtained [57, 69]. A direct t-quark identification at hadron colliders has also been sought without success. Because $(t\bar{t})$ production is strongly dependent on $\sqrt{s}$ at hadron colliders (see section 6.1), the best limit so far has been obtained by the CDF experiment, at $\sqrt{s} = 1.8$ TeV :

$$\begin{aligned} &\text{UA1} : m_t > 60 \text{ GeV} \quad (95\% \text{ CL}) \quad [70] \\ &\text{UA2} : m_t > 67 \text{ GeV} \quad (95\% \text{ CL}) \quad [71] \\ &\text{CDF} : m_t > 89 \text{ GeV} \quad (95\% \text{ CL}) \quad [72] \end{aligned} \tag{6.2}$$

assuming Standard Model couplings. In these notes we describe the t-quark search at CDF, and we outline prospects for extending the quoted mass limit at the CDF and DØ experiments in future runs.

Despite the fact that the t-quark has not been found, there is considerable prejudice in favour of its existence, and subsequent indirect limits on the top mass within the context of the Standard Model (or model dependent extensions thereof).

i) A measurement of the forward-backward asymmetry of the reaction
$e^+e^- \to b\bar{b}$ is expected to be of the form
$$A_{FB} = K(s, m_Z^2) \, I_3^e \cdot I_3^b / Q^b, \tag{6.3}$$

where I_3^b and Q^b are respectively the weak isospin and charge of the b-quark. I_3^b is measured to be ≈ -0.5 [73], consistent with a left-handed doublet. If the b-quark is a singlet, $I_3^b = 0$.

ii) Just as the GIM mechanism [74], introduced to explain the experimentally measured absence of flavour changing neutral currents, required the existence of the c-quark as a doublet partner of the s-quark, so a partner is needed in most models for the b-quark [75].

iii) $(\bar{B}^0\text{-}B^0)$ mixing measurements impose limits on the CKM matrix parameters (in particular the $t \to s$ and $t \to d$ quark transitions) because of the mixing diagrams (see section 7.4 and fig. 7.15). Within the context of the Standard Model with 3 families, these mixing data impose a lower limit on m_t [76].

iv) As noted in section 5.1.8, the cross-section ratio (σ_W^e/σ_Z^e) provides a limit on m_t if $m_t < m_W$ which is approximately independent of Standard Model decay couplings to the t-quark.

v) Standard Model calculations of m_W and m_Z (see section 5.1) are sensitive to the existence of the t-quark because of radiative corrections. From existing measurements of m_W at CDF and UA2, and m_Z and $\sin^2\theta_W$ at LEP and/or the CDHS and CHARM experiments, a limit $m_t = 137 \pm 40$ GeV (see section 5.5) can be estimated, again subject to the validity of the Standard Model.

6.1 t-quark production and decay

At Fermilab Tevatron energies, the production of t-quarks is dominated at all quark masses by $(t\bar{t})$ production (or more generally heavy quark pair production), with important lowest-order contributions from $q\bar{q}$ fusion and with substantial (gg) and (qg) corrections (fig. 6.1). At $\sqrt{s} = 630$ GeV, however, $W^\pm$ production with subsequent $W \to tb$ decay is the dominant t-quark production mechanism for $m_t < m_W$.

Nason et al. [77] have calculated the heavy quark production rate after including higher-order corrections. The dominant uncertainty of the calculation is from the choice of Q^2-scale, and from the choice of structure function parametrisations. They estimate an uncertainty of $\lesssim \pm 30$ % for $m_t \sim 50$ GeV and at higher mass this uncertainty is reduced.

The $\sqrt{s}$-dependence of $(\bar{t}t)$ production is shown in figure 6.2. For $m_t = 100$ GeV,

$$\sigma_{t\bar{t}}(\sqrt{s} = 1.8 \text{ TeV}) \simeq 50\, \sigma_{t\bar{t}}(\sqrt{s} = 0.63 \text{ TeV}), \tag{6.4}$$

making the Tevatron Collider a unique exploratory tool for new heavy quark states during the coming years. Assuming direct $(\bar{t}t)$ production and Standard Model decays $(t \to b + W^+)$ with the W either real or virtual depending on m_t, table 17 shows the diversity of allowed decay channels.

Table 17. Decay channels for direct $(\bar{t}t)$ production

PROCESS	Decay mode	Σ (branching ratio)	Final state (schematic only)
A	$t \to bq_1\bar{q}_2$ $\bar{t} \to \bar{b}q_3\bar{q}_4$	~ 0.44	$\bar{p}p \to$ 6-jets + X
B	$\left. \begin{array}{l} t \to b\ell\nu_\ell \\ \bar{t} \to \bar{b}q_1\bar{q}_2 \end{array} \right\}$ + c.c.	~ 0.15 (*3)	$\bar{p}p \to e^\pm\nu_e$ + 4-jets + X $\bar{p}p \to \mu^\pm\nu_\mu$ + 4-jets + X $\bar{p}p \to \tau^\pm\nu_\tau$ + 4-jets + X
C	$t \to b\ell_1\nu_\ell$ $\bar{t} \to \bar{b}\ell_2\bar{\nu}_\ell$	~ 0.012 (*3) ~ 0.025 (*3)	$\bar{p}p \to \ell_1^+\ell_1^- + \nu_\ell\bar{\nu}_\ell$ + 2-jets + X $\bar{p}p \to \ell_1^+\ell_2^- + \nu_\ell\bar{\nu}_\ell$ + 2-jets + X [ℓ - e, μ, τ]

Decays of type A are not rate limited ($\sigma_{t\bar{t}}$(b-jets) $\simeq 45(4.5)$ pb at $m_t \sim 100(150)$ GeV) prior to acceptance and efficiency cuts. However, because of

i) the very large QCD multi-jet background,
ii) the misidentification of (n-1)-jet events as n-jet events because of fluctuations in the underlying event and inadequacies of the jet algorithm,
iii) the loss of low-E_T b-jet events due to acceptance, 2-jet merging, or poor low-E_T jet identification efficiency, and
iv) limitations of the multi-jet mass reconstruction accuracy,

the sensitivity to a t-quark signal is poor.

The CDF experiment uses final states of type B,

$$\bar{p}p \to e^\pm\nu_e + \geq 2 \text{ jets} + X$$
$$\bar{p}p \to \mu^\pm\nu_\mu + \geq 2 \text{ jets} + X, \tag{6.5}$$

for their top search. As will be shown schematically in figure 6.8 below, the transverse mass distribution $m_T^{\ell\nu}$ will for $m_t < m_W$ be observed as a deviation from the expected $m_T^{\ell\nu}$ line shape for W-decays from the dominant background QCD process $\bar{p}p \to W^\pm + \geq 2$ jets + X. For $m_t \sim m_W$, these distributions become identical and discriminatory power is lost. For $m_t \geq m_W$

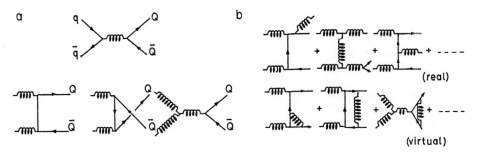

Figure 6.1. a) Typical lowest order diagrams for heavy quark production in $\bar{p}p$ collisions at CDF. b) Some typical higher-order contributions to heavy quark production.

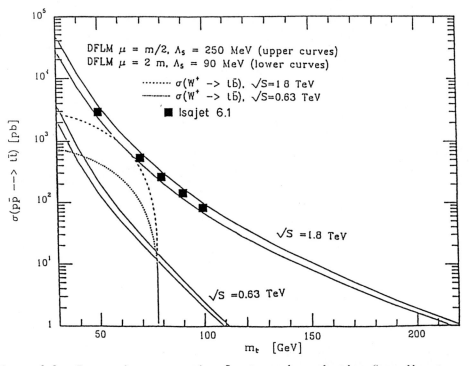

Figure 6.2. Expected cross-section for t-quark production from direct ($\bar{t}t$) pair production [77, 78], and from $W^{\pm}$ production followed by the decay $W \to t\bar{b}$. Also shown is the predicted rate for ($\bar{t}t$) production from the ISAJET generator.

additional t-quark identification criteria are required. In the case of CDF, a search is made for semileptonic decays of the associated b-jet. Multivariate analyses can also be used as noted in section 6.6. The other major background to this final state comes from direct $(\bar{b}b)$ production with the semileptonic decay of one b-quark. By requiring lepton isolation this background is small for $m_t \geq 40$ GeV.

At the expense of a low rate, dilepton final states of type C have a clean signature with little QCD background from misidentified electrons and muons. In addition to $(\bar{b}b)$ backgrounds that are rejected by isolation criteria as above, like-flavour lepton pairs (e^+e^-, $\mu^+\mu^-$) have additional backgrounds from Drell-Yan production and from direct Z production. However, unlike-flavour pairs ($e^+\mu^-$, $e^-\mu^+$) have as major additional known competing physics processes only

$$\bar{p}p \to Z + X; \quad Z \to \tau^+\tau^-; \quad \tau^\pm \to e^\pm \nu_e \nu_\tau \text{ or } \mu^\pm \nu_\mu \nu_\tau,$$
$$\bar{p}p \to W^+W^- + X; \quad W^+ \to \ell_1 \bar{\nu}_\ell; \quad W^- \to \ell_2 \bar{\nu}_\ell \tag{6.6}$$

At CDF the dilepton states (e^+e^-), ($\mu^+\mu^-$) and ($e^+\mu^-$, $e^-\mu^+$) have been used to search for the t-quark. After nominal kinematic selection efficiency and acceptance requirements are made, the ($e\mu$) rate expected at CDF for $m_t \sim 100$ GeV is approximately 20 % of the 2 pb ($e\mu$) cross-section for direct $\bar{t}t$ production. By using ($e\tau$) and ($\mu\tau$) final states, the dilepton acceptance could in principle be increased by a factor 3 (corresponding to ~ 25 GeV in the m_t limit). However, because of identification inefficiencies the real improvement of acceptance is small. Nevertheless, efforts to identify clean ($e\tau$) and ($\mu\tau$) samples should have a high priority since measured deviations from lepton decay universality may imply non-standard top decays such as $t \to H^+b$ (section 6.5).

6.2 The t-quark search at CDF using dilepton final states (type C)

The CDF analysis defines an ($e^+\mu^-$, $e^-\mu^+$) sample by requiring an identified electron and muon in the central region using lepton selection criteria similar (but not identical) to those defined in section 4. Specifically, selection criteria for electrons include :

i) a small longitudinal (< 5 %) and lateral leakage of the electromagnetic cluster into adjacent cells,
ii) strip-chamber charge distributions consistent with those expected for electrons ($\chi_\phi^2 < 10$ and $\chi_z^2 < 10$), with a good spatial match to an extrapolated high-p_T track ($|\delta x| < 15$ mm, $|\delta z| < 30$ mm),
iii) a good match (E/p < 1.5) between the measured calorimeter energy and electron momentum measurement, and
iv) a track impact within the calibrated fiducial region of the electromagnetic cell.

For muons the selection requirements are :

i) E^{em} (traversed electromagnetic calorimeter cell) < 2 GeV
   E^{had} (traversed hadronic calorimeter cell) < 6 GeV
   ($E^{em} + E^{had}$) > 0.1 GeV, and
ii) either a track segment in the muon chambers ($|\eta| < 0.65$) with a matching of the extrapolated track ($\delta\phi < 10$ cm), or alternatively a high-p_T central track with an isolation requirement $E_T < 5$ GeV in a cone $R_c = 0.4$.

Using selections $p_T^\mu > 15$ GeV and $E_T^e > 15$ GeV, figure 6.3 shows the ($e\mu$) acceptance and efficiency as a function of m_t. Figure 6.4 shows the

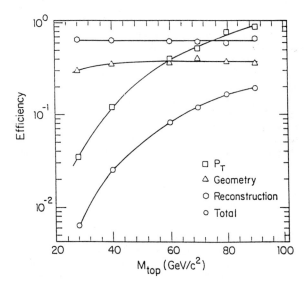

Figure 6.3. Estimated (eμ) efficiency and acceptance as a function of m_t, in analyses of CDF data.

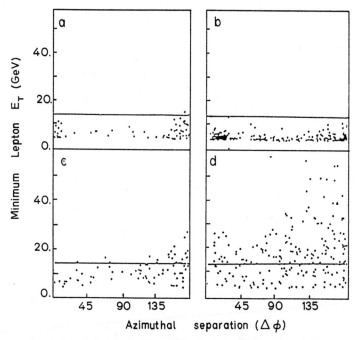

Figure 6.4. Scatter plots of the minimum lepton transverse energy vs. the azimuthal eμ angle difference for **a)** data (4.4 pb^{-1}), **b)** $\bar{b}b$ Monte Carlo (0.64 pb^{-1}), **c)** t-quark Monte Carlo with m_t = 28 GeV (2.1 pb^{-1}) and, **d)** idem with m_t = 70 GeV (79 pb^{-1}).

distribution of $p_T^{min} = \min(p_T^e, p_T^\mu)$ versus the azimuthal separation of the electron and muon. Also shown is the expectation for $(\bar{b}b)$ production and for direct $(\bar{t}t)$ production assuming either $m_t = 28$ GeV or $m_t = 70$ GeV. One $(e^+\mu^-)$ event exists, with $p_T^\mu = 42.5$ GeV and $E_T^e = 31.7$ GeV, with expected backgrounds of 0.7 events from $Z \to \tau^+\tau^+$ production, 0.2 events from WW or WZ production and 0.2 events from $(\bar{b}b)$ production. Uncertainties of the expected $(e\mu)$ production rate include :

i) the t-quark production and fragmentation,
ii) the estimated electron and muon efficiency and acceptance, and
iii) the integrated luminosity.

The nett uncertainty varies between ± 45 % at $m_t = 28$ GeV and ± 19 % for $m_t \geq 70$ GeV.

By including the (e^+e^-) and $(\mu^+\mu^-)$ channels, a 60 – 70 % increase in acceptance is obtainable. Similar electron and muon identification selections are used, with the addition of an isolation cut of $E_T < 5$ GeV within a cone $R_c = 0.4$ for electron candidates. After removing Z and Drell-Yan events with kinematic requirements

$75 < m^{\ell\ell} < 105$ GeV,
$\not{E}_T > 15$ GeV, and
$20^0 < \delta\phi^{\ell\ell} < 160^0$,   (6.7)

figure 6.5 shows the $E_T^{\ell_1}$ vs. $E_T^{\ell_2}$ distribution expected for $(\bar{t}t)$ production with $m_t = 90$ GeV. No event satisfies these kinematic selections.

Combining the $(e^+\mu^-, e^-\mu^+)$, (e^+e^-) and $(\mu^+\mu^-)$ channels, table 18 compares the number of expected and observed events at CDF.

Table 18. Expected and observed lepton pair events at CDF

Channel $(\ell_1^+ \ell_2^-)$	$m_t = 80$ GeV	$m_t = 90$ GeV	Observed events
$e^\pm\mu^\mp$	4.6	3.0	1
e^+e^-	1.4	0.7	0
$\mu^+\mu^-$	1.7	1.1	0
Total	7.7 (± 20 %)	4.8 (± 20 %)	1

The 95 % CL upper limit on the $(\bar{t}t)$ production was evaluated by finding the mean of a parent Poisson distribution that, when scaled by the measured acceptance and convoluted with a gaussian distribution representing the total $(\bar{t}t)$ acceptance uncertainty, yields a 5 % probability of observing ≤ 1 event. The resulting cross-section limit is compared with the predictions of Altarelli et al. [78] in figure 6.6. The measured upper limit on the t-quark mass from dilepton channels is

$m_t > 84$ GeV (95 % CL).   (6.8)

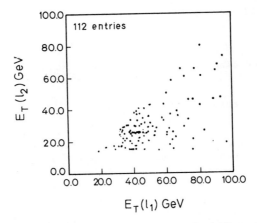

Figure 6.5. The distribution of $E_T(e_1)$ vs $E_T(e_2)$ for electron pairs following selections described in the text, for Monte Carlo ($\bar{t}t$) events generated with m_t = 90 GeV. No event satisfies these criteria.

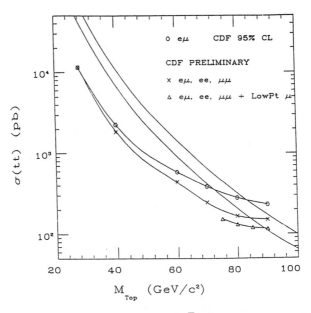

Figure 6.6. Upper limit on the measured ($\bar{t}t$) cross-section obtained from lepton pairs as a function of m_t, with superimposed Standard Model expectations. Also shown is the limit obtained when b-quark tagging is included as described in section 6.3.

6.3 The CDF top search using events $\bar{p}p \to \ell^{\pm}\nu_e$ + jets (type B)

At CDF, an event sample used to search for the top quark is

$$\bar{p}p \to \ell^{\pm}\nu_{\ell} + (\geq 2 \text{ jets}) + X; \quad \ell = e, \mu, \qquad (6.9)$$

with the electron (muon) selected in the pseudorapidity ranges $|\eta| \leq 1.1$ and $|\eta| \leq 0.6$ respectively. The identification selection criteria are similar to those already described. Additional kinematic requirements include

i) $E_T^{\ell} > 15$ GeV and $\not{E}_T > 15$ GeV, and
ii) $E_T^{jet} \geq 10$ GeV in the central rapidity region ($|\eta^J| \leq 2$) for the two leading jets.

As already noted, at least 4 jets should in principle be reconstructed for direct $(\bar{t}t)$ production. However, a requirement $E_T^J \geq 10$ GeV is necessary to unambiguously identify the jet, and the quoted kinematic selections with ≥ 2 jets are only 50 % (80 %) efficient for direct $(\bar{t}t)$ production at $m_t = 40$ GeV (80 GeV). A total of 512 electron events satisfy reaction (6.9) with $E_T^e > 15$ GeV after the removal of conversions and identified electron pairs with mass $m^{ee} > 70$ GeV. The distribution of E_T^e vs. $\not{E}_T$ is shown in figure 6.7. In following analyses two selections shown in figure 6.7 are used. A total of 104 events satisfy $E_T^e > 20$ GeV and $\not{E}_T > 20$ GeV. Similarly, 87 muon events satisfy reaction (6.9) with $p_T^{\mu} > 20$ GeV and $\not{E}_T > 20$ GeV. After accounting for the differing identification efficiency and acceptance, there is excellent agreement between the electron and muon data samples. Figure 6.8 shows the $m_T^{e\nu}$ projection of the 104 electron events noted above, with superimposed Standard Model expectations for background (W + 2 jet) production and for direct $(\bar{t}t)$ production assuming $m_t = 70$ GeV. Also shown are data using identical identification criteria for the reaction

$$\bar{p}p \to \ell^{\pm}\nu_{\ell} + (1 \text{ jet}) + X; \quad \ell = e, \mu \qquad (6.10)$$

with superimposed QCD expectations as discussed in section 4. The data are well described by standard QCD W-production processes, after detector effects are taken into account.

The published CDF results have used only the electron data, and estimate a possible t-quark contribution using a fit

$$\frac{dN}{dm_T^{e\nu}} = \alpha \, T(m_T^{e\nu}) + \beta \, W(m_T^{e\nu}) \qquad (6.11)$$

where T is the predicted $m_T^{e\nu}$ distribution for $(\bar{t}t)$ events and W is that for (W + $\geq$ 2 jet) events. The parameters α and β are normalized to 1.0 for expected Standard Model rates, and the distribution T is taken from the ISAJET Monte Carlo program that has been shown (fig. 6.2) to agree with the calculations of Nasen et al. [77]. Fitted values of α and consequent upper limits to the $(\bar{t}t)$ cross-section are shown in table 19. Systematic uncertainties contributing to the normalisation and/or the $m_T^{e\nu}$ shape include :

i) the jet and lepton energy scale and resolution, and contributions from the underlying event,
ii) uncertainties of the model used for top quark production and fragmentation, including uncertainties of initial state radiation which will complicate the event topology,
iii) uncertainties of lepton and jet identification efficiency and acceptance, and
iv) the luminosity uncertainty.

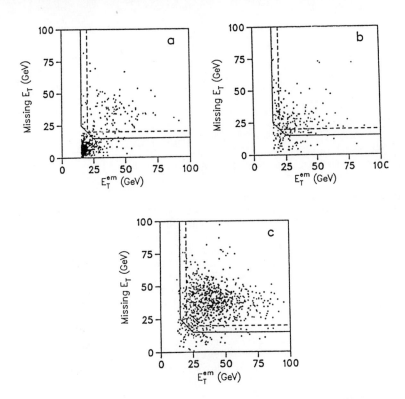

Figure 6.7. Scatter plot of $\not{E}_T$ vs E_T^e for events with an electron and two or more jets. **a)** Data, with the contours indicating tight and loose kinematic selections, **b)** Monte Carlo $\bar{t}t$ events with $m_t = 70$ GeV/c², **c)** Monte Carlo (W + 2 jet) events.

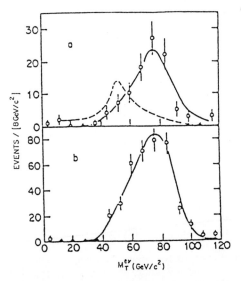

Figure 6.8. **a)** The $m_T^{e\nu}$ distribution for (electron + ≥ 2 jet) data from CDF shown as points, W + 2 jet expectation (solid curve), and expected $\bar{t}t$ production with $M_{top} = 70$ GeV/c² (dashed curve). **b)** The $m_T^{e\nu}$ distribution for (electron + 1 jet) data with the normalised (W + 1 jet) expectation.

The effect on α of the uncertainties (i) is evaluated by introducing a one standard deviation change in each contributing quantity and recalculating T and W prior to refitting α. The total uncertainty on α is then obtained by combining the individual uncertainties in quadrature. Normalisation uncertainties (items (ii) to (iv)) are separately evaluated and are listed in table 19.

A comparison of the estimated cross-section limit (95 % CL) with the predicted ($\bar{t}t$) limit is published for electron data. It excludes the mass range.

$$40 < m_t < 77 \text{ GeV}. \qquad (6.12)$$

Table 19. The number of predicted ($\bar{t}t$) events and fitted $\bar{t}t$ contribution to the electron + ≥ 2 jet rate, with 95 % CL upper limits on the $\bar{t}t$ production cross-section. Also shown are systematic uncertainties on α, and on the overall acceptance and production rate ($\delta\varepsilon$).

m_t GeV	$n_{\bar{t}t}$ predicted	α ± δα (stat)	δα(sys)	δε(sys)	$\sigma_{\bar{t}t}$ (pb)
40	130	0.07 ± .05	0.02	0.33	< 2410
50	123	0.06 ± .05	0.03	0.25	< 648
60	101	0.11 ± .08	0.04	0.22	< 408
70	43	0.00 $^{+.12}_{-.0}$	0.11	0.19	< 266
80	32	0.00 $^{+.27}_{-.0}$	0.17	0.17	< 281

The inclusion of muon data does not by itself significantly improve this limit because of the lack of sensitivity when $m_t \sim m_W$. For $m_t > m_W$ additional identification criteria are obviously needed (see also section 6.5).

The CDF Collaboration has so far pursued this approach by requiring that of the 191 single lepton events satisfying $\not{E}_T > 20$ GeV and $p_T^\ell > 20$ GeV, an additional low-p_T muon characteristic of semileptonic b-decay should be identified in some events. Following the previous nomenclature, the following requirements are made :

i) an identified muon stub matching a central track of $p_T > 2$ GeV within multiple scattering limits of (± 5σ) when extrapolated to the muon stub ($|\delta x| < \min(15, 60/p_T)$ cms),
ii) a requirement (at the expense of acceptance) on the angle δθ between the muon direction and the reconstructed jet direction of δθ > 0.5, to reject punch-through background from jets.

No event satisfies these criteria. To estimate the expected rate, a b-quark semileptonic branching ratio of BR = 0.102 ± 0.002 ± 0.007 is used, together with a fragmentation model in good agreement with identified b-decays from CLEO [79]. After taking account of known detector effects, a total of 2.3 events (1.6 events) would have been expected from direct ($\bar{t}t$) production with m_t = 80 GeV (90 GeV). The normalisation uncertainty on this estimate is approximately ± 25 %, comprising detector and luminosity uncertainties as shown in table 20.

Table 20. Systematic uncertainties in the expected rate of single lepton events with an associated low-p_T muon from semileptonic b-decay, as observed in the CDF detector.

Integrated Luminosity[+]	15 %
Monte Carlo statistics	12 %
B semileptonic BR	9.4 %
Jet E_T scale	5 %
Top fragmentation[*]	5 %
B fragmentation	5 %
Selection (p_T spectrum)[*]	5 %
Initial state radiation	4.5 %
High p_T lepton efficiency[*]	4 %
Acceptance	2 %
Total	25 %

[*] Uncertainties that are correlated with dilepton search
[+] The luminosity uncertainty has since been updated to ± 7 %

6.4 Combined limits on m_t from the Fermilab and CERN Colliders

The CDF experiment has combined their data from dilepton final states with results from the previous section on single leptons with an associated low-p_T muon. The 95 % confidence level upper limit is evaluated by finding the mean of a parent distribution which, when combined with gaussian correlated and uncorrelated uncertainties on the ($\bar{t}t$) rate, yields with 95 % probability ≤ 1 event satisfying selection criteria for one of the final states of the full sample. They obtain (see fig. 6.6)

$$m_t \text{ (CDF)} > 89 \text{ GeV (95 \% CL)} \tag{6.13}$$

The UA2 experiment [71] has performed an analysis of (e + ≥ 1 jet) events (equation 6.10) assuming t-quark production from the reactions

$$\bar{p}p \to \bar{t}t + X, \quad \text{and}$$

$$\bar{p}p \to W^\pm + X; \quad W \to \bar{t}b, \tag{6.14}$$

which is similar to that described in section 6.3 for the CDF experiment. They obtain

$$m_t \text{(UA2)} > 69 \text{ GeV (95 \% CL)}. \tag{6.15}$$

Finally, the UA1 experiment [70] have used their data for isolated muon production with associated jet activity to infer a limit

$$m_t \text{(UA1)} > 60 \text{ GeV (95 \% CL)}. \tag{6.16}$$

6.5 Indirect top quark mass determinations

In section 5.1.8, figure 5.6 showed the measured cross-section ratio (σ_W^e/σ_Z^e), and the dependence of this ratio on the t-quark mass. The CDF Collaboration is able to place a limit

$m_t > 41$ GeV (90 % CL)

$m_t > 35$ GeV (95 % CL) \hfill (6.17)

on the mass that is independent of the Standard Model decay couplings. In particular, it is independent of extensions to the Standard Model such as charged Higgs decays (t → H⁺b) that are kinematically allowed.

Because the measured value of ($\rho = m_W^2/m_Z^2 \cos^2\theta_W$) is close to one as expected in the Minimal Standard Model [43], most natural extensions to the Standard Model involve Higgs doublets, with a two-doublet model being the minimal extension for supersymmetric models. As a result the single scalar Higgs H⁰ is replaced by two charged Higgs scalars H±, two neutral Higgs scalars h⁰ and H⁰, and a pseudoscalar Higgs A. The actual couplings of the charged Higgs are extremely model dependent [80]. Depending on the mass $m_H^\pm$ and the ratio of the vacuum expectation values of the Higgs doublets, $\tan\beta = v_1/v_2$, the dominant top decay modes may be t → H⁺b, with subsequent decays H⁺ → cs and/or H⁺ → $\tau^+\nu_\tau$. As a result, the observation of single-$\tau^\pm$ or τ-pair final states, and as well e-τ and μ-τ final states, is important.

Unfortunately, the m_t limit from the cross-section ratio (σ_W^e/σ_Z^e) is sensitive only to values $m_t < m_W$.

From a measurement of the W and Z masses, the UA2 Collaboration [61] places a limit on m_t within the context of the Standard Model, from the dependence of radiative corrections to m_W and m_Z from the top quark and Higgs, of

$m_t > 76$ GeV if $m_H > 24$ GeV (90 % CL)
$m_t < 272$ GeV if $m_H < 1$ TeV (90 % CL). \hfill (6.18)

The corresponding CDF limit is

$m_t < 230$ GeV (95 % CL) \hfill (6.19)

and as noted in section 5.5, the combination of all existing Standard Model data suggests

$m_t \simeq 137 \pm 40$ GeV \hfill (6.20)

6.6 Extending the t-top quark mass limit

For the 1991 Tevatron Collider run, the CDF and DØ experiments are each expected to accumulate an integrated luminosity of at least 25 pb⁻¹. From a simple scaling of existing CDF dilepton results, a limit $m_t > 120$ GeV (95 % CL) should be attainable for each experiment. However, the DØ experiment has excellent electron and muon acceptance and is well optimised for the high-mass t-quark search. The CDF experiment intends to improve its electron and muon acceptance by almost a factor 2 by (fig. 2.11) :

i) improving the electron trigger in the range $1.1 < |\eta| < 2$,
ii) extending the muon coverage in the range $0.6 < |\eta| < 1.1$, and
iii) adding iron between the calorimeter and muon chambers in the central region.

Therefore, a limit $m_t > 140$ GeV should be achievable by each experiment, subject to expected Standard Model ($\bar{t}t$) production and decay rates.

The uncertainty of measured Standard Model processes satisfying $\not{E}_T > 40$ GeV is $\sim \pm 3$ pb at CDF. Therefore, it is attractive and important to search for at least one lepton decay (and if possible include τ-decays) with large $\not{E}_T$ and with associated jet activity. Such searches are important because they are less sensitive to non-standard top decay scenarios, such as the previously mentioned $H^+ \rightarrow \tau \nu_\tau$ decay. The actual m_t parameter space that can be excluded from these analyses is very model dependent.

As already noted in section 6.3, the sensitivity of single lepton signatures to top quark masses $m_t > m_W$ is poor unless additional t-quark signatures are used. So far, CDF has attempted to tag b-quark semileptonic decays. Other techniques might include

i) the tagging of b-quarks using the silicon microstrip vertex detector being installed at CDF (see fig. 2.11), or
ii) the use of a multivariate parameter analysis for variables such as p_T^W, m^{JJ} from the decay W → jet+jet, and the orientation in space of the leading jets. Preliminary analyses of this type [81] at CDF are encouraging.

As noted in section 7, the CDF central tracking chamber is already of sufficient precision to statistically measure an offset vertex for J/Ψ particles resulting from b-decay. With the implementation of the CDF silicon vertex detector in 1991, a transverse track offset from a primary vertex should be measurable on an event-by-event basis with a precision of $\sigma \lesssim 20$ μm for $p_T \gtrsim 5$ GeV where multiple scattering effects become small. Figure 6.9 shows the expected b-tagging efficiency for a secondary vertex measurement involving ≥ 3 tracks, and exceeding a transverse distance of (3σ) from the primary vertex. For a tag of at least one b-quark, $\varepsilon \sim 50$ % for large m_t. This implies that mass limits $m_t \sim 150$ GeV might be achievable from single lepton final states during the next CDF run.

The above scenario is a "non-discovery" limit. If indeed $m_t \sim$ 120-140 GeV with expected decay modes, its positive identification will be extremely problematic. All possible decay channels including if possible τ-decay modes must be utilised to confirm its existence.

7. HEAVY QUARK PHYSICS : b-QUARK PRODUCTION AND DECAY AT HADRON COLLIDERS

Because of the complexity of b-quark meson or baryon final states at hadron colliders and in particular

i) the desirability of tagging heavy quark decays by either lepton identification in the vicinity of jet activity (that is semileptonic b-quark decay as schematised in figure 7.1a), or by J/Ψ production as shown in fig 7.1b, and
ii) the need to separate b-quark and c-quark decays,

it was until recently accepted that b-quark studies at hadron colliders were extremely difficult. Although the b-quarks are less easy to tag than in (e^+e^-) machines (for example the decays $Y(4S) \rightarrow b\bar{b}$ at CLEO and ARGUS [73] or the decay $Z \rightarrow b\bar{b}$ at LEP), the $(b\bar{b})$ production rate is large. Approximately 10^6 B-mesons are produced with $p_T^B > 10$ GeV in the central rapidity range $|\eta| < 1.5$ per pb^{-1} of delivered luminosity at the Tevatron Collider [77]. Typical production diagrams have already been shown in figure 6.1.

The initial UA1 studies of b-quark production and decay [82] were aimed at an evaluation of possible b-quark backgrounds to t-quark signatures at low t-quark masses ($m_t \sim 35$ GeV). An initial measurement of the inclusive muon spectrum (section 7.1) lead to an evaluation of the b-quark production cross-section, and to the use of ($\mu^+\mu^-$) final states in a pioneering measurement of ($\bar{B}^0-B^0$) mixing (section 7.4). Essential to these studies was the possibility of muon identification in the vicinity of jet activity.

More recently, and still rapidly evolving, the CDF experiment has taken advantage of good non-isolated muon and electron identification capabilities in the central detector region, and of outstanding charged track pattern recognition and track momentum resolution. They have extended the UA1 studies to $\sqrt{s} = 1.8$ TeV, and have in addition succeeded in fully reconstructing B-meson final states (see sections 7.1, 7.2).

The UA1 and CDF experiments tag b-quark production from a background of multi-jet QCD processes involving light quarks by identification of semileptonic b-decays, or by the identification of J/Ψ final states resulting from b-decay. Following figure 7.1a, a background of c-quark decays should be reduced (at the expense of efficiency) by the mass reconstruction in the vicinity of the identified $e^\pm$ or $\mu^\pm$ of a D or D* meson. However, because of the existence of an additional (unidentified) neutrino, a full mass reconstruction is not possible. On the contrary (fig. 7.1b), a J/Ψ resulting from the decay of a B-meson should generally have an associated strange particle enabling an exclusive B-meson mass reconstruction. Of course, as noted below, the acceptance for full mass reconstruction is small and with existing detectors the event rates are low despite the large ($\bar{b}b$) production cross-section.

The production of b-quarks is dominated at collider energies by direct ($\bar{b}b$) production (fig. 6.2). In addition to (α_s^2) contributions of flavour excitation and flavour creation, there are substantial higher order contributions from gluon splitting and from (qg) terms. As already noted, an $O(\alpha_s^3)$ evaluation has been made [77], but because of the small value of $x_b = 2m_b/\sqrt{s}$, $O(\alpha_s^3)$ corrections are large (this implies an increased sensitivity to the factorisation scale chosen). Because of this, and because of important (gg) subprocesses in the small-x range where the gluon densities are poorly known, large theoretical uncertainties exist. Therefore b-quark production measurements potentially provide important constraints on the structure function parametrisations, and on the contribution of different higher-order production processes.

Because ($m_b/\sqrt{s}$) is small, the rapidity distribution of b-quark production extends to large values and (fig. 7.2) B-decay products are peaked at small angles. Therefore, any dedicated B-experiment should be optimised to detect both small-angle and large-angle decay products. This is the case for the proposed $\bar{B}^0 - B^0$ mixing experiment P 238 [83] at the CERN Collider. Also, to fully exploit existing detectors (for example CDF), it will be necessary to

i) trigger on single leptons ($p_T^\ell \sim 8-10$ GeV/c) and on dileptons ($p_T^\ell \gtrsim 3$ GeV) with good efficiency, and
ii) to take advantage of the b-tag capability of a silicon vertex detector, if possible at the trigger level. In the case of CDF, this will of course be an evolutionary development program for the machine, the detector, and the subsequent analysis.

7.1 b-quark production characteristics

Figure 7.3 shows the semi-inclusive p_T^μ spectrum for UA1 data [84]

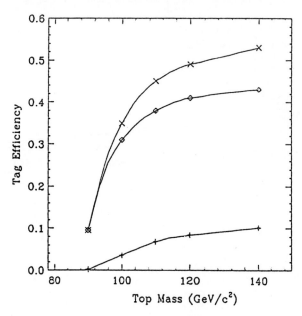

Figure 6.9. CDF efficiency for detecting $(\bar{t}t)$ through b-quark tagging as a function of m_t, for the tagging of one ($\diamond$), at least one (x) or both b-quarks (+).

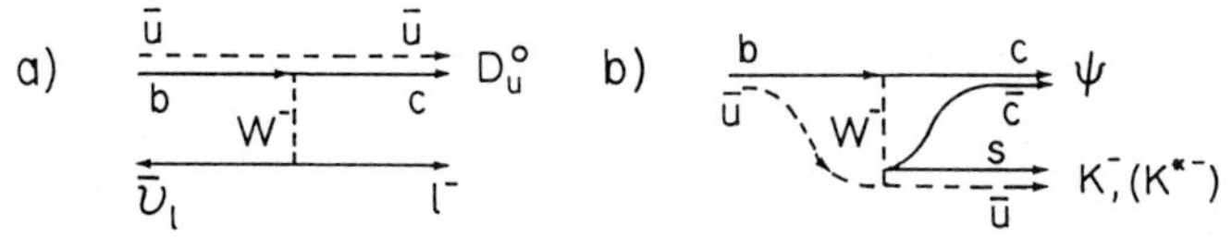

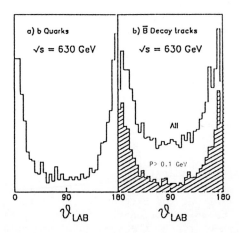

Figure 7.1. Heavy quark decay, the schematic examples shown being the decays $B_u^- \to D^0 \ell^- \nu_\ell$ and $B_u^- \to J/\Psi K^-$.

Figure 7.2. a) θ_{lab} distribution for b-quarks produced via gluon-gluon fusion using the PYTHIA event generator at $\sqrt{s} = 630$ GeV, b) θ_{lab} for the decay tracks of the B meson. The shaded portion is the distribution of decay tracks from events, for which all decay tracks have momenta exceeding 100 MeV. From reference [83].

satisfying $p_T^\mu > 10$ GeV in the central rapidity region $|\eta| < 1.5$. The identified muon is not required to be isolated, but the following kinematic requirements are made to enhance the $(\bar{b}b)$ signal and to remove $W \to \mu\nu$ decays :

i) $m_T^{\mu\nu} < 25$ GeV, and
ii) the existence of at least one reconstructed jet of $E_T^J > 10$ GeV in the rapidity range $|\eta| < 2.5$.

A comparison is made with ISAJET expectations for $(\bar{b}b)$ and $(\bar{c}c)$ production, following a full UA1 detector simulation including the above kinematic cuts. In making this comparison the ISAJET c-quark and b-quark semileptonic decays were tuned to agree with CLEO data on the lepton spectrum [79], and the D^0/D^{0*} production ratio in B-meson decays was taken to be in the ratio 28:72. Also included in the simulation were small contaminations from J/Ψ, Y, W/Z and Drell-Yan pair production, and a large background contribution from $K^\pm \to \mu^\pm\nu_\mu$ and $\pi^\pm \to \mu^\pm\nu_\mu$ decays. This latter background has been estimated from an evaluation of in-flight decays for hadrons in a jet that are subsequently reconstructed as muon tracks.

The fraction of $\bar{b}b$ events in each bin of p_T^μ has been estimated from the distribution of muon p_T relative to the jet axis, p_T^{rel}. The data are shown in figure 7.4 with the expected shape for $\bar{c}c$ decays, $K/\pi \to \mu\nu$ decay background, and $\bar{b}b$ decays superimposed. No discriminatory power exists between the shape of the p_T^{rel} distribution for c-quark and K/π decays, but the harder spectral shape of b-decays allows a measure of the fractional $(\bar{b}b)$ contribution. Averaged over p_T^μ,

$$\text{FRAC } (\bar{b}b) = 0.33 \pm 0.05 \text{ (stat)} \pm 0.03 \text{ (sys)}. \qquad (7.1)$$

Following a deconvolution of the corrected p_T^μ spectrum, the (integral) p_T^b spectrum is shown in figure 7.5. Within large statistical uncertainties, the agreement with the QCD calculations of Nason et al. [77] is excellent.

The CDF experiment has not yet published a measurement of the $(\bar{b}b)$ cross-section, but this work is in progress. Because of their ability to identify electrons in the vicinity of jets using the very localised isolation requirement that only one charged track points to the reconstructed electromagnetic cluster, CDF can measure both the non-isolated p_T^e and p_T^μ spectra. Figure 7.6 shows the p_T^e spectrum ($7 < p_T^e < 60$ GeV), using approximately the selections of section 5 and in addition

i) the requirement of one and only one charged track pointing to the electromagnetic cluster, and
ii) the removal of conversion electrons (two reconstructed tracks having a minimum separation of < 2 mm and an opening angle $\delta|\cot\theta| < 0.06$).

The background to figure 7.6 is dominated at low p_T^e by conversions (12±7% for $p_T^e > 15$ GeV)) and by charged hadrons faking the electron signature (15±15%). After the removal of $W^\pm$ and Z candidates, the p_T^e spectrum is as shown in figure 7.6 in good agreement with ISAJET predictions for the p_T^e spectrum from $(\bar{b}b)$ plus $(\bar{c}c)$ production.

Excluding the J/Ψ event sample discussed in section 7.2, the UA1 Collaboration [82] has compared the event characteristics of non-isolated $(\mu\mu)$ pairs with QCD expectations for $(\bar{b}b)$ production followed by the semileptonic decay of both b-quarks. Depending on the kinematic selections used, these data are sensitive to higher-order $(\bar{b}b)$ contributions, especially gluon splitting. Selecting 399 events from data collected until 1985 satisfying

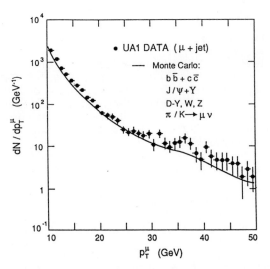

Figure 7.3. Inclusive muon spectrum from the UA1 experiment. Superimposed are the summed ISAJET expectations for ($\bar{b}b$ and $\bar{c}c$) final states, plus competing processes such as J/Ψ and W/Z production.

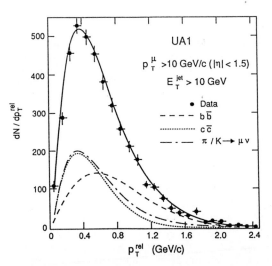

Figure 7.4. The distribution of p_T^μ with respect to the jet axis, with superimposed ISAJET estimations for $\bar{b}b$ and $\bar{c}c$ background, plus the background from π/K decay.

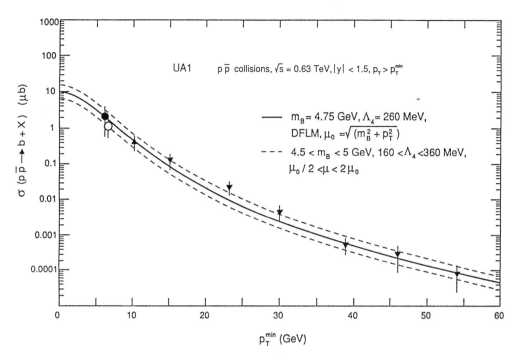

Figure 7.5. Integral cross-section, $\sigma(\bar{p}p \to b + X)$ at the UA1 experiment. The superimposed theoretical expectations are from Nason et al., for a range of QCD scales.

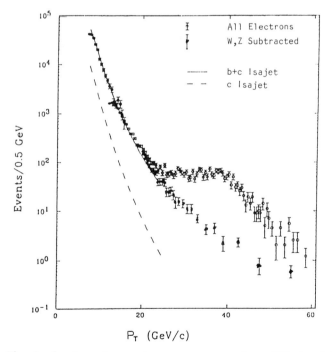

Figure 7.6. The inclusive p_T^e spectrum as measured at CDF (preliminary) for $p_T^e > 7$ GeV, after substraction of background, but before acceptance and efficiency corrections. The superimposed ISAJET expectation has been modified by the CDF efficiency and acceptance, and is arbitrary normalised.

i) $p_T^\mu > 3$ GeV,
ii) $m^{\mu\mu} > 6$ GeV, and
iii) a non-isolation requirement $S = [\Sigma\, E_T(\mu_1)]^2 + [\Sigma\, E_T(\mu_2)]^2 > 9$ in a cone $R_c = 0.7$ about each muon,

figure 7.7a shows the angle $\delta\phi^{\mu\mu}$. Whereas for leading order contributions such as flavour creation $\delta\phi^{\mu\mu} \sim 180^0$, $\delta\phi^{\mu\mu} \sim 0^0$ for gluon splitting and non-leading contributions are required for $\delta\phi^{\mu\mu} < 100^0$.

Using a sample satisfying

i) $p_T(\mu_1) > 7$ GeV, $p_T(\mu_2) > 3$ GeV,
ii) at least one jet satisfying $E_T^J > 12$ GeV, and
iii) $\Sigma E_T > 6$ GeV in a cone $R_c = 0.7$ about μ_1,

the gluon splitting is emphasised. Figure 7.7b shows good agreement with ISAJET calculations that include higher-order contributions.

Associated with semileptonic B-meson decay, an enhancement of charmed meson production is expected from the flavour-changing decay (b → cW + c.c.). Further, from the additive quark model, it is evident that the strange particle decay products of the charmed meson must have a charge correlation to the decay lepton. The full decay chain $B^- \to e^- \bar{\nu}_e D^0$; $D^0 \to K^- \pi^+$ is schematised in figure 7.8, and the quark content does not allow $(K^+ \pi^-)$ final states. Other typical allowed decays are listed below.

$$
\begin{aligned}
B^- &\to \ell^- \bar{\nu}_\ell\, D^0 \;^{1)} &;& \quad D^0 \to K^- \pi^+ \\
\bar{B}^0 &\to \ell^- \bar{\nu}_\ell\, D^+ &;& \quad D^+ \to \bar{K}^0 \pi^+ \\
&\to \ell^- \bar{\nu}_\ell\, D^0 \pi^+ &;& \quad D^0 \to K^- \pi^+ \\
B^0 &\to \ell^+ \nu_\ell\, D^- &;& \quad D^- \to K^0 \pi^- \\
&\to \ell^+ \nu_\ell\, \bar{D}^0 \pi^- &;& \quad \bar{D}^0 \to K^+ \pi^- \\
B^+ &\to \ell^+ \nu_\ell\, \bar{D}^0 &;& \quad \bar{D}^0 \to K^+ \pi^-
\end{aligned}
\qquad (7.2)
$$

The CDF Collaboration [85] has searched for $B^\pm$ decays from an identification of $(K\pi)$ pairs characteristic of D^0 or D^{0*} decay. Selecting events having an identified $e^\pm$ satisfying $11 < p_T < 30$ GeV in a central rapidity range $|\eta| = 0.6$, all correct-sign $(K\pi)$ combinations were selected in a cone of radius $R_c = 0.6$ about the $e^\pm$. The $m(K\pi)$ distribution shown in figure 7.9 has a measured signal consistent with the D^0 mass, and estimated to be 75 ± 12 events. A total of 72 ± 20 events is expected (statistical uncertainties only). The same mass distribution for wrong-sign events shows no similar $(K\pi)$ enhancement (a small contamination of wrong-sign pairs is expected to result from higher order gluon-splitting contributions). Since all opposite-sign track combinations must be considered and CDF has no particle identification capability, the background remains substantial. Further the low-mass background is increased because of $(K\pi\pi)$ decays of the D^0 or D^{0*} mesons.

In the 1991 CDF run, an improved electron and muon acceptance at the trigger and analysis levels and an additional factor 5 in the integrated luminosity will enable detailed comparisons of lepton production from B-meson decays with higher order QCD calculations. In 1991 and beyond,

---

[1]) alternatively $D^{0*}(2010)$ can be produced with subsequent decay products $D^{0*}(2010) \to D^0 \pi^0$, $D^{0*}(2010) \to D^0 \gamma$, etc.

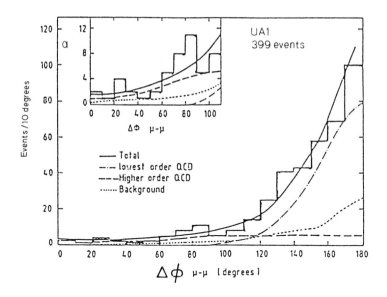
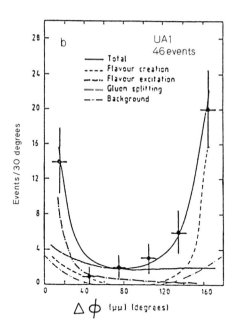

Figure 7.7. a) Azimuthal angle difference between the muons from non-isolated muon pairs. The superimposed curves are normalised ISAJET predictions normalised to 399 events (including 116 background events). Of these events 142 events are like-sign pairs. UA1 experiment. b) As above for a sample of 46 high-p_T ($\mu\mu$) pairs with associated jet activity (see text). UA1 experiment.

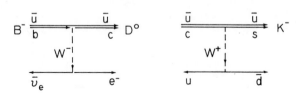

Figure 7.8. Typical quark decay chain for the decay $B_u^- \to e^- \bar{\nu}_e D^0$; $D^0 \to K^- \pi^+$.

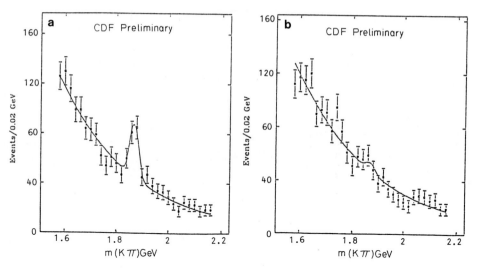

Figure 7.9. The reconstructed mass m(Kπ) for all (Kπ) pair combinations within a cone around an identified high-p_T electron, as defined in the text. **a)** Correct-sign pairs. **b)** Wrong-sign pairs.

the tagging of b-quark decays using the silicon vertex detector will be
a priority at CDF because of the large production cross-section. This is
evident from the following pessimistic bench-mark assumptions :

i) $\sigma_{bb}^-(p_T^b > 10 \text{ GeV}) \simeq 1.5 \mu b$ for $|\eta| < 1.5$,
ii) BR(≥ 1 electron) = 0.24
     BR(≥ 1 muon) $\simeq 0.24$,
iii) probability that $p_T^\ell > 7$ GeV $\simeq 0.2$,
iv) reconstruction and trigger efficiency $\simeq 0.3$ (electrons)
                                            $\simeq 0.7$ (muons)

Assuming (again pessimistically) a b-quark tagging efficiency of 0.2
(see figure 6.9), a total of 5.10^5 reconstructed electrons and 10^6 reconstructed muons would be expected from b-decays for an integrated
luminosity of 100 pb^{-1}.

7.2 Tagging b-decays and the reconstruction of exclusive B-decays

Both of the UA1 [84] and CDF [85] experiments have collected large
dimuon event samples using a trigger requirement of two muons satisfying
$p_T^\mu \gtrsim 3$ GeV. The ($J/\Psi \to \mu^+\mu^-$) sample from each experiment is shown in
figures 7.10a and 7.10b. The CDF experiment also measures a clean
$\Psi'(3685) \to \mu^+\mu^-$ signal (fig. 7.10c). Because of the trigger, the J/Ψ
sample is obviously kinematically biased towards high-p_T J/Ψ production,
as shown in figure 7.11. Because of the small background contamination,
identification criteria can be relaxed to select events with good
efficiency (apart from trigger efficiency uncertainties that limit the
accuracy of a cross-section evaluation).

The CDF experiment has also reconstructed decays $J/\Psi \to e^+e^-$, but
because of a high p_T^e trigger threshold during the 1989 run the p_T^Ψ distribution is peaked at higher values with a consequent loss of statistics. This will be remedied for the 1991 run.

Only a fraction of J/Ψ production results from B-particle decay and
kinematic selections are needed to enhance the signal. Expected rates
from ISAJET are shown in figure 7.12 for the following production mechanisms :

i)   radiative decays of high-p_T χ states produced by gluon fusion,
ii) direct J/Ψ and Ψ' production, and
iii) b-quark decays into J/Ψ.

Subsequent kinematic discriminations that will enhance b-quark contributions include

a) isolation selections (expected in processes (i) and (ii) above),
b) the selection of high-p_T J/Ψ events (already biased by the trigger),
c) a measurement of vertex displacement for b-decays (only available from 1991 on an event-by-event basis), and
d) an excess of strange particle production in association with J/Ψ production from b-quark decays.

Item (d) invites the possibility of a full reconstruction of
B-mesons (this was not possible using a semileptonic decay tag because of
the undetected decay neutrino). Typical such exclusive states might include (see fig. 7.1).

$$B_u^\pm \to J/\Psi \ K^\pm$$
$$B_d^0 \to J/\Psi \ K^{*0} \text{ and } B_d^0 \to J/\Psi \ K^0$$
$$B_s^0 \to J/\Psi \ \phi. \tag{7.3}$$

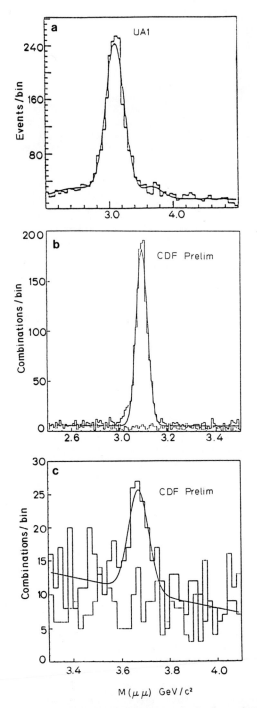

Figure 7.10. The dimuon mass distribution m($\mu\mu$) for **a)** UA1 experiment, **b)** CDF experiment (both $\mu^+\mu^-$ and $\mu^\pm\mu^\pm$ contributions shown), **c)** CDF experiment in the region of the Ψ' (3685).

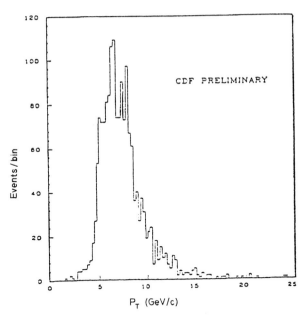

Figure 7.11. The p_T distribution of reconstructed $J/\Psi \to \mu^+\mu^-$ decays from the CDF experiment.

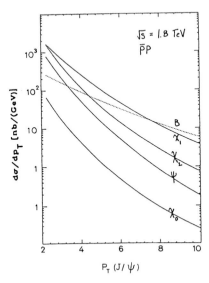

Figure 7.12. The p_T-distribution as calculated by ISAJET, for the J/Ψ events resulting from B-decays, and competing J/Ψ production processes $gg \to \chi g$; $\chi \to J/\Psi + \gamma$ and $gg \to J/\Psi + g$.

These searches are so far hampered by limited statistics, and in these notes only the decays $B^{\pm} \to J/\Psi K^{\pm}$ and $B_d^0 \to J/\Psi^* K^{*0}$ are considered. The following initial analysis has been made by CDF to optimise any possible B-meson signal.

i) The J/Ψ mass is constrained to $m_\Psi = 3.097$ GeV, and the decay muons plus the candidate $K^{\pm}$ tracks are constrained to a common vertex (not necessarily the primary vertex).
ii) A requirement $p_T^\Psi > 5$ GeV is made and the $K^{\pm}$ candidate is required to be within a cone $R_c = 1.0$ about the reconstructed J/Ψ.
iii) The $K^{\pm}$ candidate is required to satisfy $p_T^K > 2.5$ GeV, to reduce combinatorial background. The $m(\Psi K^{\pm})$ spectrum (fig. 7.13) shows a marginal enhancement at the expected mass for $B^{\pm}(5280)$. The increasing backgrounds at lower $m(\Psi K)$ values result in part from K^*-decays of the B-meson
iv) In the case of $(K^-\pi^+)$ candidates having a reconstructed mass within $\pm$ 50 MeV of the $K^*(890)$, the $m(\Psi K^-\pi^+)$ mass spectrum has been added to figure 7.13, increasing the significance of the reconstructed B-meson peak (the data described here is updated to August 1990).

The identification and study of expected B_s^0 and B_d^0 meson states in the 1991 CDF run is a major priority, especially for lifetime and mixing studies as discussed below in section 7.4.

7.3 Rare or forbidden B-meson decays

Using a signal region $(5.1 < m^{\mu\mu} < 5.7$ GeV) for the $B^0(5280)$, and requiring $p_T^{\mu\mu} > 7$ GeV to efficiently reject background from Drell-Yan and cascade decays, no evidence of flavour changing neutral current decays $B^0 \to \mu^+\mu^-$ is observed in the UA1 detector [84]. A similar analysis has been made at CDF [85]; the measured limits are :

$$\text{BR}(B^0 \to \mu^+\mu^-) < 9.10^{-6} \quad (\text{UA1; 90 \% CL})$$
$$< 3.10^{-6} \quad (\text{CDF; 90 \% CL}). \tag{7.4}$$

With increased sensitivity, rare decays such as $B^0 \to \mu^+\mu^-$ or $B^0 \to K^0\mu^+\mu^-$ will become accessible.

7.4 $\bar{B}^0 - B^0$ mixing studies [76]

Meson transitions $\bar{M}^0 - M^0$ (for example $\bar{K}^0 - K^0$ or $\bar{B}^0 - B^0$) are allowed by the weak interaction box diagram of figure 7.14. Considering the case of $\bar{B}^0 - B^0$ mixing, $\bar{B}^0$ and B^0 are combinations of mass eigenstates B_H and B_L which may have different masses ($\delta m = m_H - m_L$) and decay widths :

$$B^0 = (B_H + B_L)/\sqrt{2}$$
$$\bar{B}^0 = (B_H - B_L)/\sqrt{2}. \tag{7.5}$$

Assuming $\Gamma_H = \Gamma_L = \Gamma$ because of the large B^0-decay phase space, the probability of measuring a $\bar{B}^0$ (B^0) at any time, given a B^0 state at time $t = 0$ is (see fig. 7.15)

$$I(B^0) = e^{-\Gamma t}[1 + \cos(\delta m t)]/2$$
$$I(\bar{B}^0) = e^{-\Gamma t}[1 - \cos(\delta m t)]/2 \tag{7.6}$$

Integrating equations (7.6) over time the rates of occurrence of no oscillation (B^0) and oscillation ($\bar{B}^0$) are :

$$I_{NOSC} = \frac{2 + x^2}{2(1 + x^2)} = (1 - \chi)$$

$$I_{OSC} = \frac{x^2}{2(1 + x^2)} = \chi$$

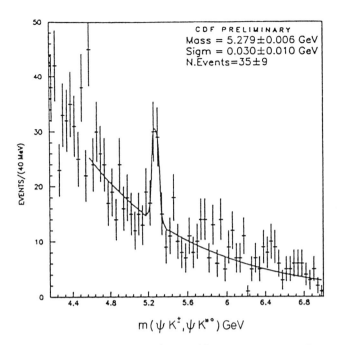

Figure 7.13. The distribution m($\Psi K^\pm + \Psi K^{*0}$) showing an enhancement consistent with B(5280).

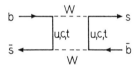

Figure 7.14. The lowest-order diagram allowing $\bar{B}^0_s - B^0_s$ transitions.

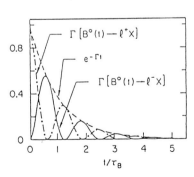

Figure 7.15. Proper time evolution of semileptonic B^0 decays with $\delta m/\Gamma = 5$.

171

with $x = \delta m/\Gamma$. For large x, $I_{OSC} \simeq I_{NOOSC} \simeq 0.5$ as expected. The long lifetime of B^0-decay makes the observation of oscillations and/or mixing favorable. For $\bar{D}^0 - D^0$ oscillations, however, δm is small and the lifetime is short. As a result, the oscillation period is long compared with the decay lifetime.

So far it has not been possible to directly measure the time evolution of $(\bar{B}^0-B^0)$ oscillations, but initial measurements at UA1, followed by measurements from Y(4S) decay at ARGUS [86] and CLEO [87], and most recently LEP [57, 88], have measured the time-integrated rates of equation (7.7). However, while only B_d^0 mixing is kinematically allowed at ARGUS and CLEO, both B_s^0 and B_d^0 production are allowed at UA1 and LEP, and each may have different x-values. The mass difference δm can be evaluated from the box diagram and is sensitive to the V_{td} and V_{ts} elements of the CKM matrix. Neglecting kinematic factors and QCD corrections it can be shown that

$$\delta m_d \simeq K \left[(V_{cb}^* V_{cd})^2 \, m_c^2/m_W^2 + (V_{tb}^* V_{td})^2 \, m_t^2/m_W^2 \right]$$

$$\delta m_s \simeq K \left[(V_{cb}^* V_{cs})^2 \, m_c^2/m_W^2 + (V_{tb}^* V_{ts})^2 \, m_t^2/m_W^2 \right]. \tag{7.8}$$

From existing unitarity constraints on the CKM matrix, it can be shown that this implies

$$\frac{x_d}{x_s} \sim \frac{(V_{td})^2}{(V_{ts})^2} \leq 0.21 \tag{7.9}$$

Then if a measure of χ is made at a collider (either e^+e^- or $\bar{p}p$),

$$\chi_{obs} = K_s \chi_s + K_d \chi_d, \tag{7.10}$$

where K_s and K_d are free parameters representing the relative production contributions of B_s^0 and B_d^0 mesons in $(b\bar{b})$ production at collider energies. From measurements of high-p_T K^+-production at the ISR, guesstimates of K_s and K_d are $K_s \simeq 0.18$ and $K_d \simeq 0.36$ at ISR energies. Because of this uncertainty, separate χ_s and χ_d measurements would be desirable.

The UA1 Collaboration has measured the ratio

$$R = \frac{N(\mu^+\mu^+) + N(\mu^-\mu^-)}{N(\mu^+\mu^-)}, \tag{7.11}$$

which in the absence of secondary lepton decays ($b \to c \to u$) can be shown to be

$$R \simeq \frac{2\chi (1-\chi)}{(1-\chi)^2 + \chi^2} \tag{7.12}$$

The previously defined sample of 399 non-isolated $(\mu\mu)$ events from section 7.1 is used. Of these events, 257 events are of opposite sign and for each sign category the decay background is estimate to be 58 events. The opposite sign category has an additional background of (15 ± 7) events from contaminating Drell-Yan and Y contributions. From this they measure

$$R = 0.45 \pm 0.08 \text{ (stat)} \pm 0.04 \text{ (sys), giving}$$
$$\chi = 0.158 \pm 0.059 \text{ (sys + stat)} \tag{7.13}$$

after allowing for secondary lepton decays of one or both leptons. This latter contribution is estimated from a maximum likelihood fit to the two-dimensional dimuon p_T-distributions.

Analyses of ARGUS and CLEO data from $\bar{b}b$ production at the Y(4S) resonance give

$$\chi_d = 0.154 \pm 0.056 \text{ (CLEO) [87]}$$
$$= 0.170 \pm 0.054 \text{ (ARGUS) [86]} \quad (7.14)$$

Figure 7.16 combines existing measurements from UA1, CLEO and ARGUS and shows the superimposed constraints provided by existing measurements or limits of the CKM matrix elements, assuming the Standard Model with three families. Taken at face value, almost maximal B_s^0-mixing must occur. Theoretically, χ_s is estimated to be in the range 0.1 to 0.5, whereas existing limits on V_{td} and V_{ts} from (7.9) imply $\chi_d \lesssim 0.2$.

What are the perspectives for $\bar{B}^0 - B^0$-mixing studies ? They fall into the four categories of improving current measurements without a change of technique, of having separate measurement of χ_d and χ_s, of following the time development of $\bar{B}^0 - B^0$ oscillations, and finally the goal of obtaining adequate sensitivity to identify possible CP-violation effects.

i) It will be difficult using present techniques to reduce the systematic uncertainty for a χ-measurement at UA1, CDF or DØ to $\lesssim 0.03$, because of uncertainties of the decay background, secondary decays, etc. In addition, using current techniques the relative contributions of χ_s and χ_d cannot be directly determined. Data from CLEO and ARGUS are expected to constrain χ_d to within an uncertainty of $\sim \pm 0.04$. LEP studies suggest that with 10^7 Z events, χ should be measurable to within an uncertainty of $\sim \pm 0.01$ if the correction for secondary decays is adequately controlled.

ii) Both CDF and the LEP detectors have (or will have) silicon microstrip detectors with adequate spatial resolution to trace the time development of $(\bar{B}^0 - B^0)$ oscillations from a measurement of the b-decay vertex. The CDF experiment has made a careful study [89] of its capability of evaluating $(x_s = \delta m/\Gamma)$ to within ± 2 units from a measurement of the dilepton decay vertices in events $\bar{p}p \to \ell^+\ell^- + X$ (recall from the equation (7.9) that $x_s \gg x_d$). The events are divided into same-sign and opposite-sign categories and for each event the ratio $Ln = L(\text{decay})/p_B$ of the B-decay length to the B-momentum is evaluated. The result is shown in figure 7.17a for the (realistic) case of $x_s = 5$, using ISAJET events satisfying $p_T^\ell > 10$ GeV. Figure 7.17b shows same distribution after replacing p_B by the lepton momentum and including resolution effects, etc. It is estimated that to resolve x_s to within ± 2 units up to a value $x_s = 10$, approximately 10^4 vertex-measured pairs are needed. Taking into account the known or expected detection efficiencies, acceptance and background at CDF, that corresponds to an integrated luminosity of approximately 70pb^{-1}.

iii) With less efficiency but more accuracy, it should be possible to tag the $\bar{B}_s^0$ (B_s^0) from its semileptonic decay and to subsequently measure the time development of the opposite-side B_s^0 ($\bar{B}_s^0$). Suitable B_s^0 decays may include for example

$$B_s^0 \to J/\Psi \, \phi, \quad \text{or}$$

$$B_s^0 \to D_s^{*-}\pi^-\pi^+\pi^-; \; D_s^{*-} \to D_s^- \gamma; \; D_s \to \phi\pi^-; \; \phi \to K^+K^-. \quad (7.15)$$

Such studies are being considered by the CDF experiment and by the LEP experiments (~ 100 events are expected from 10^6 Z-events). As already noted in fig.7.2, however, at collider energies the B-decay products are strongly enhanced in the forward direction. Schlein and others [83] have proposed the P238 experiment, a forward spectrometer system capable of reconstructing B_s^0-decays with a larger acceptance than either CDF or DØ. Tests to demonstrate the feasibility of this experiment (and in particular its ability to efficiently trigger on $(\bar{B}B)$ production) have been approved at the CERN Collider. Because of the high average B-momentum, a measurement of the proper

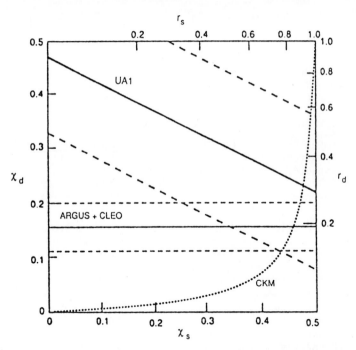

Figure 7.16. $\bar{B}^0 - B^0$ oscillations in the $\chi_s - \chi_d$ plane showing one s.d. bands. Constraints on χ_s and χ_d from the CKM matrix elements with 3 generations, are shown as a solid line. From reference [76].

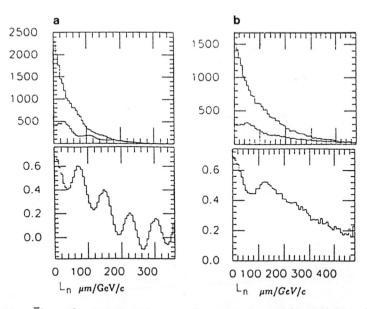

Figure 7.17. $\bar{B}^0 - B^0$ mixing with $x_s = 5$ versus $L_n = L(decay)/p_B$, for the cases of same-sign events (lower curve), opposite-sign events (upper curve) and diff/sum. a) Ln for an ideal detector. b) Ln after including resolution effects resulting from the use of the lepton momentum instead of p_B to evaluate Ln.

decay time is more accurate than in the central region as measured by CDF. The P238 proposal estimates that approximately 10^4 $\bar{B}_s^0$ or B_s^0 events could be reconstructed for an integrated luminosity of 50 pb^{-1}. An associated lepton or kaon tag would reduce this number by at least a factor 10.

iv) Estimates of the sensitivity to reach estimated CP-violating effects in for example the decay $B_d^0 \rightarrow J/\Psi \, K_S^0$ are beyond the scope of these notes [76]. At current collider energies, an integrated luminosity $\gtrsim 1$ fb^{-1} is required. If the Fermilab Collider program develops as planned, this might be achievable at CDF or D0 in the late 1990's. If not measurable at these experiments, a dedicated experiment to measure CP-violation should be considered for the LHC or SSC machines.

8. SEARCHING BEYOND THE STANDARD MODEL

So far in these notes, the principal Standard Model tests undertaken at hadron colliders have been discussed. No significant evidence exists for any deviation from Standard Model predictions. The future emphasis of hadron collider studies will be (in the era of LEP and LEP2) :

i) the identification and study of the top quark and detailed studies of b-quark production and decay, as discussed in sections 6 and 7,
ii) improved measurements of gauge boson production to allow meaningful higher order QCD comparisons, and
iii) a continued comparison of jet and gauge boson production with QCD at very high Q^2 values, to search for deviations from the Standard Model.

Two fundamental aspects of the Standard Electroweak Model have not yet been tested : gauge self-couplings, and some kind of mass generation mechanism that should result in the observation of one or more Higgs particles. Prospects for their observation are very briefly discussed in this section.

Theoretically, it is expected that at some as-yet unknown mass scale, O(1 TeV), new physics will manifest itself. Such new physics may result in

i) new particle production, as for example in supersymmetric models (SUSY) or composite models of various types such as Technicolor,
ii) the production of additional gauge bosons, to be expected in many extensions to the Standard Model,
iii) deviations at high transverse momentum or mass scales from QCD or Standard Model behaviour, as expected if quarks and/or leptons are composite.

Searching for evidence of new physics processes is now an important priority at the Tevatron Collider. Similar but largely complementary searches are also underway at LEP.

i) New particles having only strong-interaction couplings, for example the gluons $\tilde{g}$ of supersymmetric (SUSY) models, are more easily accessible by direct production at hadron colliders.
ii) Particles having electroweak couplings, for example new heavy leptons $L^{\pm}$, are most easily observable at (e^+e^-) machines if kinematically accessible. However, their direct production at hadron colliders is possible over an extended kinematic range, though at the expense of rate.

iii) If kinematically accessible, new gauge bosons can be produced at both (e^+e^-) and hadron colliders; searches at (e^+e^-) colliders will be constrained by $\sqrt{s}$, while those at hadron colliders are constrained by rate.

The very brief notes of this section are not exhaustive, and are intended mainly to show the complementarity of searches for new physics at LEP and hadron colliders. The few topics chosen indicate our own prejudice of the most useful future searches at hadron colliders. Recent reviews by Ellis et al. [90] and Pauss [91] are strongly recommended for a more complete assessment of discovery prospects.

8.1 Triple gauge coupling [92]

Figure 8.1 shows the approximate cross-section for the boson pair production and figures 8.2a and 8.2b show the expected lowest-order diagrams for (W^+W^-), ($W^\pm Z$) and ($W^\pm\gamma$) production. Because of cancellations between different diagrams, the cross-sections are low.

It is unlikely that the existence of trilinear coupling terms such as ($W^+W^-Z^0$) involving W^+W^- production can be demonstrated at the CDF or DØ experiments prior to LEP2 operation. Although for an integrated luminosity of 100 pb^{-1}, approximately 750 W^+W^- pairs are produced, the events are difficult to isolate. The signal of approximately 150 events expected for the process

$$\bar{p}p \rightarrow W^+W^- + X; \quad W_1^\pm \rightarrow \ell^\pm\nu_\ell; \quad W_2^\pm \rightarrow q_1\bar{q}_2 \tag{8.1}$$

will be masked by irreducible physics backgrounds of

i) (W + ≥ 2 jet) events from standard QCD processes, and
ii) (W^+W^-) pairs from ($\bar{t}t$) decay if $100 \lesssim m_t \lesssim 150$ GeV.

In the dilepton decay mode, the signal must be separated from Z-decays, from Drell-Yan background in the case of (e^+e^-) and ($\mu^+\mu^-$) pairs, and from ($\bar{t}t$) decays in the case of ($e\mu$) pairs.

However, the direct observation of ($W\gamma$) or ($Z\gamma$) final states is feasible, because of the larger production cross-section. Selecting $E^\gamma > 10$ GeV, a total of approximately 1650 (1200) events from $W^\pm\gamma$ ($Z^0\gamma$) production are expected prior to the leptonic W/Z decay for an integrated luminosity of 100 pb^{-1}. However, good photon identification is required at small angles, and an acceptance requirement $20^\circ < \theta_\gamma < 160^\circ$ reduces the event rate by a factor 0.5. This means that at least a total cross-section measurement can be made. The ($W\gamma$) coupling probes the anomalous magnetic moment of the $W^\pm$, for which precise predictions exist. A measure of the ($d\sigma/d\cos\theta_\gamma$) distribution, shown in figure 8.3 before and after resolution effects and an E^γ cut are included, will eventually provide a direct measurement of the magnetic moment.

8.2 Higgs searches

A search has been made at CDF [93] for a light Standard Model (H^0) Higgs produced via the bremsstrahlung process, as shown in figure 8.4. The CDF experiment was able to exclude

$$2m_\mu < m_H < 818 \text{ MeV}$$
$$846 < m_H < 2m_K \text{ MeV (90 \% CL)} \tag{8.2}$$

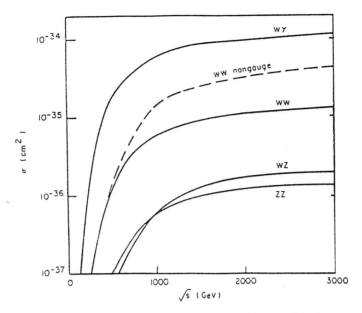

Figure 8.1. The cross-section for gauge boson pair production as a function of $\sqrt{s}$. Not included is the (W^+W^-) production rate from $(\bar{t}t)$ production if $m_t > m_w$. From Humpert [92].

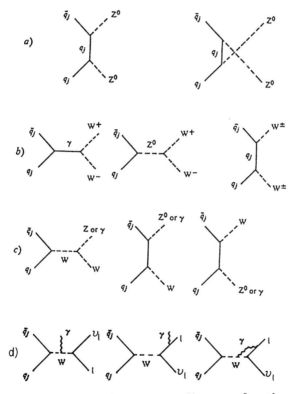

Figure 8.2. Typical lowest-order Feynman diagrams for the production of gauge boson pairs: a) $\bar{q}_j q_j \to Z^0 Z^0$, b) $\bar{q}_j q_j \to W^+W^-$, c) $\bar{q}_j q_j \to WZ^0/\gamma$. In the case of $(W + \gamma)$ production competing on-shell radiative terms must be included (d).

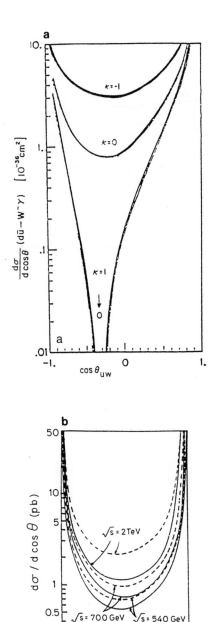

Figure 8.3. The distribution (a) dN/dcosθ at the parton level in the (Wγ) rest frame (b) dN/dcosθ after resolution smearing in pp rest frame, and application of a selection Eγ > 10 GeV. The cases K = 1 as expected from the Standard Model, and K = -1 are shown.

Since this search, each of the LEP experiments have published far more stringent limits for the Standard Model Higgs H^0. The exclusion region is [57, 94]

$$m_H < 41.6 \text{ GeV}. \tag{8.3}$$

Because of the small Higgs production cross-section (see fig. 8.5), direct Higgs production [95] is not expected to be observable at the Tevatron Collider.

The minimal SUSY extension of the Standard Model allows in addition to the light h^0 and H^0, two charged Higgs, $H^\pm$ and a pseudo scalar particle A. The charged Higgs, $H^\pm$, have already been noted in section 6 as a possible decay product for the t-quark. The LEP experiments presently require [57, 96] at 95 % CL.

$$m_H^\pm > 42.0 \text{ GeV}, \tag{8.4}$$

assuming BR ($H^\pm \to \tau\nu$) = BR ($H^\pm \to cs$) = $\cdot 5$

The minimal SUSY model also requires that the neutral Higgs h^0 is of mass $m_h \sim m_Z$; such a particle has not yet been observed, and according to some SUSY models, its observation should be possible at LEP2.

8.3 Additional gauge bosons

In almost any extension of the Standard Model, for example from SUSY models or from composite models, additional gauge bosons are expected with masses and coupling strengths appropriate to the given model. The most stringent preliminary limits on direct W' or Z' production are from the CDF experiment [85].

The preliminary (e^+e^-) mass spectrum from CDF is shown in figure 8.6a for electron pairs satisfying

i) $P_T^{e1} > 15$ GeV, $P_T^{e2} > 7$ GeV

ii) $|\eta^{e1}| < 1.1$ and $|\eta^{e2}| < 2.4$, and

iii) $m^{ee} > 30$ GeV $\tag{8.5}$

One (e^+e^-) pair is measured with a mass m^{ee} = 189 GeV, corresponding to a production cross-section of (0.7 ± 0.7) pb. This compares with an expected Drell-Yan cross-section

$$\sigma_{DY}(m^{ee} > 150 \text{ GeV}) = 1 \text{ pb} \tag{8.6}$$

From the lack of events satisfying $m^{ee} > 200$ GeV, CDF infers a cross-section limit

$$\sigma \cdot \text{BR} (m^{ee} > 200 \text{ GeV}) < 1.6 \text{ pb } (95 \text{ \% CL}), \tag{8.7}$$

which, assuming standard couplings and the EHLQ structure functions (the most pessimistic) with an overall normalisation to the measured Z cross-section, corresponds to a limit

$$m(Z') > 380 \text{ GeV } (95 \text{ \% CL}). \tag{8.8}$$

This kinematic range is available only to the Tevatron Collider, the advantage of higher $\sqrt{s}$ compared with the CERN Collider being evident in figure 8.7, where expected (10 event) discovery limits are shown.

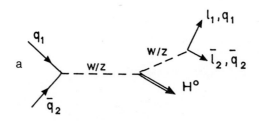
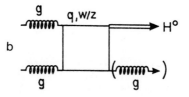

Figure 8.4. a) Bremsstrahlung production of the Standard Model Higgs from $W^{\pm}$. b) A typical lowest-order production diagram for H^0 from gluon fusion.

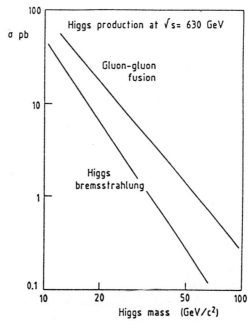

Figure 8.5. The cross-section as a function of m_H for Standard Model Higgs production.

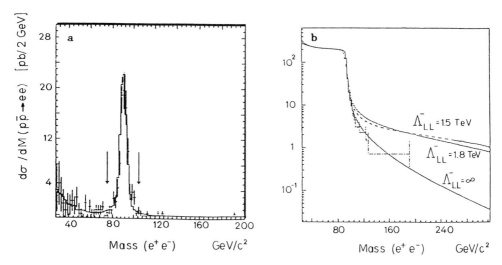

Figure 8.6. a) Preliminary CDF data showing the $m(e^+e^-)$ distribution with superimposed ISAJET expectations normalised at m_Z. b) The same data shown as an integral spectrum, with the expected rate if a composite term with scale Λ is included.

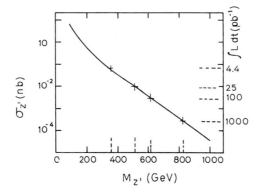

Figure 8.7. The expected dependence of $\sigma_{Z'}$ on $m_{Z'}$ at $\sqrt{s} = 1.8$ TeV. Superimposed are expected $m_{Z'}$ limits to be obtained with increasing data samples from the Tevatron Collider, normalised to the measured limit of $m_{Z'} > 380$ GeV (95 % CL), and assuming Standard Model couplings.

181

For an integrated luminosity of 100 pb^{-1}, a limit of $m(Z') \gtrsim 600$ GeV (95 % CL) should be attainable.

Again assuming Standard Model couplings and the EHLQ structure functions, a preliminary limit

$$m(W') > 480 \text{ GeV (95 \% CL)}. \tag{8.9}$$

has been presented by the CDF Collaboration.

8.4 Heavy lepton searches

The UA1 Collaboration [97] searched for heavy leptons ($L^{\pm}$) in decays $\bar{p}p \to W^{\pm} + X$; $W^{\pm} \to L^{\pm}\nu_L$; $L^{\pm} \to q\bar{q} + \nu_L$. By searching for events of large $\not{E}_T$ with associated jet activity, they obtained a limit assuming Standard Model couplings of

$$m_L > 41 \text{ GeV (90 \% CL)}, \tag{8.10}$$

Again, the event cleanliness and high statistics at LEP allows an improved limit. Using the decay mode $L^{\pm} \to \nu_L W^*$, ALEPH [98] exclude all m_L with 95 % CL if $m(\nu_L) < 42.7$ GeV, and the OPAL experiment [99] places a limit $m_L > 44.3$ GeV. Already these limits are effectively constrained by the limited LEP energy. In addition to these direct searches, an indirect upper limit can be placed on m_L from analyses of electroweak radiative corrections.

From their measurements of Γ_z [57], the LEP experiments have also placed limits on the mass of a heavy 4th neutrino. Assuming stable Dirac neutrinos, the best limit, from the L3 experiment [100], is $m_\nu > 42.8$ GeV (95 % CL). In the case of Majorana neutrinos, this limit becomes $m_\nu > 34.8$ GeV. The similar limit (in the context of SUSY models) of the neutralino is $m(\tilde{\nu}) > 27.8$ GeV.

8.5 Quark and lepton substructure

There is at present no theoretical need to require the compositeness of quarks, leptons or gauge bosons. There are, though, a number of theoretical motivations, including for example an explanation of the proliferation of quark and lepton flavours, and secondly the natural consequence in Technicolor models of composite scalar Higgs particles. The idea of composite models may include just the gauge bosons as in some models, or it may extend to the quarks and/or leptons.

A consequence of compositeness in the case of quark and leptons is the existence of contact interactions between (identical) fermions and in many models contact terms between leptons and quarks ($\ell\ell qq$). This has the effect of an increased production cross-section for jet-production ($qqqq$ contact interactions), or lepton pair production ($\ell\ell qq$ contact interactions) (see fig. 8.8). Current limits [101] are shown in table 21. To compare the existing collider data with composite models, the generalised Lagrangian of Eichten et al. and of Peskin [102],

$$L = g_{eff}^2 \left[\frac{\eta_{LL}}{2\Lambda_L^2} (\bar{\Psi}_L \gamma^\mu \Psi_L)(\bar{\Psi}_L \gamma_\mu \Psi_L) + \frac{\eta_{RR}}{2\Lambda_R^2} (\bar{\Psi}_R \gamma^\mu \Psi_R)(\bar{\Psi}_R \gamma_\mu \Psi_R) \right.$$
$$\left. + \frac{\eta_{LR}}{2\Lambda_{LR}^2} (\bar{\Psi}_L \gamma^\mu \Psi_R)(\bar{\Psi}_R \gamma_\mu \Psi_L) + \frac{\eta_{RL}}{2\Lambda_{RL}^2} (\bar{\Psi}_R \gamma^\mu \Psi_L)(\bar{\Psi}_L \gamma_\mu \Psi_R) \right] \tag{8.11}$$

with Λ representing the composite scale, and $\eta_{ij} = \pm 1$, is used.

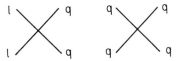

Figure 8.8. Contact terms included in the Lagrangian of Eichten et al. [102].

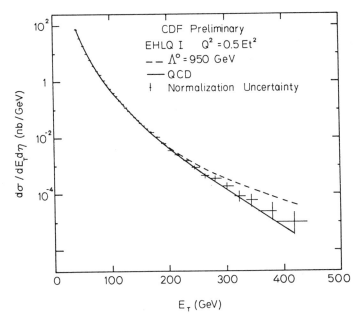

Figure 8.9. Preliminary inclusive E_T^J distribution for jet production at CDF. QCD predictions are shown, with and without composite contact terms of scale Λ.

Table 21. Existing limits on the composite scale for lepton and quark couplings [101].

Coupling	Λ^+ (TeV)	Λ^- (TeV)	Expt. limit
eeee	1.4	3.3	Tasso
ee$\mu\mu$	4.4	2.1	Jade
ee$\tau\tau$	2.2	3.2	Jade
qqqq	.95	.95	CDF (prelim)
eeqq	.9	1.5	Venus

The UA2 and CDF experiments have compared the large-p_T jet cross-section (or equivalently the 2-jet mass spectrum dN/dm^{JJ}) with expectations of QCD, together with a quark-quark contact amplitude of the form above but compressed to a single term with $\eta = +1$. Taking account of theoretical and experimental uncertainties, UA2 has published a 95 % confidence limit [103].

$$\Lambda^{qq} > 370 \text{ GeV} \tag{8.12}$$

CDF [85] have presented a conservative preliminary 95 % confidence limit based on the inclusive jet transverse energy distribution of figure 8.9, of

$$\Lambda^{qq} > 950 \text{ GeV} \tag{8.13}$$

Both UA2 and CDF are expected to publish more stringent limits soon. The UA1 Collaboration [104] used deviations of the 2-jet centre-of-mass angular distribution $d\sigma/d\cos\theta^*_{JJ}$ to set a limit on Λ^{qq}, the $\cos\theta^*$ distribution for 2-jet production being more isotropic. They set a 95 % confidence limit

$$\Lambda^{qq} > 410 \text{ GeV} \tag{8.14}$$

The measurements are energy-limited and the higher limit from CDF reflects solely the increased available $\sqrt{s}$.

The CDF Collaboration have also shown preliminary limits for the (eeqq) contact term, and a similar analysis is in progress for the ($\mu\mu$qq) contact term. They have compared the (integral) (e^+e^-) mass distribution of figure 8.6b with the expected rate using the left-handed Lagrangian of (8.11) and $\eta_{LL} = \pm 1$. The expectation for several values of Λ are superimposed on the figure. The preliminary CDF limits are [85]

$$\Lambda^{eq}(\eta_{LL} = +1) > 1.7 \text{ TeV } (95 \% \text{ CL})$$

$$\Lambda^{eq}(\eta_{LL} = -1) > 2.1 \text{ TeV } (95 \% \text{ CL}) \tag{8.15}$$

A consequence of many composite models [105] is the existence of "leptoquark" states, which are strongly interacting colour triplets that decay into a lepton-quark pair. Leptoquarks decaying into a quark + neutrino generate a missing E_T signature, while for lepton-quark decays an analysis of the Drell-Yan mode may be sensitive. At hadron colliders, leptoquarks should be pair-produced. Many possible theoretical implementations of leptoquark states exist. In most models, the couplings conserve generation resulting in decays ($Q = -\frac{1}{3}, +\frac{2}{3}$)

$LQ_1 \to d\nu_e$, ue^- or $u\bar{\nu}_e$, de^+

$LQ_2 \to s\nu_\mu$, $c\mu^-$ or $c\bar{\nu}_\mu$, $s\mu^+$

$LQ_3 \to b\nu_\tau$, $t\tau^-$ or $t\bar{\nu}_\tau$, $b\tau^+$ (8.16)

The CDF experiment is expected to be sensitive to leptoquark production of both types in excess of 100 GeV, but no results have been presented. The UA1 Collaboration [90] have placed limits on second generation leptoquark masses, using their $\not{E}_T$ data sample. They assume branching ratios [BR(LQ $\to \mu^+$s) + BR(LQ $\to$ c$\bar{s}$)] and apart from a limited region in the range $21 \leq m_{LQ} \leq 25$ GeV, they obtain

$m_{LQ} > 33$ GeV (95 % CL) (8.17)

Another consequence of many composite models is the existence of excited leptons. Using event candidates

$\bar{p}p \to W^\pm + X$; $W^\pm \to e^{\pm *}\nu_e$; $e^{\pm *} \to e^\pm \gamma$,

the UA2 and UA1 experiments have searched for evidence of excited lepton production [106]. The sensitivity of this search is limited by the $W^\pm$ mass. Therefore, the recent data from LEP for the decay $\ell^{\pm *} \to \ell^\pm \gamma$ are now far more stringent, ALEPH [107] results being with 95 % CL

$m_{\ell^*} > 45.6$ GeV, or

$m_{\ell^*} > 89$ GeV (8.19)

The first result assumes ($\ell^{*+}\ell^{*-}$) production, and the latter result assumes ($\ell^*\ell$) production with a coupling λ/m_{ℓ^*} where $\lambda = 1$.

8.6 Supersymmetric SUSY models [90, 108]

At each of the collider experiments, there have been extensive searches for evidence of new particle production, using signatures expected for supersymmetric particles. Supersymmetric theories are of interest because they provide an elegant way of cancelling mass divergences caused by radiative corrections due to scalar bosons of the Standdard Model. Since the mass divergences of bosons and fermions are of opposite sign, the terms will cancel if bosons and fermions occur in pairs with identical couplings. It is therefore postulated that to every particle there exists a supersymmetric partner (called an s-particle or SSP) as shown in table 22.

The charged SSP bosons may mix to give $\tilde{\chi}^\pm$ (charginos), and similarly for the neutral bosons $\tilde{\chi}^0$ (neutralinos). The couplings of the SSP's are analogous to those of natural particles, but with different helicity assignments.

In the lack of any experimental evidence constraining the above description, supersymmetric models cover an enormous parameter space meaning than in any confrontation with data more explicit models must be used. It is therefore natural to introduce a new quantum number, the R-parity, defined by

$$R = (-1)^{3B+L+2S} \quad (8.20)$$

which is (+ 1) for natural particles an (− 1) for SSP's. In 8.19, B is the baryon number, L is the lepton number, and the particle spin is denoted by S. If R is conserved (and this is conventionally assumed but is certainly not essential)

i) SSP's are produced in pairs,
ii) any SSP decays into some other SSP, and
iii) the lightest SSP (LSP) must be stable.

Table 22. Spin assignment of the supersymmetric particles.

Particle	Spin	s-Particle	Spin
quark $q_{L,R}$	½	s-quark $\tilde{q}_{L,R}$	0
lepton $\ell_{L,R}$	½	s-lepton $\tilde{\ell}_{L,R}$	0
photon γ	1	photino $\tilde{\gamma}$	½
gluon g	1	gluino $\tilde{g}$	½
W	1	wino $\tilde{W}$	½
Z	1	zino $\tilde{Z}$	½
Higgs H	0	s-higgs $\tilde{H}$	½
graviton G	2	gravitino $\tilde{G}$	3/2

At existing hadron colliders SSP's may be produced in two ways.

i) SSP leptons can be produced as pairs in W and Z decay, that is for example $Z \to \tilde{q}\tilde{q}$, $Z \to \tilde{\ell}^+\tilde{\ell}^-$ and $Z \to \tilde{W}^+\tilde{W}^-$ if kinematically accessible. Assuming subsequent decays $\tilde{\ell} \to \ell\tilde{\gamma}$ and $\tilde{W} \to \tilde{\ell}\nu$ with an undetected $\tilde{\gamma}$ as the LSP, one expects dilepton states with associated $\not{E}_T$. Such SSP decays are also accessible from Z decays at LEP. Prior to LEP, the most stringent limits were from UA2 [109]. However, LEP data [57, 110] have now excluded these decays within the kinematic limit imposed by m_Z (i.e. m ~ 44.5 GeV).

ii) The direct production of SSP pairs is kinematically possible over an extended mass range at hadron colliders, but in this case cross-section limitations restrict such searches to strongly-interacting SSP's, that is s-quarks and gluinos. The following discussion is restricted to such searches.

In most model comparisons, it is presently (and in some cases unjustifiably) assumed that

i) the LSP is the photino, and $m(\tilde{\gamma}) \simeq 0$
ii) the existing s-quarks are mass degenerate,
iii) all couplings are strong,
vi) the $\tilde{g}$ and $\tilde{q}$ fragmentations are exactly as for g and q fragmentations at equivalent mass, and
v) cascade decays are ignored (cascade decays have now been considered for some studies [111]).

Two possible decay chains are generally considered. If $m(\tilde{q}) > m(\tilde{g})$, it is assumed that $(\tilde{g} \to q\bar{q}\tilde{\gamma})$ and $(\tilde{q} \to q\tilde{g}$ or $qg)$. If $m(\tilde{q}) < m(\tilde{g})$, it is assumed that $(\tilde{g} \to \tilde{q}q$ or $q\tilde{q})$, and that $(\tilde{q} \to q\tilde{\gamma})$. Production is dominated at hadron colliders by the production of $\tilde{g}\tilde{g}$, $\tilde{g}\tilde{q}$ or $\tilde{q}\tilde{q}$ pairs, and in all model variations, the final state will be characterised by hadronic jet(s) production in association with $\not{E}_T$ from the outgoing $\tilde{\gamma}'s$.

The most stringent limits for $\tilde{g}$ and $\tilde{q}$ production are from recent UA2 [112] and preliminary CDF results [114]. The CDF data were selected as follows :

i) $\not{E}_T$ > 40 GeV, and $\sigma(\not{E}_T)$ > 2.8 standard deviations,
ii) at least one well-measured jet satisfying E_T^J > 15 GeV and $|\eta|$ < 1, and with at least one additional well-measured jet satisfying E_T^J > 15 GeV in the rapidity range $|\eta|$ < 3.5, and
iii) the removal of events with signatures characteristic of leptonic W and Z decays, and as well unwanted backgrounds such as multiple vertex events and cosmic or beam-gas events.

A total of 98 events are shown in figure 8.10 and Standard Model sources or background are expected to contribute 86.4 events (table 23). Making a more stringent selection, CDF measure (table 24)

Table 23. Contributions to the preliminary CDF $\not{E}_T$-signal from Standard Model sources or background.

DATA	98 events
W + ≥ 2 JET ↳ $e\nu$	6.4
W + ≥ 2 JET ↳ $\mu\nu$	16.6
W + ≥ 1 JET ↳ $\tau\nu$	30.7
Z + ≥ 2 JET ↳ $\nu\bar{\nu}$	32.7
QCD MISIDENT. MISMEAS.	4 ± 4
TOTAL (exc.QCD)	86.4 ± 14.1 (stat) ± 11.6 (sys)

Table 24. Contributions to the preliminary/CDF E_T-signal from the Standard Model or background events

Selection	Data	Estimated background
$\not{E}_T$ > 100 GeV; ≥ 2 jets	3	1.3 ± 1.3
$\not{E}_T$ > 40 GeV; ≥ 4 jets	2	1.3 ± 1.3

The data have been compared with an ISAJET model for $\tilde{g}$ and $\tilde{q}$ production at different $m(\tilde{q})$ and $m(\tilde{g})$ values, using the (conservative) EHLQ structure function parametrisations. No cascade decays were implemented. The

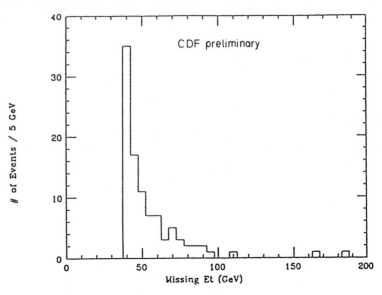

Figure 8.10. The distribution in $\not{E}_T$ of 98 events from the CDF experiment satisfying $\not{E}_T > 40$ GeV, subject to criteria discussed in the text.

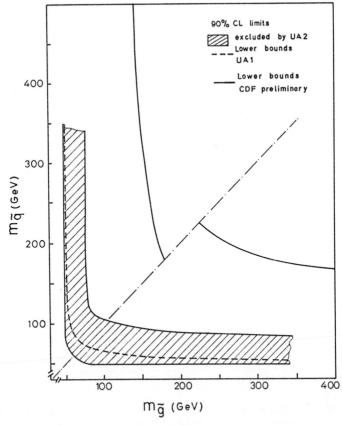

8.11. Excluded 90 % CL limits for $m(\tilde{g})$ and $m(\tilde{q})$ from UA1 and UA2 [112, 113], and CDF (preliminary) [114].

model used required ≥ 4 jets for $m(\tilde{q}) > m(\tilde{g})$, and $\not{E}_T > 100$ GeV for the case $m(\tilde{g}) > m(\tilde{q})$. Resulting 90 % CL limits on $m(\tilde{q})$ and $m(\tilde{g})$ after taking account of background and luminosity uncertainties are shown in figure 8.11, taking values $m(\tilde{q}) > 150$ GeV and $m(\tilde{g}) > 150$ GeV. The discontinuity at $m(\tilde{q}) = m(\tilde{g})$ is an inadequacy of the ISAJET model, and the results are preliminary. Recent studies including cascade decays suggest that the quoted mass limits should be reduced by $\sim 10 - 15$ GeV.

The UA2 analysis used more stringent selection criteria :

i) E_T(jet 1) ≥ 25 GeV and E_T(jet 2) ≥ 15 GeV with azimuthal opening angle $\delta\phi_{12} < 160^0$ (that is not back-to-back), and
ii) $E_T > 40$ GeV with $\delta\phi$ (E_T^{jet}, $\not{E}_T$) $< 140^0$ with no additional jet activity in the direction of the $\not{E}_T$ vector.

No event survived these cuts. The UA2 model used the cross-section calculations of Dawson et al. [115], and as for the CDF evaluations they used EHLQ structure function parametrisations. The resulting mass limits are shown in figure 8.11.

Existing $\tilde{g}$ and $\tilde{q}$ analyses should be modified to include a realistic implementation of cascade decays. The quoted preliminary CDF limits are likely to improve following more detailed background studies, and a comparison with models taking the cases $m(\tilde{g}) \geq m(\tilde{q})$ and $m(\tilde{g}) \leq m(q)$ consistently. The aim of future analyses with data of the 1991 Tevatron run and beyond should be to extend the tested model parameter space.

9. CONCLUDING REMARKS

Hadron colliders are essentially exploratory machines, and with the completion of the UA1 and UA2 experiments, Fermilab will have from 1991 a unique exploratory tool with two outstanding general-purpose detectors (CDF and DØ) for the coming decade.

These lectures have attempted from an experimental viewpoint to survey the contribution of the CERN and Fermilab Colliders to both exploratory and precision measurements relevant to the Standard Model, and to indicate interesting future directions of activity. Since many of us will be involved in the future CDF and DØ programs, it seems sensible to conclude with a shopping list of priorities for these experiments.

The most important exploratory tasks include

i) an unambiguous identification of the t-quark and an accurate evaluation of its mass (thereby constraining the range of possible Standard Model Higgs masses),
ii) a positive and direct identification of the (WWγ) coupling from a measurement of the ($W^{\pm} + \gamma$) production rate and angular distribution,
iii) extending the mass scale limit for possible composite particle production, and
iv) extending mass or cross-section limits on the production of supersymmetric particles, additional gauge bosons and other manifestations of extensions to (or departures from) the Standard Model.

It should be noted that item (iii) above implies precision comparisons of the data with QCD, and item (iv) implies considerably improved model comparisons (for example the implementation of cascade decays in supersymmetric models). Major experimental efforts are also required at CDF and DØ in the context of (i) to (iv) above to improve the identification capability for τ-leptons and direct photons. Studies to optimise missing-E_T measurements at CDF and DØ are also essential, especially in the presence of multiple interactions in a $\bar{p}p$ bunch crossing (as expected with increasing Tevatron luminosity).

Precision QCD comparisons continue to be important at CDF and DO, and comparisons with NLO QCD evaluations are necessary. In particular measurements of the p_T-dependence of $W^\pm$, Z, γ and hadron jet production at the largest p_T-values and a comparison with QCD models, are important. Equally important, though more mundane, are possible constraints that collider data might provide for the measurement of parton distributions within the nucleon. Useful measurements include

i) $W^\pm$ asymmetry data that provide a constraint on the relative u-quark and d-quark nucleon content, and

ii) data on low-E_T direct photon production and two-photon production to constrain parametrisations of the gluon content of the nucleon.

It is very likely that at LEP2 the W mass will be measured more accurately than at CDF and DØ. Nevertheless, a continued emphasis on accurate W mass measurements is essential. Experimentally, this implies the need for accurate measurements of missing-E_T.

Finally, the pioneering B-physics studies at UA1 and CDF have demonstrated that accurate magnetic spectrometers can be used to great advantage in a hadron collider environment. CDF have demonstrated their ability to reliably reconstruct exclusive B-meson decays, and their current upgrade plans emphasise the possibility of high-statistics B-physics studies. These studies include the identification of rare decays for sensitive tests of the Standard Model, a measurement of the time-evolution of $B^0 - \bar{B}^0$ oscillations, and even the possibility of measuring CP- violation in B-meson decays if planned machine upgrades are implemented.

ACKNOWLEDGEMENTS

I would like to thank Professor and Mrs. T. Ferbel for their kindness and hospitality at this School, and in particular for the informal but lively and stimulating atmosphere that they succeeded in creating. Thanks are equally due to the participants of the School; their interest and enthusiasm made it a real pleasure to give these lectures.

I am grateful to the following people for discussions and/or material : N. Ellis, K. Eggert and F. Pauss from the UA1 experiment, P. Lubrano, K.-H. Meier and H. Plothow-Besch from the UA2 experiment, and A. Baden, A. Barbaro-Galtieri, T. Fuess, M. Gold, J. Huth, R. Kephart, C. Newman-Holmes, R. Plunkett, P. Tipton, M. Shapiro, M. Shochet and A. Tollestrup from the CDF experiment.

Finally, many thanks are due to Mme U. Fischer and M. F. Maschiocchi for their help in preparing this text.

References

1. S.L. Glashow, Nuc. Phys. **22** (1961) 579.
   S. Weinberg, Phys. Rev. Lett. **19** (1967) 1264.
   A. Salam, in "Elementary Particle Theory", ed. W. Svartholm, Almquist and Wiksell, Stockholm (1968).

2. C. Rubbia et al., Proc. Int. Conf. on Neutrinos, Aachen, 1976. ed. H. Faissner et al., (Kreweg and Braunschweig), 683.

3. G. Altarelli et al., Z. Phys. **C27** (1985) 617.
   For more recent evaluations including partial $O(\alpha_s^3)$ corrections, see for example
   T. Matsuura et al., Nucl. Phys. **B319** (1989) 570.
   K. Ellis, Proceeding 8th Topical Workshop in $\bar{p}p$ Collider Physics, Castiglione, Italy (1989). To be published.

4. G.J. Budker, At. Energ. **22** (1967) 346.

5. See D. Möhl et al., Phys. Rep. **58** (1980) 73.
   D. Möhl, CERN report 84–15 (1984).

6. L. Evans et al., in Proton-Antiproton Collider Physics, eds. G. Altarelli and L. di Lella, World Scientific, 1.
   Fermilab Accelerator Upgrade, Fermilab report, April 1989.

7. UA1 Collaboration, A. Astbury et al., CERN/SPSC/78-06 (1978).
   See also ref. [42], and references therein.

8. UA2 Collaboration, B. Mansoulié, Proc. Moriand Workshop, La Plagne, France, 1983 (Editions Frontières, Gif-sur-Yvette) 609.

9. CDF Collaboration, F. Abe et al., Nucl. Inst. Methods **A271** (1988) 387, and references therein.

10. DØ Design Report, Fermilab, 1983.

11. UA4 Collaboration, M. Bozzo et al., Phys. Lett. **B147** (1984) 392.
    D. Bernard et al., Phys. Lett. **B198** (1987) 583.

12. E710 Collaboration, N. Amos et al., Phys. Lett. **B243** (1990) 158.
    R. Rubinstein, Fermilab-Conf-90/160-E (1990).

13. UA5 Collaboration, G. Alner et al., Phys. Lett. **B138** (1984) 304.

14. For a review of soft collisions at the CERN $\bar{p}p$ Collider, see D.R. Ward, in Proton-Antiproton Collider Physics, eds. G. Altarelli and L. di Lella, World Scientific, 85.

15. M. Shapiro, Proceedings 17th SLAC Summer Inst., July 1989, SLAC-PUB-5219 (1990).

16. PYTHIA event generator, H.-U. Bengtsson and T. Sjostrand, LU TP 87-3 (1987).

17. I. Hinchcliffe, Papageno Monte Carlo, unpublished.

18. UA2 Collaboration, C. Booth, Proc. of the Topical Workshop on Proton-Antiproton Collider Physics, Aachen, 1986, eds. K. Eggert et al., World Scientific, 381.

19. UA1 Collaboration, C. Albajar et al., Phys. Lett. **B185** (1987) 233 and Addendum, Phys. Lett. 191B (1987) 462.

20. For a review of the Physics of hadronic jets at the CERN Collider prior to 1987 (including fragmentation) see R.K. Ellis and W.G. Scott, in Proton-Antiproton Collider Physics, ed. G. Altarelli and L. di Lella, World Scientific, 131.

21. CDF Collaboration, F. Abe et al., Phys. Rev. Lett. **65** (1990) 968.

22. For a recent review, see J. Morfin, Fermilab preprint, Fermilab-Conf-90/155 (1990).
    Frequently used structure function parametrisations include
    DFLM : M. Diemoz et al., Z. Phys. **C39** (1988) 21.
    MRS : A. Martin et al., Phys. Lett. **B206** (1988) 327.
    Mod. Phys. Lett. **A4** (1989) 1135.
    Phys. Rev. **D37** (1988) 1161.
    HMRS : P. Harriman et al., RAL preprint RAL-90-018 (1990).
    DO   : D. Duke et al., Phys. Rev. **D30** (1984) 49.
    EHLQ : E. Eichten et al., Rev. Mod. Phys. **56** (1984) 579.

23. S. Ellis, Z. Kunszt and D. Soper, Phys. Rev. Lett. **64** (1990) 2121.

24. UA1 Collaboration, G. Arnison et al., Phys. Lett. **B158** (1985) 494.

25. UA2 Collaboration, P. Lubrano, Proc. 4th Rencontre de physique de la Vallée d'Aoste, La Thuile, 1990. To be published.

26. UA2 Collaboration, J. Appel et al., Z. Phys. **C30** (1986) 341.

27. Z. Kunszt and W. Stirling, Phys. Lett. **B171** (1986) 307.
    DTP preprint DTP /87-16.

28. S. Parke and T. Taylor, Phys. Rev. Lett. **56** (1986) 2459.

29. H. Kuijf, PhD thesis, Univ. Leiden, in preparation. Private communication to UA2 Collaboration.

30. UA1 Collaboration, G. Arnison et al., Phys. Lett. **B177** (1986), 244.

31. UA2 Collaboration, R. Ansari et al., Phys. Lett. **B215** (1988) 175.

32. CDF Collaboration, D. Brown and J. Huth, private communication.

33. P. Aurenche et al., Nucl. Phys. **B297** (1988) 661.
    P. Aurenche et al., Fermilab preprint PUB-89/226-T (1989), and references therein.

34. R. Ansari et al., Z. Phys. **C41** (1988) 395.

35. CDF Collaboration, presented by A.M. Harris, Fermilab preprint CONF-90/118-E.

36. E. Eichten et al., Rev. Mod. Phys. **58** (1986) 1065.

37. J. Collins and D. Soper, Phys. Rev. **D16** (1977) 2219.

38. E.L. Berger et al., Nucl. Phys. **B239** (1984) 52.

39. G. Altarelli and L. di Lella, in "Proton-Antiproton Collider Physics", World Scientific (Singapore) 177.

40. UA2 Collaboration, J. Alitti et al., CERN-EP/90-20 (1990) to be published in Z. Phys. C.

41. CDF Collaboration, to be published.

42. UA1 Collaboration, C. Alabajar et al., Z. Phys. C44 (1989) 15.

43. H. Plothow, The Standard Model Parameters, CERN preprint CERN-PPE 3/90-168 (1990).

44. UA2 Collaboration, J. Alitti et al., CERN preprint CERN-EP/90-52 (1990), to be published in Z. Phys. C.

45. CDF Collaboration, to be published.

46. G. Altarelli et al., Nucl. Phys. B246 (1984) 12, and ref. [39].

47. P. Arnold et al., Nucl. Phys. B319 (1989) 37 and B330 (1990) 284.
P. Arnold et al., Phys. Rev. D40 (1989) 613.

48. F. Berends et al., Phys. Lett. B224 (1989) 237.
See also A. Bawa et al., Phys. Lett. B203 (1988) 172.

49. S. Ellis et al., Phys. Lett. B154 (1985) 435.
See also R. Kleiss et al., Nucl. Phys. B262 (1985) 235.

50. E. Berger et al., Phys. Rev. D40 (1989) 83.

51. CDF Collaboration, J. Hauser, Proc. of Fermilab Structure Function Workshop, April 1990. Fermilab preprint FERMILAB-Conf-90/109-E.

52. For pedagogical review and derivation see :
C. Jarlskog, Proc. CERN School of Physics, CERN 82-04 (1982) 63.
I. Aitchison and A. Hey "Gauge Theories in Particle Physics", edition 2 (Adam and Hilger).

53. N. Cabibbo, Phys. Rev. Lett. 10 (1963) 531.
M. Kobayashi and K. Maskawa, Prog. Theor. Phys. 49 (1973) 652.

54. CDF Collaboration, F. Abe et al., Phys. Rev. Lett. 63 (1989) 720.
F. Abe et al., Phys. Rev. Lett. 65 (1990) 2243.

55. "Review of Particle Properties", Particle Data Group, Phys. Lett. B239 (1990).

56. UA2 Collaboration, J. Alitti et al., Phys. Lett. B241 (1990) 150.

57. F. Dydak, Invited talk at International High Energy Physics Conference, Singapore 1990, to be published.

58. UA1 Collaboration, C. Albajar et al., CERN preprint CERN-PPE/90-141 (1990).

59. CDF Collaboration, F. Abe et al., Phys. Rev. Lett. 64 (1990) 157.

60. A. Martin et al., Phys. Lett. B189 (1987) 220, and references therein.
A. Martin et al., Phys. Lett. B228 (1989) 149.

61. UA2 Collaboration, J. Alitti et al., CERN preprint CERN-PPE/90-105 (1990). Submitted to Z. Phys. C.

62. R. Brown et al., Nucl. Phys. **B75** (1974) 112.

63. CDF Collaboration, F. Abe et al., to be published.

64. W. Marciano et al., Phys. Rev. **D22** (1980) 2695, and references therein.

65. G. Fogli et al., Z. Phys. **C40** (1988) 379.
    P. Langacker et al., Phys. Rev. **D36** (1987) 2191.

66. P. Langacker, Proc. 1989 SLAC Summer Institute, to be published.
    P. Langacker, Phys. Rev. Lett. **63** (1989) 1920.
    ALEPH Collaboration, D. Decamps et al., CERN-PPE/90-104 (1990), Submitted to Z. Phys. C.

67. CDHS Collab., H. Abramowicz et al., Phys. Rev. Lett. **57** (1986) 298.
    A. Blondel et al., Z. Phys. **C45** (1990) 361.

68. CHARM Collaboration, J. Allaby et al., Z. Phys. **C36** (1987) 611.
    Z. Phys. **C41** (1989) 567.

69. ALEPH Collaboration, D. Decamp et al., Phys. Lett. **B236** (1990) 511, and reference [66].
    DELPHI Collaboration, P. Abreu et al., Phys. Lett. **B242** (1990) 536.
    OPAL Collaboration, M. Akrawy et al., Phys. Lett. **B236** (1990) 364.

70. UA1 Collaboration, C. Albajar et al., Z. Phys. **C48** (1990) 1.

71. UA2 Collaboration, T. Akesson et al., Z. Phys. **C46** (1990) 179.

72. CDF Collaboration, F. Abe et al., Phys. Rev. Lett. **63** (1990) 142.
    F. Abe et al., Phys. Rev. Lett. **63** (1990) 147.
    F. Abe et al., Fermilab-PUB-90/137, submitted to Phys. Rev.
    CDF Collaboration presented by K. Sliwa, Fermilab-CONF-90/93-E, to be published in Proc. XXVth Rencontres de Moriond, France, March 1990.

73. See B. Naroska, Phys. Rep. **148** (1987) 68, and references therein.

74. S. Glashow et al., Phys. Rev. **D2** (1970) 1285.

75. G. Kane et al., Nucl. Phys. **B195** (1982) 29.
    A. Bean et al., Phys. Rev. **D35** (1987) 3533.

76. K. Eggert et al., Proc. 7th Topical Workshop on $p\bar{p}$ Collider Physics, Fermilab, 1988. Ed. R. Raja et al., World Scientific, 599.

77. P. Nason et al., Nucl. Phys. **B303** (1988) 603.
    See also R. K. Ellis, Fermilab-Conf-89/168 (1989).

78. G. Altarelli et al., Nucl. Phys. **B308** (1988) 724.

79. CLEO Collaboration. M. Artuso et al., Phys. Rev. Lett. **62** (1989) 2233.

80. See for example : S. Glashow et al., Phys. Lett. **B196** (1987) 233.
    V. Barger et al., Phys. Rev. **D40** (1989) 2875.
    J. Gunion, UCD preprint UCD-89-10 (1989).

81. CDF Collaboration. Private communication, H. Grassmann and T. Westhusing.

82. See M. Della Negra et al., Proton-Antiproton Collider Physics, ed. G. Altarelli and L. di Lella, World Scientific, p. 225.

83. Proposal P238, CERN-SPSC/88-33, and CERN-SPSC/88-43.
P. Schlein, Proc. 8th Workshop on $p\bar{p}$ Collider Physics, Castiglione, 1989. Eds. Belletini and Scribano, World Scientific, 612.

84. A. Morsch, Proc. XXVth Rencontres de Moriond, Les Arcs, France, (1990), to be published.
S. McMahon, same proceedings.

85. L. Pondrom, Proc. XVth International Conference on High Energy Physics, Singapore 1990, to be published.

86. ARGUS Collaboration, H. Albrecht et al., Phys. Lett. B192 (1987) 246.

87. CLEO Collaboration, A. Jawahery, Proc. XXIVth International Conference on High Energy Physics, Munich 1988, eds. R. Kotthaus et al., (Springer Verlag.)
CLEO Collaboration, A. Bean et al., Phys. Rev. Lett. 58 (1987) 183.

88. L3 Collaboration. B. Adeva et al., L3 preprint 020 (1990), submitted to Phys. Lett.

89. M. Gold and C. Haber, Proc. Workshop "Physics at Fermilab in 1990's", Breckenridge, 1989, ed. D. Green et al., World Scientific, 283.

90. J. Ellis and F. Pauss, "Proton-Antiproton Collider Physics", eds. G. Altarelli and L. di Lella, World Scientific, 269.

91. F. Pauss, Proc. Cargèse Summer Inst. on Particle Physics, 1989, Plenum Press.

92. R. Brown et al., Phys. Rev. D19 (1979) 922.
K. Hagwara et al., Phys. Rev. D41 (1990) 2113.
U. Baur et al., Phys. Rev. D41 (1990) 1476.
B. Humpert, Phys. Lett. B135 (1984) 179.
For an excellent summary, see also D. Zeppenfeld, Proc. 8th Topical Workshop on $p\bar{p}$ Collider Physics, Castiglione, 1989, Eds. Belletini and Scribano, World Scientific, 326.

93. CDF Collaboration, F. Abe et al., Phys. Rev. D41 (1990) 1717.

94. ALEPH Collaboration, D. Decamp et al., Phys. Lett. B246 (1990) 306.
D. Decamp et al., Phys. Lett. B241 (1990) 141.
DELPHI Collaboration, P. Abren et al., Phys. Lett. B245 (1990) 276.
L3 Collaboration, B. Adeva et al., Phys. Lett. B248 (1990) 203.
OPAL Collaboration, M. Akrawy et al., Phys. Lett. B236 (1990) 224.
M. Akrawy et al., Z. Phys. C, to be published.

95. See for example H. Georgi et al., Phys. Rev. Lett. 40 (1978) 692.

96. ALEPH Collaboration, D. Decamp et al., Phys. Lett. B241 (1990) 633.
DELPHI Collaboration, P. Abren et al., Phys. Lett. B241 (1990) 449.
L3 Collaboration, B. Adeva et al., L3 preprint 018, submitted to Phys. Lett.
OPAL Collaboration, M. Akrawy et al., Phys. Lett. B242 (1990) 299.

97. UA1 Collaboration, C. Albajar et al., Phys. Lett. **B185** (1987) 241.

98. ALEPH Collaboration, D. Decamp et al., Phys. Lett. **B236** (1990) 511.

99. OPAL Collaboration, M. Akrawy et al., Phys. Lett. **B240** (1990) 250 and CERN preprint CERN-EP/90-72 (1990).

100. L3 Collaboration, B. Adeva et al., L3 preprint 016, submitted to Phys. Lett.

101. TASSO Collaboration, W. Braunschweig et al., Z. Phys. **C37** (1988) 171.
JADE Collaboration, see reference [73].
VENUS Collaboration, K. Abe et al., Phys. Lett. **B232** (1989) 425.

102. E. Eichten et al., Phys. Rev. Lett. **50** (1983) 811.
M. Peskin, Proc. 1981 International Symposium on Lepton and Photon Interactions at High Energies, ed. W. Pfeil, 880.

103. UA2 Collaboration, P. Bagnaia et al., Phys. Lett. **B129** (1983) 130.

104. UA1 Collaboration, G. Arnison et al., Phys. Lett. **B147** (1984) 241.

105. For a review, see B. Schremp, Proc. 6th Topical Workshop on $\bar{p}p$ Collider Physics, Aachen, 1986, eds. K. Eggert et al., World Scientific, 642.

106. UA1 Collaboration, G. Arnison et al., Phys. Lett. **B135** (1984) 250.
UA2 Collaboration, R. Ansari et al., Phys. Lett. **B195** (1987) 613.

107. ALEPH Collaboration, D. Decamp et al., Phys. Lett. **B236** (1990) 501.

108. See for example H. Haber and G. Kane, Phys. Rep. 117 (1985) 75.

109. UA2 Collaboration, T. Akesson et al., Phys. Lett. **B238** (1990) 443.

110. **ALEPH Collaboration**, D. Decamp et al., Phys. Lett. **B236** (1990) 86.
D. Decamp et al., CERN preprint CERN-EP/90-63 (1990).
DELPHI Collaboration, P. Abren et al., CERN preprint CERN-EP/90-79 (1990). P. Abren et al., CERN preprint CERN-EP/90-80 (1990).
L3 Collaboration, B. Adeva et al., Phys. Lett. **B233** (1989) 530.
OPAL Collaboration, M. Akrawy et al., Phys. Lett. **B240** (1990) 261.
M. Akrawy et al., CERN preprint CERN-PPE/90-95 (1990).

111. H. Baer et al., Phys. Rev. Lett. **63** (1989) 352.

112. UA2 Collaboration, J. Alitti et al., Phys. Lett. **B235** (1990) 363.

113. UA1 Collaboration, C. Albajar et al., Phys. Lett. **B198** (1987) 261.

114. CDF Collaboration, F. Abe et al., Phys. Rev. Lett. **62** (1989) 1825.
P. Hu, thesis, Rutgers University (1990).

115. S. Dawson et al., Phys. Rev. **D31** (1985) 1581.

A PERSPECTIVE ON MESON SPECTROSCOPY*

D. W. G. S. Leith

Stanford Linear Accelerator Center
Stanford University, Stanford, California 94309, USA

1. INTRODUCTION

We will visit the field of meson spectroscopy during these lectures, with the main emphasis on the systematic progress in the area of light quark spectroscopy. To begin with, we will review some basic concepts and definitions, and generally set the stage for a discussion of the data in Section 2. Section 3 will examine the experimental data on strange meson resonances—the K^* states—where the most systematic and most complete spectroscopic data exist. In Section 4 we review the data on strangeonium resonances—the $(s\bar{s})$ states. The experimental situation with respect to glueballs and exotic mesons is examined in Section 5, and the story on heavy quark spectroscopy is briefly reviewed in Section 6. Finally, in Section 7, we summarize the experimental situation and draw some conclusions on the topic of meson spectroscopy.

2. PREFACE

a) <u>What Are Mesons?</u>

We believe that regular mesonic matter is made up of a quark and an anti-quark pair bound together, while baryonic matter is a composite of three quarks. However, our theory of the strong interaction—quantum chromodynamics (QCD)—with its non-abelian character for the quark and gluon couplings, would also expect bound states of two, or more, gluons in a state of matter called gluonium—or less colorfully, just "glueballs." Other possible forms of mesonic matter may exist—in which a gluon binds to a quark-antiquark pair forming a kind of matter which would be a hybrid of the quark and gluon states discussed above. Finally, there are expectations that multiquark states, or quark molecules, may exist. These various possible kinds of mesonic matter are summarized in Table 1, and discussed further in Ref. 1.

What would be the signal that we have found evidence for the existence of the more exotic forms of matter, beyond the classical $q\bar{q}$ pairs? A very direct indication—a so called "smoking gun"—would be the discovery of a meson with internal quantum numbers which would *not* be allowed by $q\bar{q}$ matter, and only permitted by one of the other kinds of matter listed in Table 1. We will revisit this again in Section 5 and even

---

* Work supported by Department of Energy contract DE-AC03-76SF00515.

Table 1. The various forms of mesonic matter.

Types of Mesonic Matter	Composition
Quark-antiquark	$q\bar{q}$
Glueball	gg, ggg
Hybrid	$q\bar{q}g$
Multi-quark	$q\bar{q}q\bar{q}$

later, in Section 7. Another strategy would be to programmatically study the mesons of a specific type and find *all* of the candidate states required in the $q\bar{q}$ family and have other states left over—with properties not expected, or required, by the $q\bar{q}$ picture. Such states would then be candidates for the other, exotic form of mesons. We will consider this approach frequently in Sections 3 and 4, when we review the K^* and $s\bar{s}$ meson situation.

b) What Are the Basic Building Blocks of Matter?

The quarks are the basic building blocks that make up the hadrons. We believe the quarks come in three generations, or families, with electric charge of 2/3 or $-1/3$ and with intrinsic spin of 1/2. The masses of the quarks are thought to be around 4 MeV and 7 MeV for the u, d quarks in the first generation, 150 MeV and 1.5 GeV for the strange and charm quarks of the second generation. In the third generation, the bottom quark has a mass of 4.7 GeV, and the sixth quark—the top quark—is, as yet, undiscovered. The limits on the mass of the top quark are that it is greater than 41 GeV, but less than 250 GeV.* Table 2 lists the six quarks and indicates some of the properties of the individual quark flavors.

The quarks are bound by the strong interaction which is mediated by the gluon. This quantum of the strong force has spin 1, charge zero and mass zero.

c) What Are the Forces Experienced by the Building Blocks?

The forces that we must consider are the strong interaction, the electromagnetic interaction, the weak interaction and gravity.

The strong interaction is mediated by the gluons, which act on the color charge of the quarks. The strength of the force between two quarks at a separation of 10^{-15} cm is about 25 times stronger than the electromagnetic force.

The electromagnetic interaction is mediated by the photon, which acts on all electrically charged bodies. We take its strength as our normalization, (i.e., unity).

Table 2. Properties of the quarks.

	d	u	s	c	b	t
Electric charge	$-1/3$	$+2/3$	$-1/3$	$+2/3$	$-1/3$	$+2/3$
Isospin—Third Component	$-1/2$	$+1/2$	0	0	0	0
Strangeness	0	0	-1	0	0	0
Charm	0	0	0	$+1$	0	0
Bottomness	0	0	0	0	-1	0
Topness	0	0	0	0	0	$+1$

⋆ Since then, the summer conference results claim that $m_t > 89$ GeV.

The weak interaction is mediated by the heavy vector bosons—the W^+, W^- and the Z^0, and acts on the quarks and leptons. The strength of this force between two quarks at a separation of 10^{-15} cm is about 80% of the electromagnetic force, but it is very short range. For example, the force between two protons in a nucleus is about 10^{-7} of the electromagnetic force!

In the Standard Model—the powerful paradigm that attempts to describe our world—these two forces are described as separate realizations of one force in nature—the electro-weak force.

Finally, gravity we suspect is mediated by an as yet unobserved particle called the graviton, and acts on all particles with mass. The strength of the force between two quarks at a distance of 10^{-15} cm is $\sim 10^{-41}$ of the electromagnetic force. Very weak indeed!

As an example of the relative strength of these forces, we consider two processes with comparable available energy, but one proceeding via the strong interaction and the other, a weak decay process. The strong decay of the $K^*(890)$ and the weak decay of the K^0 are the examples, with 280 MeV available in the first, and 206 MeV available in the second, process. The lifetime of the first decay is $\sim 10^{-23}$ secs, while the $K_1^0 \to \pi^+\pi^-$ has a lifetime of $\sim 10^{-10}$ secs!

d) What Are the Useful Quantum Numbers for Mesons?

First, we list and define a number of generally useful quantum numbers:

- Baryon number, B = number of baryons − number of antibaryons

  = 0 for mesons.

- Hypercharge, $Y = \quad S \quad + \quad B$

  (strangeness)(baryon number)

  = S for mesons.

- Strangeness, S = number of strange quarks − number of strange antiquarks.

  = 3 for Ω baryon

  = 1 for K^- meson

  = 0 for π^- meson

Similarly, there are related quantum numbers for charm, beauty and topness for mesons.

- Isotopic Spin, I = describes the grouping of particles into charge multiplets.

  The projection of this quantum number, I, along the charge axis is called I_3, or I_z, and is related to the value of the electric charge of the particle by

  $$I_z = Q - \frac{Y}{2}.$$

- Charge Conjugation, C

  The operation of C takes a particle into its anti-particle

  $$C|B, Y, I, I_z\rangle = \eta_I | - B, -Y, I, -I_z\rangle$$

  where η_I is a phase factor.

  A second operation of C should lead back to the initial state. This requires that $\eta_I^2 = 1$.

From this we see that a meson with
$$B = Y = I_Z = 0$$
may be an eigenstate of C, with eigenvalue ± 1.

The commutation relations are:
$$CI_X = I_X C$$
$$CI_Y = -I_Y C$$
$$CI_Z = -I_Z C$$

- G–Parity, G

For particles that are charged, (i.e., not eigenstates of C), it is useful to define a new quantum number—G-parity—such that
$$G = CR_X = C \; exp(i\,\pi\,I_X)$$
where R_X rotates the state by 180°, around the x–axis. We then have:
$$(CR_X)I_X = I_X(CR_X)$$
$$(CR_X)I_Y = I_Y(CR_X)$$
$$(CR_X)I_Z = I_Z(CR_X)$$
so that a meson of any charge may be simultaneously an eigenstate of G, I^2 and I_Z.

In general,
$$G|B,\,Y,\,I,\,I_Z\rangle = \eta_I \; i^{2I}|-B,\,-Y,\,I,\,I_Z\rangle$$

The value of the G parity must be determined from experiment. The eigenvalues of G are $G = \pm 1$. The eigenstates of G must have $B = S = 0$. [This is so, since for a particle with baryon number B, charge Q and strangeness S, an operation by C leads to the antiparticle state with quantum numbers $-B$, $-Q$ and $-S$. For an operation of G on such a particle, its charge would be changed by the rotation but not its strangeness nor baryon number. Therefore, only particles with $B = S = 0$ can be eigenstates of G.]

$$Q = 0 \text{ for an eigenstate of C}$$
$$B = S = 0 \text{ for an eigenstate of G}$$

The pion has odd G–parity.

If a state decays into several mesons, each of which is an eigenstate of G, then the G parity of the state is the product of the G parities of the individual mesons.

If a neutral meson has definite isospin and C parity, i.e., is an eigenstate of both, then
$$G = (-1)^I C \; .$$

- Parity, P

    A function is said to have even parity if it is invariant under the reflection in the origin of all of the space coordinates.
    $$f(-\vec{r}) = f(\vec{r}) \ .$$
    It is said to be odd, if
    $$f(-\vec{r}) = -f(\vec{r}) \ .$$
    In a reaction or a decay process, the total wave functions of the initial and final states have definite parity. We attribute an intrinsic parity to each particle involved (leptons excepted).

    Finally, for a meson, consisting of a bound state of a quark and an antiquark, the internal quantum numbers may be calculated from:

    $$J = L + S$$

    $$P = (-1)^{L+1}$$

    $$C = (-1)^{L+S}$$

    $$G = (-1)^{L+S-I}$$

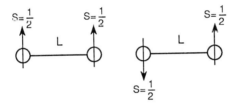

Fig. 1. The spin parallel and the spin anti-parallel quark pair configurations of a classical meson.

In Table 3 we list the internal quantum numbers G, C and P for several example mesons.

Table 3. The P, G and C quantum numbers for some example mesons.

Particle	P	G-Parity	C-Parity
$\pi(140)$	−1	−1	+1
$\eta(550)$	−1	+1	+1
$\rho(770)$	−1	+1	−1
$w(783)$	−1	−1	−1
$f_2(1270)$	+1	+1	+1
$w_3(1670)$	−1	−1	−1

Table 4. Conservation laws.

Conservation Laws	Respected in the following interactions:
Conservation of energy, E	S, EM, W
Conservation of three momentum, p	S, EM, W
Conservation of angular momentum, J	S, EM, W
Conservation of charge, Q	S, EM, W
Conservation of baryon number B	S, EM, W
Conservation of lepton number L	S, EM, W
Conservation of isotopic spin, I	S
Conservation of parity, P	S, EM
Conservation of charge conjugation, C	S, EM
Conservation of G–parity, G	S
Conservation of strangeness, S	S, EM
Invariance under CPT	S, EM, W
Invariance under T	S, EM, W

e) **What Are the Conservation Laws and Selection Rules?**

In Table 4 we list the important conservation laws that can help us understand the production and decay of meson states, and that can help provide insight as to the quantum numbers belonging to a specific meson state.

An example of how the conservation laws just mentioned may be used to construct a table of the various allowed and forbidden two–body decays for non–strange mesons, is shown in Table 5.[2] The first column lists the possible spin, parity and charge conjugation quantum numbers of meson states. Subsequent columns list various two–body processes (with the corresponding threshold energy) that may be potential decay modes. The appearance of a symbol in the table indicates that the decay mode in question is forbidden by the corresponding selection rule, but allowed by all "stronger" selection rules. The order in which they are applied is J, S (for symmetry not spin as is usually the case in these lectures) P, C, I, G.

Suppression of the decay rate for an electromagnetic process is indicated by α or α^2 in the column entry. Decays which are forbidden by I and by G can occur by emission and reabsorption of a virtual photon, and the rate is approximately suppressed by α^2 compared to the fully allowed strong decays which appear as blanks in the table.

An example of the application of these selection rules is given in Table 6, for three possible mesonic states—a state with I^G, J^{PC} of $0^{-+}1^{-}$ —a pion–like object; a $1^{--}1^{+}$ or a rho–like object; and a $0^{++}0^{+}$—a scalar object. In the table, these three states are examined for a set of possible decay modes. Often the spin–parity of a new meson may be indicated by the nature of the decay modes.

f) **Angular Distribution**

Normally, in particle–particle scattering processes, the first few excited states formed in the scattering stand out as peaks in the cross-section, but for higher energies no clear structure is observed. This usually does *not* mean that higher mass

Table 5. Allowed and forbidden two-body decay modes.

ALLOWED AND FORBIDDEN TWO-BODY DECAY MODES

$J^{P(C)}$	I^G	$\gamma\gamma$ 0	$\gamma\pi$ 140	$\pi^0\pi^0$ 270	$\pi\pi$ 280	$\gamma\eta$ 550	$\pi\eta$ 690	$\gamma\rho$ 750	$\gamma\omega(\phi)$ 780(1020)	$\pi^0\rho^0$ 885
$0^{+(+)}$	0^+	α^2	J			J	IG	α	α	P
	1^-	α^2	J	IG	IG	J		α	α	P
	2^+	α^2	J			J	IG	α	α	P
$0^{+(-)}$	0^-	C	J	C	C	J	C	C	C	P
	1^+	C	J	C	$(C)I$	J	$(C)G$	$(C)\alpha$	C	P
	2^-	C	J	C	$(C)G$	J	$(C)I$	$(C)\alpha$	C	P
$0^{-(+)}$	0^+	α^2	J	P	P	J	P	α	α	C
	1^-	α^2	J	P	P	J	P	α	α	C
	2^+	α^2	J	P	P	J	P	α	α	C
$0^{-(-)}$	0^-	C	J	P	P	J	P	C	C	
	1^+	C	J	P	P	J	P	$(C)\alpha$	C	IG
	2^-	C	J	P	P	J	P	$(C)\alpha$	C	
$\text{even}^{+(+)}$	0^+	α^2	C			C	IG	α	α	C
	1^-	α^2	$(C)\alpha$	IG	IG	C		α	α	C
	2^+	α^2	$(C)\alpha$			C	IG	α	α	C
$\text{even}^{+(-)}$	0^-	C	α	C	C	α	C	C	C	
	1^+	C	α	C	$(C)I$	α	$(C)G$	$(C)\alpha$	C	IG
	2^-	C	α	C	$(C)G$	α	$(C)I$	$(C)\alpha$	C	
$\text{even}^{-(+)}$	0^+	α^2	C	P	P	C	P	α	α	C
	1^-	α^2	$(C)\alpha$	P	P	C	P	α	α	C
	2^+	α	$(C)\alpha$	P	P	C	P	α	α	C
$\text{even}^{-(-)}$	0^-	C	α	P	P	α	P	C	C	
	1^+	C	α	P	P	α	P	$(C)\alpha$	C	IG
	2^-	C	α	P	P	α	P	$(C)\alpha$	C	
$\text{odd}^{+(+)}$	0^+	(S)	C	S	P	C	P	α	α	C
	1^-	(S)	$(C)\alpha$	S	P	C	P	α	α	C
	2^+	(S)	$(C)\alpha$	S	P	C	P	α	α	C
$\text{odd}^{+(-)}$	0^-	(S)	α	S	P	α	P	C	C	
	1^+	(S)	α	S	P	α	P	$(C)\alpha$	C	IG
	2^-	(S)	α	S	P	α	P	$(C)\alpha$	C	
$\text{odd}^{-(+)}$	0^+	$(S)P$	C	S	C	C	IG	α	α	C
	1^-	$(S)P$	$(C)\alpha$	S	$(C)G$	C		α	α	C
	2^+	$(S)P$	$(C)\alpha$	S	$(C)I$	C	IG	α	α	C
$\text{odd}^{-(-)}$	0^-	$(S)P$	α	S	IG	α	C	C	C	
	1^+	$(S)P$	α	S		α	$(C)G$	$(C)\alpha$	C	IG
	2^-	$(S)P$	α	S	IG	α	$(C)I$	$(C)\alpha$	C	

(Continued)

Table 5 (Continued)
Allowed and Forbidden Two-Body Decay Modes

$J^{P(C)}$	I^G	$\pi\rho$ 890	$\pi\omega(\phi)$ 920(1160)	$K\bar{K}$ 990	$K_1^0 K_1^0$ $K_2^0 K_2^0$ 995	$K_1^0 K_2^0$ 995	$\eta\eta$ 1100	γf 1250	$\eta\rho$ 1300	$\eta\omega(\phi)$ 1330(1570)
$0^{+(+)}$	0^+	P	P			C		C	P	P
	1^-	P	P			C	IG	C	P	P
	2^+	P	P	I	I	C	I	C	P	P
$0^{+(-)}$	0^-	P	P	C	C		C	α	P	P
	1^+	P	P	(C)I*	C		C	α	P	P
	2^-	P	P	(C)I	C	I	C	α	P	P
$0^{-(+)}$	0^+	G	C	P	P	P	P	C	C	C
	1^-		(C)G	P	P	P	P	C	(C)G	C
	2^+	G	(C)I	P	P	P	P	C	(C)I	C
$0^{-(-)}$	0^-		IG	P	P	P	P	α	IG	
	1^+	G		P	P	P	P	α		IG
	2^-		IG	P	P	P	P	α	IG	I
even$^{+(+)}$	0^+	G	C			C		C	C	C
	1^-		(C)G			C	IG	C	(C)G	C
	2^+	G	(C)I	I	I	C	I	C	(C)I	C
even$^{+(-)}$	0^-		IG	C	C		C	α	IG	
	1^+	G		(C)I*	C		C	α		IG
	2^-		IG	(C)I	C	I	C	α	IG	I
even$^{-(+)}$	0^+	G	C	P	P	P	P	C	C	C
	1^-		(C)G	P	P	P	P	C	(C)G	C
	2^+	G	(C)I	P	P	P	P	C	(C)I	C
even$^{-(-)}$	0^-		IG	P	P	P	P	α	IG	
	1^+	G		P	P	P	P	α		IG
	2^-		IG	P	P	P	P	α	IG	I
odd$^{+(+)}$	0^+	G	C	P	S	P	S	C	C	C
	1^-		(C)G	P	S	P	S	C	(C)G	C
	2^+	G	(C)I	P	S	P	S	C	(C)I	C
odd$^{+(-)}$	0^-		IG	P	S	P	S	α	IG	
	1^+	G		P	S	P	S	α		IG
	2^-		IG	P	S	P	S	α	IG	I
odd$^{-(+)}$	0^+	G	C	C	S	C	S	C	C	C
	1^-		(C)G	(C)I*	S	C	S	C	(C)G	C
	2^+	G	(C)I	(C)I	S	C	S	C	(C)I	C
odd$^{-(-)}$	0^-		IG		S		S	α	IG	
	1^+	G			S		S	α		IG
	2^-		IG	I	S	I	S	α	IG	I

$J^{P(C)}$	I^G	$K\overline{K}^*$ $\overline{K}K^*$ 1380	πf 1390	$\rho^0\rho^0$ 1500	$\rho\rho$ 1500	$\rho\omega(\phi)$ 1530(1770)	$\omega\omega(\phi)$ 1560(1800)	$K^*\overline{K}^*$ 1775	ηf 1800	$N\overline{N}$ 1880
$0^{+(+)}$	0^+	P	P			IG			P	
	1^-	P	P	IG	IG		IG		P	
	2^+	P	P			IG	I	I	P	I
$0^{+(-)}$	0^-	P	P	C	C	C	C	C	P	C
	1^+	P	P	C	$(C)I$	$(C)G$	C	$(C)I^*$	P	$(C)I^*$
	2^-	P	P	C	$(C)G$	$(C)I$	C	$(C)I$	P	$(C)I$
$0^{-(+)}$	0^+		IG			IG				
	1^-			IG	IG		IG		IG	
	2^+	I	IG			IG	I	I	I	I
$0^{-(-)}$	0^-		C	C	C	C	C	C	C	C
	1^+		$(C)G$	C	$(C)I$	$(C)G$	C	$(C)I^*$	C	$(C)I^*$
	2^-	I	$(C)I$	C	$(C)G$	$(C)I$	C	$(C)I$	C	$(C)I$
even$^{+(+)}$	0^+		IG			IG				
	1^-			IG	G		IG		IG	
	2^+	I	IG			IG	I	I	I	I
even$^{+(-)}$	0^-		C	C	G	C	C		C	C
	1^+		$(C)G$	C		$(C)G$	C		C	$(C)I^*$
	2^-	I	$(C)I$	C	G	$(C)I$	C	I	C	$(C)I$
even$^{-(+)}$	0^+		IG			IG				
	1^-			IG	G		IG		IG	
	2^+	I	IG			IG	I	I	I	I
even$^{-(-)}$	0^-		C	C	G	C	C		C	
	1^+		$(C)G$	C		$(C)G$	C		C	
	2^-	I	$(C)I$	C	G	$(C)I$	C	I	C	I
odd$^{+(+)}$	0^+		IG			IG				
	1^-			IG	G		IG		IG	
	2^+	I	IG			IG	I	I	I	I
odd$^{+(-)}$	0^-		C	C	G	C	C		C	
	1^+		$(C)G$	C		$(C)G$	C		C	
	2^-	I	$(C)I$	C	G	$(C)I$	C	I	C	I
odd$^{-(+)}$	0^+		IG			IG				C
	1^-			IG	G		IG		IG	$(C)I^*$
	2^+	I	IG			IG	I	I	I	$(C)I$
odd$^{-(-)}$	0^-		C	C	G	C	C		C	
	1^+		$(C)G$	C		$(C)G$	C		C	
	2^-	I	$(C)I$	C	G	$(C)I$	C	I	C	I

Table 6. Example of use of selection rules.

Decay Mode	$1^-, 0^{-+}$	$1^+, 1^{--}$	$0^+, 0^{++}$
$\gamma\pi$	J	α	J
$\pi^+\pi^-$	P	✓	✓
$\pi\eta$	P	G	I, G
$\gamma\rho$	α	C	α
$K\bar{K}$	P	✓	✓
ηw	C	I, G	P
$K\bar{K}^*$	✓	✓	P
$\rho\rho$	I, G	C	✓
$K^*\bar{K}^*$	✓	✓	✓
$N\bar{N}$	✓	✓	✓

resonances are not being formed, but rather that there are a substantial number of overlapping resonances being produced and the scattering cross–section becomes featureless, as the resonant structures blend into one another.

The best way to identify, and then to study, these higher–lying resonances is to investigate the scattering angular distributions. In the simplest case of two-body scattering, see Fig. 2, one can observe the angular structure in terms of the energy dependence of the moments (the coefficients t_L^M) of the spherical harmonics in the t–channel helicity

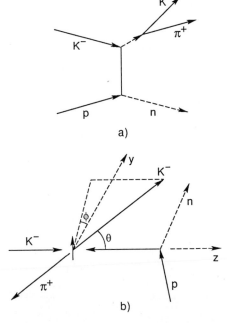

Fig. 2. An example of a two–body scattering process. In Fig. 2a the $K\pi \to K\pi$ scattering process is indicated, while Fig. 2b defines the t-channel helicity angles θ and ϕ in the $K\pi$ center-of-mass.

frame describing the angular distribution. The production angular distribution I_{prod} of the $K^-\pi^+$ system, for example, can be expanded as

$$I_{prod}(M_{t\pi}, t, \Omega) = \frac{1}{\sqrt{4\pi}} \sum_{L, M\geq 0} t_L^M (M_{K\pi}, t', \Omega) (2 - \delta_{M0}) \operatorname{Re}[Y_L^M(\Omega)]$$

where Ω stands for $(\cos\theta_{GJ}, \phi_{TY})$, the polar and azimuthal angles of the $K\pi$ scattering in the helicity frame, and $\delta_{MM'}$ is Krönecker's delta. The normalization condition is such that t_0^0 corresponds to the acceptance corrected number of events.

The production angular distribution, I_{prod}, is obtained from the observed distribution, I_{obs}, corrected for the spectrometer acceptance A,

$$I_{obs}(M_{K\pi}, t', \Omega) = A(M_{K\pi}, t', \Omega) \cdot I_{prod}(M_{K\pi}, t', \Omega) \quad .$$

In practice, the acceptance function A is also expanded in spherical harmonic moments for use in the fit to the observed data. The acceptance corrected moments are obtained by an extended maximum likelihood method. The number of moments required to fit the $K\pi$ angular distribution is found to increase with increasing $K\pi$ mass—indicating increasing complexity of the angular distribution as the energy in the scattering process increases.

Resonances appear in moments up to an L value of $2J$ (i.e., a $J = 1$ resonance would show structure in $L = 0$, 1 and 2 moments, but not in $L = 3$ or greater).

For $J \leq 3$, one can now relate these moments, t_L^M, to the scattering amplitudes as follows:

$$t_0^0 = |S|^2 + |P_0|^2 + |P_-|^2 + |P_+|^2|D_0|^2 + |D_0|^2 + |D_-|^2 + |D_+|^2 + |F_0|^2 + |F_-|^2 + |F_+|^2$$

$$t_1^0 = 2|S||P_0|\cos(\varphi_s - \varphi_{P_0}) + \frac{4}{\sqrt{5}}|P_0||D_0|\cos(\varphi_{P_0} - \varphi_{D_0})$$

$$+ \frac{2\sqrt{3}}{\sqrt{5}}(|P_+||D_+|\cos(\varphi_{P_+} - \varphi_{D_+}) + |P_-||D_-|\cos(\varphi_{P_-} - \varphi_{D_-}))$$

$$+ \frac{6\sqrt{3}}{\sqrt{35}}|D_0||F_0|\cos(\varphi_{D_0} - \varphi_{F_0}) + \frac{4\sqrt{6}}{\sqrt{35}}(|D_+||F_+|\cos(\varphi_{D_+} - \varphi_{F_+})$$

$$+ |D_-||F_-|\cos(\varphi_{D_-} - \varphi_{F_-}))$$

$$t_1^1 = \sqrt{2}|S||P_-|\cos(\varphi_s - \varphi_{P_-}) + \sqrt{\frac{6}{5}}|P_0||D_-|\cos(\varphi_{P_0} - \varphi_{D_-})$$

$$- \sqrt{\frac{2}{5}}|P_-||D_0|\cos(\varphi_{P_-} - \varphi_{D_0}) + \frac{6}{\sqrt{35}}|D_0||F_-|\cos(\varphi_{D_0} - \varphi_{F_-})$$

$$- \frac{3\sqrt{2}}{\sqrt{35}}|D_-||F_0|\cos(\varphi_{D_-} - \varphi_{F_0})$$

$$t_2^0 = 2|S||D_0|\cos(\varphi_s - \varphi_{D_0}) + \frac{2}{\sqrt{5}}|P_0|^2 - \frac{1}{\sqrt{5}}(|P_+|^2 + |P_-|^2) + \frac{2\sqrt{5}}{7}|D_0|^2$$

$$+ \frac{\sqrt{5}}{7}(|D_+|^2 + |D_-|^2) + \frac{1}{\sqrt{5}}(|F_+|^2 + |F_-|^2) + \frac{4}{3\sqrt{5}}|F_0|^2$$

$$+ 6\sqrt{\frac{3}{35}}|P_0||F_0|\cos(\varphi_{P_0} - \varphi_{F_0}) + 6\sqrt{\frac{2}{35}}(|P_+||F_+|\cos(\varphi_{P_+} - \varphi_{F_+})$$

$$+ |P_-||F_-|\cos(\varphi_{P_-} - \varphi_{F_-}))$$

$$t_2^1 = \frac{\sqrt{6}}{\sqrt{5}}|P_0||P_-|\cos(\varphi_{P_0} - \varphi_{P_-}) + \frac{\sqrt{10}}{7}|D_0||D_-|\cos(\varphi_{D_0} - \varphi_{D_-})$$

$$+ \sqrt{2}|S||D_-|\cos(\varphi_s - \varphi_{D_-}) + \frac{4\sqrt{3}}{\sqrt{35}}|P_0||F_-|\cos(\varphi_{P_0} - \varphi_{F_-})$$

$$+ \frac{2\sqrt{5}}{15}|F_0||F_-|\cos(\varphi_{F_0} - \varphi_{F_-}) - \frac{3\sqrt{2}}{\sqrt{35}}|P_-||F_0|\cos(\varphi_{P_-} - \varphi_{F_0})$$

$$t_2^2 = -\frac{\sqrt{3}}{\sqrt{10}}(|P_+|^2 - |P_-|^2) - \frac{\sqrt{15}}{7\sqrt{2}}|(D_+|^2 - |D_-|^2)$$

$$+ \sqrt{\frac{3}{35}}(|P_+||F_+|\cos(\varphi_{P_+} - \varphi_{F_+}) - |P_-||F_-|\cos(\varphi_{P_-} - \varphi_{F_-}))$$

$$- \sqrt{\frac{2}{15}}(|F_+|^2 - |F_-|^2)$$

$$t_3^0 = \frac{6\sqrt{3}}{\sqrt{35}}|P_0||D_0|\cos(\varphi_{P_0} - \varphi_{D_0}) - \frac{6}{\sqrt{35}}(|P_+||D_+|\cos(\varphi_{P_+} - \varphi_{D_+}) + |P_-||D_-|$$

$$\cos(\varphi_{P_-} - \varphi_{D_-})) + 2|S||F_0|\cos(\varphi_s - \varphi_{F_0}) + \frac{8\sqrt{5}}{15}|D_0||F_0|\cos(\varphi_{D_0} - \varphi_{F_0})$$

$$+ \frac{2\sqrt{10}}{15}(|D_+||F_+|\cos(\varphi_{D_+} - \varphi_{F_+}) + |D_-||F_-|\cos(\varphi_{D_-} - \varphi_{F_-}))$$

$$t_3^1 = \frac{4\sqrt{3}}{\sqrt{35}}|P_0||D_-|\cos(\varphi_{P_0} - \varphi_{D_-}) + \frac{6}{\sqrt{35}}|P_-||D_0|\cos(\varphi_{P_-} - \varphi_{D_0})$$

$$+ \sqrt{2}|S||F_-|\cos(\varphi_S - \varphi_{F_-}) + \frac{2\sqrt{5}}{15}|D_-||F_0|\cos(\varphi_{D_-} - \varphi_{F_0})$$

$$+ \sqrt{\frac{2}{5}}|D_0||F_-|\cos(\varphi_{D_0} - \varphi_{F_-})$$

$$t_3^2 = -\sqrt{\frac{6}{7}}(|P_+||D_+|\cos(\varphi_{P_+} - \varphi_{D_+}) - |P_-||D_-|\cos(\varphi_{P_-} - \varphi_{D_-}))$$

$$-\frac{1}{\sqrt{3}}(|D_+||F_+|\cos(\varphi_{D_+} - \varphi_{F_+}) - |D_-||F_-|\cos(\varphi_{D_-} - \varphi_{F_-}))$$

$$t_4^0 = -\frac{4}{7}(|D_+|^2 + |D_-|^2) + \frac{6}{7}|D_0|^2 + \frac{1}{11}(|F_+|^2 + |F_-|^2) + \frac{6}{11}|F_0|^2$$

$$+ \frac{8}{\sqrt{21}}|P_0||F_0|\cos(\varphi_{P_0} - \varphi_{F_0}) - \frac{2\sqrt{2}}{\sqrt{7}}(|P_+||F_+|\cos(\varphi_{P_+} - \varphi_{F_+})$$

$$+ |P_-||F_-|\cos(\varphi_{P_-} - \varphi_{F_-}))$$

$$t_4^1 = \frac{2\sqrt{15}}{7}|D_0||D_-|\cos(\varphi_{D_0} - \varphi_{D_-}) + \frac{5\sqrt{2}}{\sqrt{35}}|P_0||F_-|\cos(\varphi_{P_0} - \varphi_{F_-})$$

$$+ \frac{2\sqrt{5}}{\sqrt{21}}|P_-||F_0|\cos(\varphi_{P_-} - \varphi_{F_0}) + \frac{\sqrt{30}}{11}|F_0||F_-|\cos(\varphi_{F_0} - \varphi_{F_-})$$

$$t_4^2 = -\frac{\sqrt{10}}{7}(|D_+|^2 - |D_-|^2) - \sqrt{\frac{5}{7}}(|P_+||F_+|\cos(\varphi_{P_+} - \varphi_{F_+})$$

$$- |P_-||F_-|\cos(\varphi_{P_-} - \varphi_{F_-})) - \frac{\sqrt{10}}{11}(|F_+|^2 - |F_-|^2)$$

$$t_5^0 = \frac{20\sqrt{5}}{3\sqrt{77}}|D_0||F_0|\cos(\varphi_{D_0} - \varphi_{F_0}) - \frac{10\sqrt{10}}{3\sqrt{77}}(|D_+||F_+|\cos(\varphi_{D_+} - \varphi_{F_+})$$

$$+ |D_-||F_-|\cos(\varphi_{D_-} - \varphi_{F_-}))$$

$$t_5^1 = \frac{10}{\sqrt{77}}|D_0||F_-|\cos(\varphi_{D_0} - \varphi_{F_-}) + \frac{20\sqrt{2}}{3\sqrt{77}}|D_-||F_0|\cos(\varphi_{D_-} - \varphi_{F_0})$$

$$t_5^2 = -\frac{5}{\sqrt{33}}(|D_+||F_+|\cos(\varphi_{D_+} - \varphi_{F_+}) - |D_-||F_-|\cos(\varphi_{D_-} - \varphi_{F_-}))$$

$$t_6^0 = \frac{-25}{11\sqrt{13}}(|F_+|^2 + |F_-|^2) + \frac{100}{33\sqrt{13}}|F_0|^2$$

$$t_6^1 = \frac{50\sqrt{7}}{33\sqrt{13}}|F_0||F_-|\cos(\varphi_{F_0} - \varphi_{F_-})$$

$$t_6^2 = \frac{-5\sqrt{35}}{11\sqrt{39}}(|F_+|^2 - |F_-|^2)$$

It is often interesting to make linear combinations of these moments in order to project out some particular bilinear combination of amplitudes. For example, when we have only S and P waves present we can define

$$\sigma_0^P = |P_0|^2 + \frac{1}{3}|S|^2 = \frac{1}{3}t_0^0 + \frac{\sqrt{5}}{3}t_2^0$$

$$\sigma_+^P = |P_+|^2 + \frac{1}{3}|S|^2 = \frac{1}{3}t_0^0 - \frac{\sqrt{5}}{6}t_2^0 - \sqrt{\frac{5}{6}}t_2^2$$

$$\sigma_-^P = |P_-|^2 + \frac{1}{3}|S|^2 = \frac{1}{3}t_0^0 - \frac{\sqrt{5}}{6}t_2^0 + \sqrt{\frac{5}{6}}t_2^2 \quad ;$$

if the S wave amplitude is small, these in effect project the P wave amplitudes directly.

Similarly, for S, P and D waves the combinations

$$\sigma_0^D = |D_0|^2 + \frac{4}{10}(|S|^2 + \sum |P|^2) = \frac{4}{10}t_0^0 + \frac{7}{10}t_4^0$$

$$\sigma_+^D = |D_+|^2 + \frac{3}{10}(|S|^2 + \sum |P|^2) = \frac{3}{10}t_0^0 - \frac{7}{20}t_4^0 - \frac{7}{2\sqrt{10}}t_4^2$$

$$\sigma_-^D = |D_-|^2 + \frac{3}{10}(|S|^2 + \sum |P|^2) = \frac{3}{10}t_0^0 - \frac{7}{20}t_4^0 + \frac{7}{2\sqrt{10}}t_4^2$$

may prove useful in regions where the D wave is dominant.

It is interesting to note how clearly the data exhibit the moment, and indeed the amplitude structure of the basic meson-meson scattering process. In Fig. 3 and Fig. 4 we show the polar and azimuthal angular distributions for $K\pi$ scattering in the region of the $K_1^*(890)$ and $K_2^*(1420)$ for two different reactions—$K^-p \to K^-\pi^+n$ and $K^-p \to K^0\pi^-p$. The basic meson-meson scattering processes for these production reactions are $K^-\pi^+ \to K^-\pi^+$ elastic scattering, and $K^-w \to \bar{K}^0\pi^-$ charge exchange scattering. One clearly sees the spin 1 and spin 2 character of the two K^* states in the $\cos\theta$ plots, and also the clear evidence of spin 0 and spin 1 in the exchange process (i.e., the π-like and w-like property of the t-channel exchange).

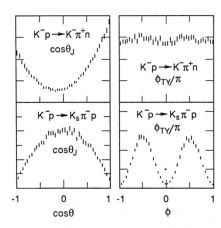

Fig. 3. Polar and azimuthal $K\pi$ angular distribution for $M_{K\pi}$ in the regions of K_1^* (890), and for the two reactions $K^-p \to K^-\pi^+n$ and $K^-p \to K_s^0\pi^-p$.

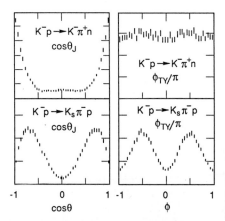

Fig. 4. Polar and azimuthal $K\pi$ angular distributions for $M_{K\pi}$ in the region of K_2^* (1420), and for the two reactions $K^-p \to K^-\pi^+n$ are $K^-p \to K_s^0\pi^-p$.

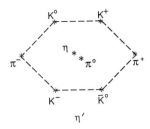

Fig. 5. SU(3) octet, singlet representation for mesons.

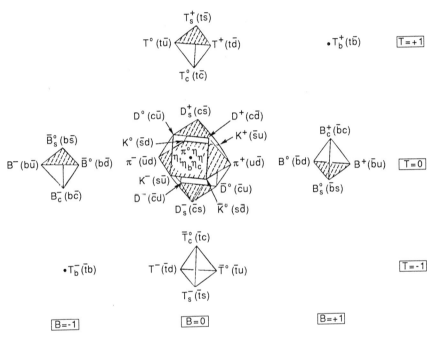

Fig. 6. The six quark representation of the meson spectrum.

h) **Classification**

The discussion of the classification of mesons was very simple before the discovery of the heavy quarks. The relevant classification scheme was SU(3) and all the mesons fit into octet or singlet configurations[3]—see Fig. 5. The singlet states were usually heavily mixed with the iso-scalar member of the octet.

Now, with the full three generation structure of the quark families, the classification configuration is a little more complicated. See Fig. 6.[4]

Let us examine empirically what we might expect by way of meson states within the $q\bar{q}$ quark model of mesons. In Table 7 we show all the quark flavor combinations that we might expect, and give some examples of such configurations found in nature.

Within a $q\bar{q}$ model, we can reconstruct the spectrum of states that we might expect, see Fig. 7. This is called a Grotrian plot, and is used by nuclear physicists and atomic

211

Table 7. Standard quark model assignments for some of the known mesons. Only the states in the $u\bar{u}$, $d\bar{d}$, $s\bar{s}$, $c\bar{c}$, and $b\bar{b}$ columns and the neutral states in the $I = 1$ column, are eigenstates of charge conjugation, C.

$u\bar{d}$, $u\bar{u}$, $d\bar{d}$	$u\bar{u}$, $d\bar{d}$, $s\bar{s}$	$c\bar{c}$	$b\bar{b}$	$\bar{s}u$, $\bar{s}d$	$c\bar{u}$, $c\bar{d}$	$c\bar{s}$	$\bar{b}u$, $\bar{b}d$
$I = 1$	$I = 0$	$I = 0$	$I = 0$	$I = 1/2$	$I = 1/2$	$I = 0$	$I = 1/2$
π	η, η'	η_c		K	D	D_s	B
ρ	ω, ϕ	J/ψ	Υ	K^*_{890}	D^*_{2010}		

physicists before them, to reconstruct the expected spectroscopy in non-relativistic systems. On the left side of the figure, we reconstruct the states with the quark spins antiparallel. For relative orbital angular momentum between the $(q\bar{q})$ pair, $L = 0, 1, 2, 3$ units we have the mesonic states with $J^P = 0^-, 1^+, 2^-, 3^+$. These correspond to the spectroscopic notations ^1S, ^1P, ^1D, ^1F for the $(q\bar{q})$ states. The states on the right-hand side of Fig. 7 are generated by the $(q\bar{q})$ system with spins parallel. For zero relative orbital angular momentum in this $(q\bar{q})$ system we have the ground state with $J^P = 1^-$. When we add one unit of angular momentum, $L = 1$, we find a triplet of levels, ^3P, corresponding to the vector addition of $L = 1$ and $S = 1$ and giving rise to states with $J^P = 2^+, 1^+$ and 0^+. Further relative angular momentum gives rise to more triplet levels with higher and higher spins. The states on the left and the "middle" levels of the triplet states are called "un-natural" spin-parity states, while the others are called "natural" spin-parity states. This comes from the fact that in pseudoscalar–pseudoscalar meson scattering, one can couple to the "natural" states, but can not access the "un-natural" states.

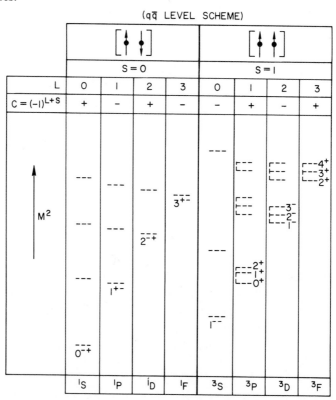

Fig. 7. A Grotrian plot; a classification scheme for the meson system.

The vertical towers under each specific spin-parity assignment represent the radial excitations of the ground state structure.

The spectroscopy develops via the excitation of the orbital quantum number, and via the excitation of the radial quantum number.

Meson states with the same spin and parity, and the same additive quantum numbers can mix. For states which are eigenstates of C, they must have the same value of C-parity to mix. For example, the axial vector nonets $J^{PC} = 1^{++}$, 1^{+-} have strong mixing of the K_1 states—the Q_A and the Q_B, which do not have definite C-parity. Again, the second excited $c\bar{c}$ state, the $J^{PC} = 1^{--}$ at 3770 MeV is surely a mixture of the second radially excited $3\,^3S_1$ and the $^3D_1$ ground state—see the Grotrian plot in Fig. 7.

And finally, the isoscalar states also mix—the singlet isoscalar and the isoscalar member of the octet. This mixing may be understood in terms of physical states η and η' and the pure octet and pure singlet states η_8, η_1.

$$\eta = \eta_8 \cos\theta - \eta_1 \sin\theta$$

$$\eta' = \eta_8 \sin\theta + \eta_1 \cos\theta$$

where θ is the mixing angle and the η_1, η_8 are defined as:

$$\eta_1 = (u\bar{u} + d\bar{d} + s\bar{s})/\sqrt{3}$$

$$\eta_8 = (u\bar{u} + d\bar{d} - 2s\bar{s})/\sqrt{6}$$

These combinations diagonalize the mass-squared matrix

$$M^2 = \begin{pmatrix} M_{11}^2 & M_{18}^2 \\ M_{18}^2 & M_{88}^2 \end{pmatrix}$$

where

$$M_{88}^2 = \frac{1}{3}\left(4\,M_K^2 - M_\pi^2\right)\ .$$

For "ideal" mixing (i.e., $\tan\theta = 1\sqrt{2}$, $\theta \sim 35°$) the mainly octet state, η, is pure $s\bar{s}$ while the mainly singlet state η' has no $s\bar{s}$ content. The natural parity nonets that are well measured seem to have nearly ideal mixing—see Table 8; this does not appear to be the case for unnatural parity nonets.

Finally, let's return to the Grotrian plots of Fig. 7, but try to incorporate not just the single flavor aspects of our first attempt. In Fig. 8, we have flipped the axes around, but it is still our Grotrian plot, and now we represent all of the light quark flavors. In Section 7, in the conclusion, we will revisit this kind of plot as we try to summarize the whole meson sector.

Table 8. Singlet–octet mixing for the well known mesons.

J^{PC}	Mesons	θ
0^{--}	π, K, η, η'	$-10°$
1^{--}	ρ, K^*, ϕ, w	$39°$
2^{++}	A_2, K^*_{1430}, f'_{1525}, f_{1270}	$28°$
3^{--}	ρ_3^{1690}, K^*_{1780}, ϕ_3^{1810}, w_3^{1670}	$29°$

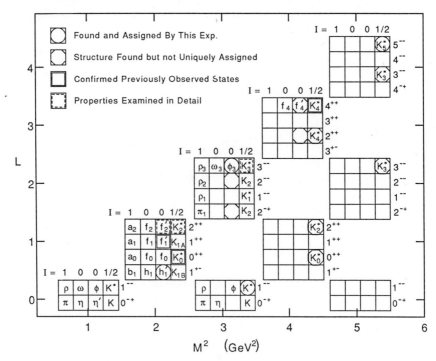

Fig. 8. A summary of the meson spectroscopy, circa 1987 as reported by S. Suzuki for the LASS experiment, Aston et al., SLAC–PUB–4340.

i) Where Do We Study Mesons?

There are many different opportunities to study the meson resonances. A list of a number of the possible processes is given in Table 9.

J) A Perspective

We have seen from the brief look at the classification section that we expect a very large number of meson states in nature, with a wide variety of flavors and specific internal quantum numbers.

The basic interest in studying the spectroscopy goes beyond the botanical classification, or the thrill of stamp collecting, but rests in the deep empirical nature of our science. We should experimentally measure enough of the spectrum to be reasonably sure that we really *do* understand the spectrum of states produced in nature, the ordering, the mass splitting, etc.

An interesting fact in planning experiments is that the states are many, they are broad and their splittings are not large. This implies that it will not always be possible to find the states as nice clean peaks in a cross-section, but that we will have to ferret out their presence by studying (in some detail) the scattering amplitudes in meson-meson scattering processes, and by measuring the strength of each amplitude through its interference patterns, in order to piece together the full story of the meson spectrum.

I will take us through just such an experimental program, looking at what has been learned of the K^* and $(s\bar{s})$ meson sectors.

From the systematics discovered there, I will try to summarize the situation for all the light quark mesons.

Table 9. Examples of the many ways that experiments study meson resonances.

Reaction	Example
Meson-meson collisions in peripheral strong interaction processes	$K^-p \to K^-\pi^+ n$
Diffractive processes, which may be analyzed in terms of meson-Pomeron scattering	$K^-p \to K\pi\pi p$
Meson production in the Coulomb field, or Primakoff Effect.	$\pi A \to \rho A$
Electron-positron annihilation	$e^+e^- \to \rho \to \pi^+\pi^-$
Central collisions in strong interaction processes	$pp \to ppX$
Photoproduction	$\gamma p \to Mp$
Decay of heavy particles, H^0, t, Z^0, $W^\pm$, upsilon.	$Z^0 \to B\bar{B}$
Jet hadronization	quark$\to$ Mesons, ...

Following that, we will briefly look at the experimental situation in heavy quark spectroscopy.

3. K^* SPECTROSCOPY

a) Introduction

In this section I will discuss the experimental program at SLAC, which has systematically studied the K^* spectroscopy using separated K^+ and K^- beams and a 4π multiparticle spectrometer, the Large Aperture Superconducting Solenoid (LASS).

To be successful, this program required the development of three quite distinct kinds of tools:

— a detector with 4π geometrical coverage, capable of effectively handling multiparticle final states, having good particle identification and operating with a trigger that provided full coverage in angle and momentum space.

— analysis programs to extract the off-mass-shell meson-meson scattering amplitudes, partial wave by partial wave.

— computer time to process and analyze an event sample of some 140 million events at summary tape level.

The tools were developed.

The detector, LASS, is shown in Fig. 9 as an artistic recreation and Fig. 10 as a line drawing. The spectrometer consists of two parts—the first is a classical forward spectrometer with tracking chambers, a large aperture bending magnet and a pressurized gas Čerenkov counter, which does a good job of measuring the fast forward particles; the second is a target spectrometer comprising a 1 m long liquid hydrogen target surrounded by cylindrical proportional chambers, and planar tracking chambers, time-of-flight hodoscope and Čerenkov hodoscope all inside a 22.4 kG superconducting solenoid magnet, which did an excellent job of measuring the slow and large angle particles produced in the Kp interactions. The chambers and the associated electronics were designed to handle the (5-10) MHz rates driven by the SLAC duty cycle.

Fig. 9. An artist's impression of the LASS spectrometer. Ref. 5.

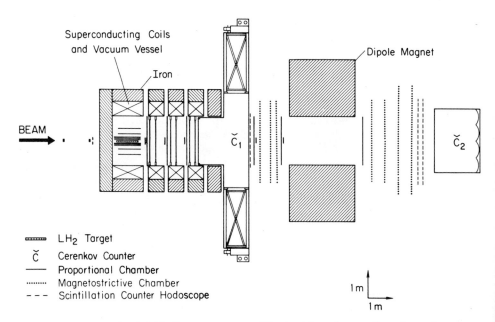

Fig. 10. The LASS Spectrometer. The incident beam direction, the solenoid/target, the Cerenkov counters, and the dipole chambers are shown. See Ref. 5 for details.

The spectrometer is fully described in Ref. 5.

The analysis programs were developed as the SLAC-LBL partial wave analysis program, and are fully described in Ref. 6.

The computer time problem was solved by use of mainframe capacity at SLAC (IBM 360) and at Nagoya University (FACOM) and through the development of special purpose processor farms—the 168/E and later the 3081/E emulators—which provided some five years of equivalent IBM 3081K processing power for the experiment. The processor farm is described in Ref. 7.

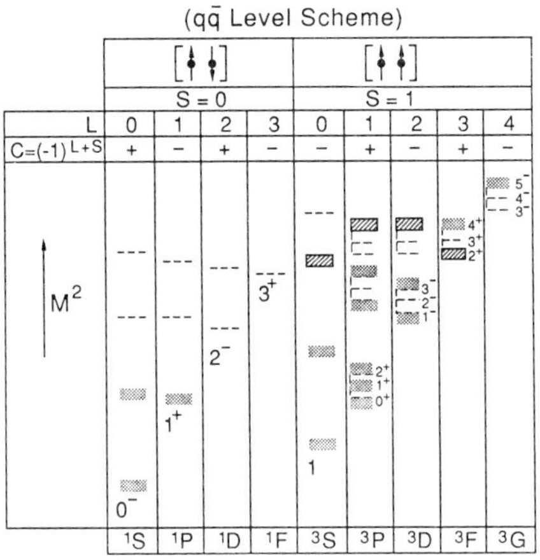

Fig. 11. The K^* states found in the SLAC/LASS experiment. The entries marked ▓ are fully confirmed observations, while those marked ▨ have been only seen in one channel and require confirmation.

The experimental data were taken in 1981 and 1982. The data was processed in 1984 and 1985—140,000,000 events on the data summary tapes.

The K^* states found in the SLAC experiment are indicated in Fig. 11. One sees the expected structure of states—the L-excitations with the triplet $(L+S, L, L-S)$ states and the radial excitation repeating the ground state structure, but with ~ 500 MeV of energy separating each level of excitation.

Let us review some of the studies that went into these discoveries.

b) <u>The Reaction $K^-p \to K^-\pi^+n$</u>

This process has a cross-section of 370 μbarns at 11 GeV/c K^- momentum, and results in a very simple event topology in the detector. The production process is peripheral, and dominated at small momentum transfers by single pion exchange.

217

The $K\pi$ final state is analyzed in the t-channel helicity frame—see Fig. 2—and the angular distribution described in terms of spherical harmonic moments, t_L^M,

$$I_{prod}(M_{K\pi}, t, \Omega) = \frac{1}{\sqrt{4\pi}} \sum_{L,M} t_L^M(M_{K\pi}, t)\, (2 - \delta_{M,0})\, Re\,[Y_L^M(\Omega)] \;;$$

Ω stands for the polar, azimuthal angles $(\cos\theta, \phi)$, and $\int I_{prod}(\Omega)d\Omega = t_0^0$.

The actual observed angular distribution, I_{obs}, is related to I_{prod} via

$$I_{obs}(M_{K\pi}, t, \Omega) = A(M_{K\pi}, t, \Omega)\, I_{prod}(M_{K\pi}, t, \Omega)$$

where A is the acceptance of the spectrometer, which is determined by Monte Carlo simulation. The moments are determined from the relationship

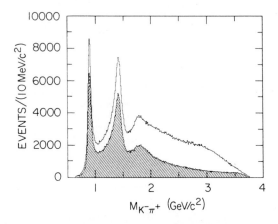

Fig. 12. The $K^-\pi^+$ invariant mass distribution for the reaction $K^-p \to K^-\pi^+n$. The outer histogram is for events with momentum transfer, t, less than 1 $(\text{GeV}/c)^2$ while the inner histogram is for events with mass $n\pi^+$ greater than 1.7 GeV/c^2. The outer histogram contains some 780,000 events.

$$I_{obs}(M_{K\pi}, t, \Omega) = A(M_{K\pi}, t, \Omega)\frac{1}{\sqrt{4\pi}} \sum t_L^M(M_{K\pi}, t)\, (2 - \delta_{M,0}) Re\,[Y_L^M(\Omega)]$$

The $K^-\pi^+$ mass spectrum from the LASS data sample is shown in Fig. 12, for momentum transfers less than 1.0 $(\text{GeV}/c)^2$. There are $\sim$ 780,000 events in this plot. The moments describing the $K^-\pi^+$ angular distributions are shown in Fig. 13. Clear indications of the first five leading orbitally excited K^* states—the $K_1^*(890)$, $K_2^*(1430)$, $K_3^*(1780)$, $K_4^*(2000)$ and $K_5^*(2380)$—can be seen in the t_2^0, t_4^0, t_6^0, t_8^0, and t_{10}^0 moments, respectively. Each state dominates the highest moment required in the relevant mass region, clearly demonstrating the spin-parity.

A fit to the moments—see Figs. 14 and 15—yields the masses and widths for the K^* states listed in Table 10.

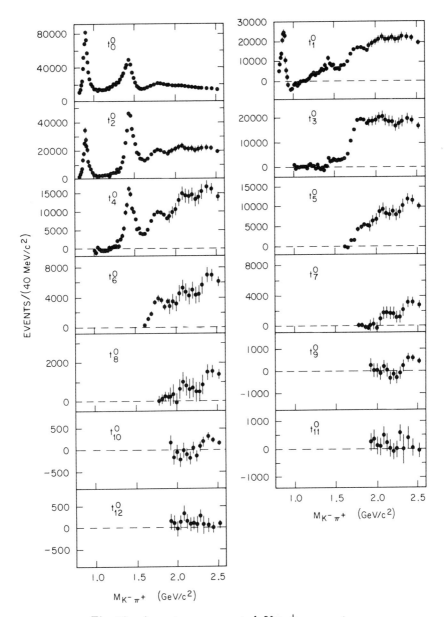

Fig. 13. Acceptance corrected $K^-\pi^+$ moments.

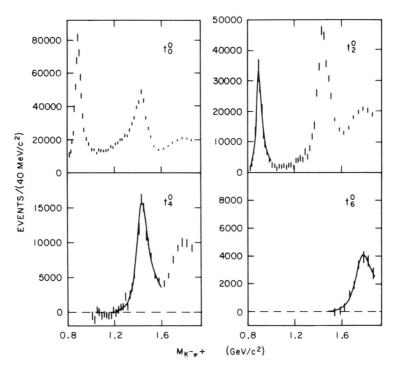

Fig. 14. A Breit–Wigner resonance fit to the $K^-\pi^+$ moments for spin 1, 2, and 3 states.

The $K^-\pi^+$ angular distributions show clearly the leading resonances. However, to reveal the underlying structure it is necessary to extract the partial wave amplitudes of the meson-meson scattering. Over 30 years ago, Chew and Low[8] suggested that it would be possible to extract the physical $\pi\pi$ scattering amplitude in a πp reaction dominated at small momentum transfers by π exchange, via an extrapolation of the differential cross section to the pion pole. Many extensions and refinements of this classic technique have been applied since, to study $\pi - \pi$ and $K - \pi$ scattering. One of the most powerful methods was developed by Estabrooks and Martin.[9] This model has been used in the analysis of our data. The $K^-\pi^+$ partial wave amplitudes—both magnitude and phase—are shown in Fig. 16 and again in the Argand diagram form in Fig. 17. A resonance is indicated by an anti-clockwise looping of the scattering amplitude on an Argand plot, with the movement of the amplitude being fastest at resonance.

The S-wave amplitude—which is the sum of the $I = 3/2$ and $I = 1/2$ scattering lies outside the unitary circle. However the $I = 1/2$ part, shown by the dotted line in Fig. 17, is consistent with a purely elastic amplitude up through 1500 MeV. (This was unraveled from the scattering amplitude plotted here, by separately measuring the $I = 3/2$ part in the process $K^-p \to K^-\pi^-\Delta^{++}$.) It exhibits a slow motion through the region 800-1300 MeV, followed by a rapid loop around 1400 MeV, and again near 1900 MeV. The P-wave amplitude shows an elastic resonance around 890 MeV, a small hesitation around 1400 MeV and a further loop near 1700 MeV and perhaps again near 2100 MeV.

The D-wave amplitude goes through its first resonance at 1420 MeV, and then has a hesitation near 1750 MeV and a second loop around 2000 MeV. The F-wave amplitude has a clear loop at 1780 MeV, the G-wave amplitude shows resonance structure at 2060 MeV and the H-wave amplitude indicates resonant behavior around 2400 MeV. It is interesting to note the decrease in size of the resonant amplitude as the spin (or L-excitation) increases.

Table 10. Resonance parameters derived from fits to the moments of the $K\pi$ angular distributions. The first number in the error is the parabolic error from the fitting, and the second number is the estimated systematic error.

Resonance	J^P	Mass (MeV)	Width (MeV)
$K^*(890)$	1^-	$897.0 \pm 0.7 \pm 0.7$	$49.9 \pm 1.7 \pm 0.8$
$K^*(1430)$	2^+	$1433.0 \pm 1.6 \pm 0.5$	$115.8 \pm 2.7 \pm 1.6$
$K^*(1780)$	3^-	$1778.1 \pm 6.4 \pm 1.3$	$185.9 \pm 23.3 \pm 12.3$
$K^*(2062)$	4^+	$2062 \pm 14 \pm 13$	$221 \pm 48 \pm 27$
$K^*(2380)$	5^-	$2382 \pm 14 \pm 19$	$178 \pm 37 \pm 32$

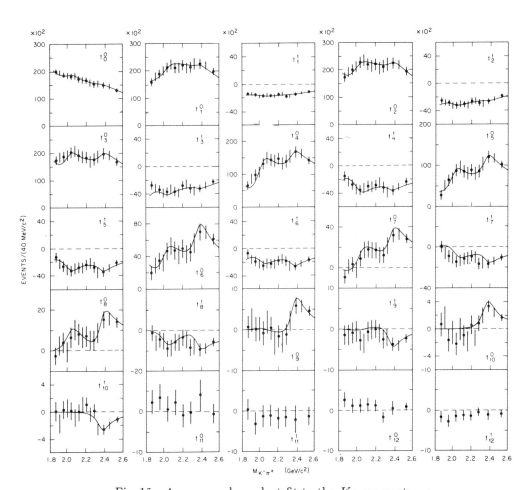

Fig. 15. An energy dependent fit to the $K\pi$ moments.

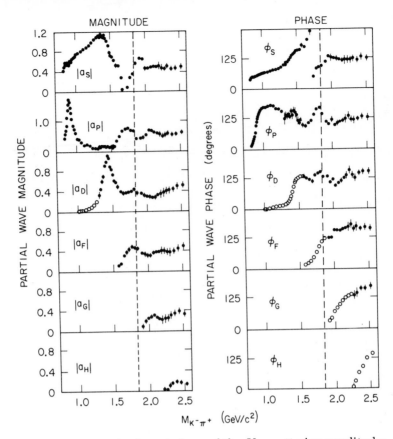

Fig. 16. The magnitude and phase of the $K\pi$ scattering amplitudes.

c) The Results of the Partial Wave Analysis of $K\pi$ Elastic Scattering

In the following sections, we discuss the important features of each partial wave amplitude and extract parameters for the resonances. In all cases, except when specifically indicated otherwise, these parameters are derived from fits using a relativistic Breit-Wigner parametrization for a spin-L resonance of the form

$$BW = \frac{\sqrt{2L+1}}{(M_R^2 - M^2) - iM_R\Gamma} \frac{\epsilon \quad M_R\Gamma}{} \quad ,$$

where M_R is the resonance mass, ϵ is the $K\pi$ elasticity, M is the invariant mass of the $K\pi$ system, and Γ is the mass dependent width given by

$$\Gamma(M) = \left(\frac{q}{q_R}\right)^{2L+1} \left(\frac{M_R}{M}\right) \frac{D_L(q_R r)}{D_L(qr)} \Gamma_R \quad ,$$

where q is the momentum in the $K\pi$ center of mass, Γ_R is the width of the resonance, q_R is q evaluated at the resonance mass, r is the interaction radius, and D_L is the barrier factor for a spin-L wave defined by Blatt and Weisskopf.[10] In addition to the statistical errors on the fit parameters, additional uncertainties also arise when different mass ranges or background forms are used in the fits and when different normalization procedures are used in the model. Estimates for the size of these model dependent variations are shown as the systematic errors in the tables.

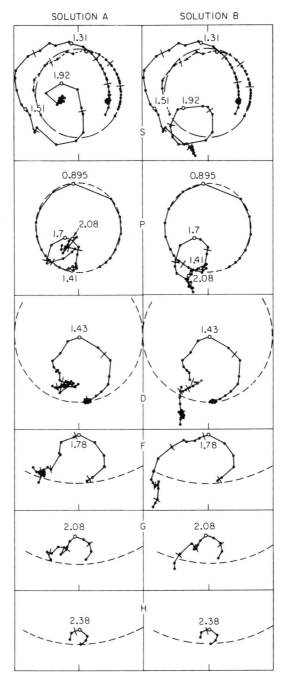

Fig. 17. Argand diagram for $K\pi$ scattering amplitudes. The dashed line represents the $I=1/2$ S-wave amplitude, determined by subtracting the $I=3/2$ component. The thick marks on all amplitudes are at 200 MeV/c^2 intervals.

1) The S-wave Amplitude. In the region below 1.3 GeV/c^2, the magnitude and phase of the S-wave amplitude rise slowly and are compatible with unitarity. The first inelastic two-body threshold, the $K\eta$, is at about 1.05 GeV/c^2, but this channel appears to be very weakly coupled to the S-wave K^* system, as is predicted by SU(3). The phase reaches 90° at 1.34 GeV/c^2. Above this mass, the motion becomes more rapid. The magnitude peaks just below 1.4 GeV/c^2, and then drops precipitously to nearly zero at 1.7 GeV/c^2, while the phase varies rapidly in this same mass region. Above 1.85 GeV/c^2, the two solutions have different detailed structure, but both show a peak in the magnitude around 1.9 GeV/c^2 with rapidly varying phase motion. The Argand diagrams also display rapid circular motion in these mass regions, which support the assertion that there are two resonant structures in the $K\pi$ S-wave, one at about 1.4 GeV/c^2 and the other around 1.9 GeV/c^2. The structure around 1.4 GeV/c^2 has been seen by many earlier experiments,[11] while the observation of the second structure around 1.9 GeV/c^2 confirms evidence provided by an earlier experiment in LASS.[12] Determination of the resonance parameters for the lower mass object is complicated by the large elastic phase shift in the low mass region and the proximity of the $K\eta'$ threshold. Historically, the mass value has been associated with the point where $\delta_S^{1/2}$ reaches 90° and, more recently, a unitary coupled-channel analysis of earlier data obtained a value which was consistent with this simple definition.[13]

To indicate the range of values allowed for the $K_0^*(1350)$ resonance, we quote parameter values from two models as shown in Table 11. In the first (Model I) the mass is taken as the point where the phase shift reaches 90° while the width is approximately determined from the speed of the phase motion there. As a second method (Model II) the data are fit to a parametrization developed by Estabrooks[14] that considers $K\eta'$ to be the only important inelastic channel. The S-wave amplitude is parametrized as the sum of an inelastic Breit-Wigner resonance and a background term parametrized as an effective range,

$$a_S = BG + BW' e^{i\phi}$$

$$BG = \sin \delta_{BG} e^{i\delta_{BG}}; \quad \cot \delta_{BG} = \frac{1}{aq} + \frac{1}{2}bq \qquad (1)$$

$$BW' = \frac{M_R \Gamma_1}{(M_R^2 - M^2) - iM_R(\Gamma_1 + \Gamma_2)}; \quad \Gamma_i = q_i \Gamma_{R,i} \quad \text{for } i = 1, 2$$

where the subscripts 1 and 2 refer to the $K\pi$ and $K\eta$ channels respectively and the elasticity ϵ is the ratio of the $K\pi$ partial width to the total. The results of the fit are shown by the curve in Fig. 18 and the parameters are shown in Table 11.

To determine the parameters of the higher mass S-wave resonance around 1.9 GeV/c^2, the data are fit to a model containing a Breit-Wigner resonance plus a simple polynomial term for the background,

$$a_S = BG + BW e^{i\phi}$$

$$BG = (a_0 + a_1 M) e^{i(\varphi_0 + \varphi_1 M)}$$

where $a_0, a_1, \varphi_0, \varphi_1$, and ϕ are parameters to be determined by the fit. The two solutions in this mass region are fit separately to the model. The results of the fits are shown as solid lines in Fig. 19, and their parameters are given in Table 12.

2) The P-Wave Amplitude. In the P-wave, as shown in Fig. 20, the well known $K^*(892)$ is evident from the classic Breit-Wigner resonance phase motion below 1.0 GeV/c^2. A fit to the P-wave phase in this region, with no background, gives the resonance parameters for the $K^*(892)$ as noted in Table 13.

Between about 1.2 and 1.55 GeV/c^2, the P-wave magnitude is small (about 0.15) and relatively flat. However there is a small, but statistically significant, increase in magnitude

Table 11. Resonance parameters for the $K_0^*(1350)$ determined using two different models as described in the text. The indicated errors are statistical and systematic, respectively.

$K_0^*(1350)$	
Model I	
Mass (MeV/c²)	1340
Width (MeV/c²)	350
Model II	
Mass (MeV/c²)	$1429 \pm 4 \pm 5$
Width (MeV/c²)	$287 \pm 10 \pm 21$
ϵ	$0.93 \pm 0.04 \pm 0.09$
ϕ (degrees)	$-16.9 \pm 12.8 \pm 9.0$
a(GeV/c)⁻¹	$4.03 \pm 1.72 \pm 0.06$
b(GeV/c)⁻¹	$1.29 \pm 0.63 \pm 0.67$
χ^2/NDF	52.4/45

around 1.4 GeV/c² which is accompanied by substantial structure in the phase. Above 1.55 GeV/c², the magnitude increases to a peak around 1.74 GeV/c² and then decreases. The peak is accompanied by rapid phase motion which leads to clear resonance-like behavior in the Argand diagram. This higher mass structure is very similar to that observed by Estabrooks et al.[15] and by Aston et al.,[12] and is interpreted as the observation of a P-wave resonance, the $K^*(1790)$, decaying into $K\pi$. A $1^- K^*$ resonance decaying into $K_s\pi^-\pi^+$ has also been observed in this mass region, and is most simply interpreted as another decay mode of this same state.[16] That analysis also presented evidence for a $K^*(892)\pi$ decay mode of a second $1^- K^*$ resonance at around 1.42 GeV/c², where the $K^-\pi^+$ coupling is very small. These conclusions are discussed in Section 3(e), below.

The $K^-\pi^+$ data in the regions of the two prominent resonances can be well represented separately by Breit-Wigner forms, with an additional background term in the high mass region. However, these fits do not reproduce the behavior of the amplitude in the 1.4 GeV/c² region. Therefore, given the evidence for two $1^- K^*$ states in the region between 1.3 and 1.8 GeV/c² decaying into $\bar{K}^0\pi^-\pi^+$, and the excellent statistics of this experiment, we performed a fit to the P-wave for the entire mass region below 1.84 GeV/c² with all three of these states to quantify the evidence for a weakly coupled $K^-\pi^+$ decay mode of the state in the 1.42 GeV/c² region. The model has the form,

$$a_P = BG + \sum_i BW_i e^{i\phi_i}$$

$$BG = (a_0 + a_1 M)e^{i(\varphi_0 + \varphi_1 M)}$$

$$BW_i = \frac{\sqrt{2L+1}\; \epsilon_i M_{R,i}\Gamma_i}{(M_{R,i}^2 - M^2) - iM_{R,i}\Gamma_i}$$

The results of the fit are shown by the curves in Fig. 20 and the parameters are summarized in Table 14. The dashed line indicates the result of the fit when only the two large resonances, the $K^*(892)$ and the $K^*(1790)$ are included. Though the data are well represented below 1.2 and above 1.6 GeV/c², the amplitude is poorly reproduced around 1.4 GeV/c². The solid line shows the results when a third resonance around 1.4 GeV/c² is included in the fit. The data are now well represented throughout the entire mass

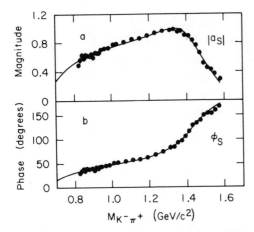

Fig. 18. The magnitude and phase of the I=1/2 S-wave $K\pi$ scattering amplitude in the mass region up to 1600 MeV.

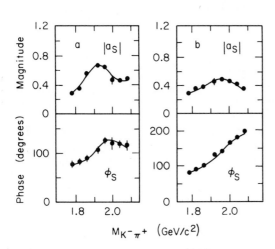

Fig. 19. The magnitude and phase of the S-wave $K\pi$ scattering amplitude between masses of 1760 MeV and 2140 MeV.

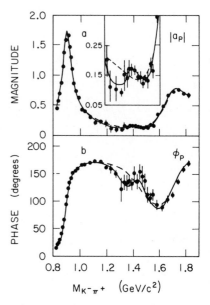

Fig. 20. The behavior of the P wave $\bar{K}\pi$ scattering amplitude up to 1.8 GeV/c^2.

Table 12. Resonance parameters for the $K_0^*(1950)$; the two ambiguous amplitudes are fitted separately using the Breit-Wigner parametrization described in the text. The indicated errors are statistical and systematic, respectively.

$K_0^*(1950)$	
Solution A	
Mass (MeV/c^2)	$1934 \pm 8 \pm 20$
Width (MeV/c^2)	$174 \pm 19 \pm 79$
ϵ	$0.55 \pm 0.08 \pm 0.12$
ϕ (degrees)	$48.3 \pm 5.6 \pm 15.5$
χ^2/NDF	$4.2/8$
Solution B	
Mass (MeV/c^2)	$1955 \pm 10 \pm 8$
Width (MeV/c^2)	$228 \pm 34 \pm 22$
ϵ	$0.48 \pm 0.02 \pm 0.02$
ϕ (degrees)	$49.1 \pm 4.5 \pm 1.1$
χ^2/NDF	$5.4/8$

Table 13. Resonance parameters for the $K^*(892)$. The indicated errors are statistical and systematic, respectively.

$K^*(892)$	
Mass (MeV/c^2)	$895.9 \pm 0.5 \pm 0.2$
Width (MeV/c^2)	$50.8 \pm 0.8 \pm 0.9$
ϵ	1.0 (Fixed)
r (GeV/c)$^{-1}$	$3.4 \pm 0.6 \pm 0.3$
χ^2/NDF	$16.9/12$

Table 14. P-wave resonance parameters for the $K^*(1410)$ and $K^*(1790)$ as determined in the model described in the text.

$K^*(1410)$	
Mass (MeV/c^2)	$1380 \pm 21 \pm 19$
Width (MeV/c^2)	$176 \pm 52 \pm 22$
ϵ	$0.066 \pm 0.010 \pm 0.008$
ϕ (degrees)	$22.9 \pm 20.6 \pm 16.9$
r (GeV/c^{-1})	2.0 (Fixed)
$K^*(1790)$	
Mass (MeV/c^2)	$1677 \pm 10 \pm 32$
Width (MeV/c^2)	$205 \pm 16 \pm 34$
ϵ	$0.388 \pm 0.014 \pm 0.022$
ϕ (degrees)	$-2.5 \pm 8.6 \pm 8.3$
r (GeV/c^{-1})	2.0 (Fixed)
χ^2/NDF	$75.4/51$

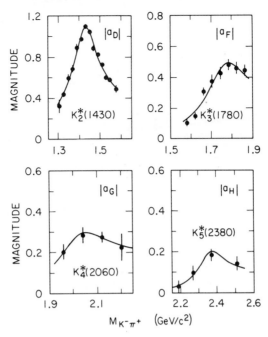

Fig. 21. The magnitudes of the leading $D, F, G,$ and H wave $K\pi$ scattering amplitudes.

range. The significance of the third resonance in this model is > 6 standard deviations, and the parameters for this resonance are compatible with those found in the three-body experiments. These resonance parameters are correlated with the background parameters, and can change significantly when different background forms are used. Several different background parametrizations have been used, in addition to the one described, and the model dependent variations in the parameters that are obtained are taken into account in the systematic errors indicated in Table 14.

The observation of the $K^*(1410)$ with a strongly suppressed $K^-\pi^+$ coupling is corroborated by the production dependence of the three-body amplitudes which indicates that the 1^- amplitudes in the $K^*(1790)$ region are steep as is expected from π exchange, while the slope of the 1^- amplitude in the $K^*(1410)$ region is much flatter, as is expected from a heavier production mechanism [see Section 3(e)]. It should be noted that the $K^-\pi^+$ decay of this state is so suppressed, it would have been essentially unobservable in earlier $K^-\pi^+$ experiments.[12,15]

3) Higher waves. The dominant structures observable in the higher spin waves arise from the leading orbitally excited K^* series. The resonance parameters for these states are extracted by fitting the partial wave magnitudes to a Breit-Wigner form. In performing these fits, the turn-on of the partial waves associated with each of these states is assumed to be resonance dominated.

The D-wave magnitude clearly shows the presence of the well established $K_2^*(1430)$. The fit to the D-wave magnitude is shown in Fig. 21, with the parameters of Table 15.

Results from the three-body channel $\bar{K}^0\pi^-\pi^+$ measured in this experiment have shown evidence for a second resonance in the D-wave around 2.0 GeV/c^2.[16] The Argand plot shows a cusp-like structure around 1.7 GeV/c^2 followed by the arc shaped path around 2.0 GeV/c^2 region which may be interpreted as arising from a similar resonance. However, the relatively small phase motion, the large background, the slow drift of this solution toward nonunitarity at high mass, and the overall phase

Table 15. Resonance parameters for the leading D, F, G, and H states as determined by Breit-Wigner fits to the high spin waves as described in the text. The indicated errors are statistical and systematic respectively. Systematic errors are not estimated for the G- and H-wave states.

	$K_2^*(1430)$
Mass (MeV/c^2)	$1431.2 \pm 1.8 \pm 0.7$
Width (MeV/c^2)	$116.5 \pm 3.6 \pm 1.7$
ϵ	$0.485 \pm 0.006 \pm 0.020$
r(GeV/c^{-1})	$2.7 \pm 1.2 \pm 0.6$
	$K_3^*(1780)$
Mass (MeV/c^2)	$1781 \pm 8 \pm 4$
Width (MeV/c^2)	$203 \pm 30 \pm 8$
ϵ	$0.187 \pm 0.008 \pm 0.008$
r(GeV/c^{-1})	2.0 (Fixed)
	$K_4^*(2060)$
Mass (MeV/c^2)	2055 ± 51
Width (MeV/c^2)	245 ± 124
ϵ	0.099 ± 0.012
r(GeV/c^{-1})	2.0 (Fixed)
	$K_5^*(2380)$
Mass (MeV/c^2)	2382 ± 34
Width (MeV/c^2)	147 ± 80
ϵ	0.061 ± 0.012
r(GeV/c^{-1})	2.0 (Fixed)

uncertainty prevent us from drawing any firm conclusions regarding such a structure. The most that can be said is that these data are not inconsistent with the existence of a D-wave resonance around 2.0 GeV/c^2.

The F-wave amplitude is required above a $K\pi$ mass of 1.58 GeV/c^2. The leading $K_3^*(1780)$ resonance can be clearly observed in the magnitude which rises steadily to a peak around 1.8 GeV/c^2. There is a slight drop just above 1.8 GeV/c^2 but the magnitude then remains rather large indicating a significant background to $K_3^*(1780)$ production above the resonance mass. The resonance fit for the leading state shown by the curve in Fig. 21 with parameters given in Table 15.

A G-wave amplitude is required above a $K\pi$ mass of 1.92 GeV/c^2. The leading $K_4^*(2060)$ can be observed in the magnitude which rises to a peak around 2.1 GeV/c^2. The fit to the G-wave magnitude is shown in Fig. 21. The resulting resonance parameters given in Table 15 agree with the results from the overall mass dependent fit to the moments presented in an earlier publication.[17]

Evidence for a leading $K_5^*(2380)$ can be seen in the H-wave magnitude that peaks at about 2.4 GeV/c^2. The fit to this magnitude uses the four independent mass bins as

Table 16. The preferred quark state assignments and the measured mass values for the states observed in this experiment; the predicted values from the model of Ref. 19 are given for comparison. The quoted errors include the statistical and systematic errors added in quadrature.

Spin Parity	Probable $\bar{q}q$ state	Measured Mass (MeV/c^2)	Predicted Mass (MeV/c^2)
0^+	1^3P_0	(I) 1340	1240
		(II) 1429±6	
	2^3P_0	(A) 1934±22	1890
		(B) 1955±13	
1^-	1^3S_1	895.9±0.6	900
	2^3S_1	1380±28	1580
	1^3D_1	1677±34	1780
2^+	1^3P_2	1431±2	1420
3^-	1^3D_3	1781±9	1790
4^+	1^3F_4	2055±51	2110
5^-	1^3G_5	2382±34	2390

shown, and the resulting parameters are given in Table 15. These also agree with the results from the mass dependent fit to the moments presented earlier.[17]

4) Discussion and Summary. Results from an energy independent partial wave analysis, of a data sample more than four times larger than those used by earlier analyses, have provided new information on the $K\pi$ resonance structure.[18] Clear resonance behavior can be seen in the partial wave amplitudes, which can be reproduced by simple Breit-Wigner modeling. Fits of the amplitudes with these models provide estimates for the masses, widths, and elasticities of the observed resonances.

The parameters of these states are summarized in Table 16 and in Fig. 10. For the most part, the observed states can be naturally assigned to levels expected in a quark model. The preferred quark model assignments are also summarized in Table 16, along with predictions for their mass values taken from the $q\bar{q}$ potential model of Godfrey and Isgur.[19]

The amplitude analysis confirms the first four leading resonances, the $K^*(892)$, the $K_2^*(1430)$, the $K_3^*(1780)$, and the $K_4^*(2060)$, and fits to these amplitudes provide new measurements for the parameters of these states. In the highest mass region analyzed, the behavior of the H-wave amplitude supports the earlier observation of a spin-5 $K_5^*(2380)$ state based on the moments fit.[17] In the quark model, these states can naturally be understood as the leading orbitally excited triplet series, the 1^3S_1, 1^3P_2, 1^3D_3, 1^3F_4, and 1^3G_5. These five states lie close to a linear Regge trajectory with a slope 0.84 $(GeV/c^2)^{-1}$, and the model predicts their masses well.

The lowest mass S-wave state, the $K_0^*(1350)$, is naturally assigned as the lowest 0^+ quark model state, the 1^3P_0. The predicted and measured masses show a substantial difference ($\sim 100 - 200$ MeV/c^2), however, the measured mass of this state depends critically on the model used for the estimation.

A second 0^+ state is observed around 1.95 GeV/c^2, confirming an observation in a previous study of the $K^-\pi^+$ channel,[12] but with a much better determination of the parameters. Within the quark model, this state can only be assigned as a radial excitation of the 0^+ member of the $L = 1$ triplet, most naturally the 2^3P_0 state. This state is one of the clearest candidates for a radial excitation of a light quark system. The measured mass value agrees well with the predictions of the model.

The two P-wave resonances observed in earlier $K^-\pi^+$ analyses,[15][12], the $K^*(892)$, and the $K^*(1790)$, are clearly seen in the amplitudes. In addition, new evidence for a structure with a small elasticity around 1.4 GeV/c^2 is provided which is most easily interpreted as a confirmation of the $K^*(1410)$ resonance seen in the results of the three-body $\bar{K}^0\pi^-\pi^+$ PWA, decaying into the $K^-\pi^+$ mode. The most recent of the three-body analyses, using data from this experiment, observed the two higher mass states at masses of 1420 MeV/c^2 and 1735 MeV/c^2 respectively,[16] in good agreement with this analysis. Even though mixing is not excluded, it is simplest to associate the higher mass state with the 1^3D_1, based primarily on the small mass splitting between this state and the $K_3^*(1780)$. The lower state is then interpreted as the first radial excitation of the $K^*(892)$. The suppression of the $K^-\pi^+$ decay mode of this lower state can be understood in some models as being a dynamical effect resulting from the presence of a node in the radial wave function.[20] The agreement of the $K^*(1410)$ with the mass predictions of the model of Ref. 19 for the 2^3S_1 is rather poor. However, other models give mass predictions for the 2^3S_1 that are consistent with the $K^*(1410)$ mass and also suggest that the 3^3S_1 state should lie well below 2.0 GeV/c^2.[20] The assignment discussed here implies that the nonet containing the 2^3S_1 states should lie in the mass region between about 1250 and 1600 MeV/c^2, and that these states will not be easily observable in the simple two-body modes. Though many previous experiments have found evidence for a candidate isovector state, the $\rho(1250)$,[2] its present status is equivocal,[21] and there are no firm candidates for the other states in this nonet. On the other hand, a recent analysis has obtained a consistent picture of several different channels by postulating two ρ' resonances with masses[22] 1.465 and 1.700 GeV/c^2.

d) The Reaction $K^-p \rightarrow \bar{K}^0\pi^-p$

When we study the reaction

$$K^-p \rightarrow \bar{K}^0\pi^-p \;,$$

we find a reaction *not* dominated by one pion exchange, but by isoscalar, natural J^P exchange. The Dalitz plot for this reaction is shown in Fig. 22, where the N^* and K^* bands are clearly visible. The decay distribution data for the K^*'s, are shown in Figs. 23 and 24, as a function of $K\pi$ mass; the cosine of the Jackson angle in the $K\pi$ center-of-mass is shown in Fig. 23, while the azimuthal Treiman–Yang angle is shown in Fig. 24. Clear indications of the spin 1, natural parity exchange are seen, as are the $J^P = 1^-$ 2^+, 3^- of the three leading K^* states at 890, 1420, and 1780 MeV/c^2.

A more quantitative analysis yields the moments of the $K\pi$ decay distribution in the three mass regions of interest. Figure 25 gives the moments in the region around 1 GeV/c^2 —the region dominated by the production and decay of the $K^*(892)$. The parameters of the $K^*(892)$ Breit–Wigner line shape, as obtained from fits to the different moments, are summarized in Table 17, and described in detail in Ref. 23. Figure 26 shows the behavior of the moments in the 1400 MeV/c^2 region, and Fig. 27 shows the moments in the 1800 MeV/c^2 region. Tables 18 and 19 give the resonance parameters from fits to these moments, as described in Ref. 23. Figure 28 shows the leading natural spin-parity amplitudes and Table 20 gives the best estimates for the parameters of the three leading K^* resonances.

Having examined the leading amplitudes, we can now explore the structure in the underlying waves. The mass dependence of the natural parity exchange amplitudes from threshold to 2 GeV/c^2 is shown in Fig. 29. The relative phase between the P- and D-waves and between the D and F-waves is shown in Fig. 30, and indicates the presence of resonant waves in addition to the leading K^*'s. This data has been fit using

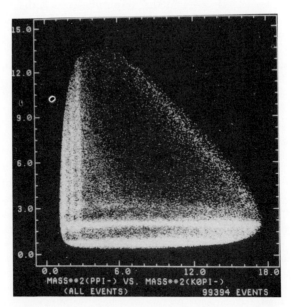

Fig. 22. Dalitz plot for the reaction $K^-p \to \bar{K}^0 p\pi^-$.

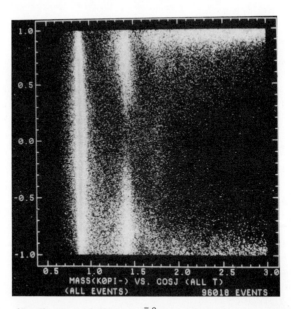

Fig. 23. The decay distribution of $K^{*-} \to \bar{K}^0 \pi^-$ is shown in a scatterplot of the cosine of the Jackson angle in the $K\pi$ center-of-mass, as a function of the $K\pi$ mass.

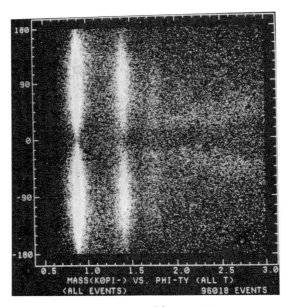

Fig. 24. The decay distribution of $K^{*-} \to \bar{K}^0 \pi^-$ is shown in a scatterplot of the azimuthal (Treiman–Yang) angle as a function of $K\pi$ mass.

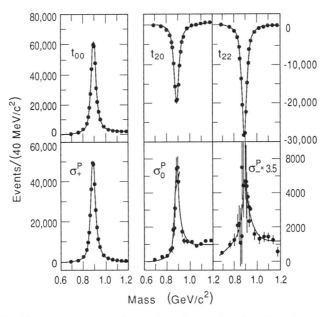

Fig. 25. The moments and projected amplitudes for the process $K^- p \to K^{*-} p$, $K^* \to \bar{K}^0 \pi^-$ in the region around $K\pi$ mass of 1 GeV.

233

Table 17. Breit–Wigner parameters for the $K^*(892)$ region. The first error is statistical, the second systematic.

Fit	Mass (MeV/c^2)	Width (MeV/c^2)	Radius (GeV^{-1})
t_{00}	$890.9 \pm 0.2 \pm 0.5$	$46.2 \pm 0.5 \pm 0.5$	$5.6 \pm 0.5 \pm 1.0$
t_{20}	$889.6 \pm 0.5 \pm 0.5$	$50.7 \pm 1.4 \pm 0.5$	$100. \pm 50. \pm 5.0$
t_{22}	$890.5 \pm 0.3 \pm 0.5$	$46.0 \pm 0.7 \pm 0.5$	$11.4 \pm 2.5 \pm 1.0$
σ_0^P	$890.3 \pm 0.2 \pm 0.5$	$50.4 \pm 1.7 \pm 0.5$	$0.0 \pm 10. \pm 1.0$
σ_+^P	$890.7 \pm 0.2 \pm 0.5$	$46.4 \pm 0.4 \pm 0.5$	$15.4 \pm 5. \pm 1.0$
σ_-^P	$898.8 \pm 7.7 \pm 0.5$	$87.9 \pm 30. \pm 3.0$	$100. \pm 50. \pm 5.0$

Table 18. Breit–Wigner parameters for the $K_2^*(1430)$ region. The first error is statistical, the second systematic.

Fit	Mass (MeV/c^2)	Width (MeV/c^2)	Radius (GeV^{-1})
t_{00}	$1419 \pm 0.8 \pm 1$	$99.3 \pm 3.0 \pm 2$	$100 \pm 78 \pm 10$
t_{22}	$1424 \pm 2.6 \pm 1$	$100.7 \pm 10.1 \pm 3$	$0 \pm 53 \pm 5$
t_{40}	$1424 \pm 3.5 \pm 1$	$101 \pm 11.3 \pm 3$	$5 \pm 5 \pm 3$
t_{42}	$1420 \pm 2 \pm 1$	$89.8 \pm 5 \pm 1$	$3.1 \pm 1.2 \pm 1$
σ_0^D	$1412 \pm 4.8 \pm 1$	$100.7 \pm 14.7 \pm 3$	$100 \pm 59 \pm 15$
σ_+^D	$1420.5 \pm 1.1 \pm 1$	$98.8 \pm 4.4 \pm 3$	$12.5 \pm 54 \pm 10$

Table 19. Breit–Wigner parameters for the $K_3^*(1780)$ region. The first error is statistical, the second systematic.

Fit	Mass (MeV/c^2)	Width (MeV/c^2)	Radius (GeV^{-1})
t_{00}	$1747 \pm 12 \pm 4$	$145 \pm 59 \pm 10$	100
t_{62}	$1738 \pm 21 \pm 5$	$195 \pm 36 \pm 15$	100
σ_0^F	$1784 \pm 43 \pm 10$	$233 \pm 360 \pm 50$	100
σ_+^F	$1741 \pm 9.8 \pm 5$	$243 \pm 60 \pm 10$	100

Table 20. The leading natural spin-parity amplitudes. A single Breit–Wigner resonance is assumed for each wave.

Resonance	Mass (GeV/c^2)	Width (GeV/c^2)	Radius (GeV^{-1})
$K^*(892)$	$0.8904 \pm 0.0002 \pm 0.0005$	$0.0452 \pm 0.001 \pm 0.002$	$12.1 \pm 3.2 \pm 3$
$K_2^*(1430)$	$1.4234 \pm 0.002 \pm 0.003$	$0.098 \pm 0.004 \pm 0.004$	$4.8 \pm 2.3 \pm 3$
$K_3^*(1780)$	$1.720 \pm 0.010 \pm 0.015$	$0.187 \pm 0.031 \pm 0.020$	$8.5 \pm 3 \pm 10$

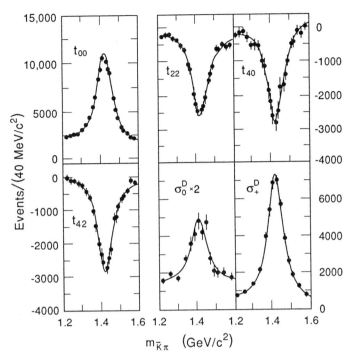

Fig. 26. The moments for the reactions $K^-p \to K^{*-}p$, $K^{*-} \to \bar{K}^0\pi^-$ in the region around $K\pi$ mass of 1400 MeV.

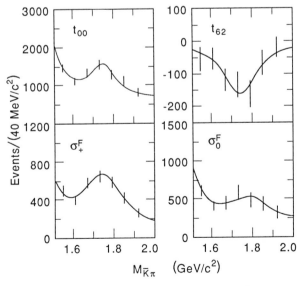

Fig. 27. The moments for the processes $K^-p \to K^{*-}p$, $K^{*-} \to \bar{K}^0\pi^-$ in the region around $K\pi$ mass of 1750 MeV.

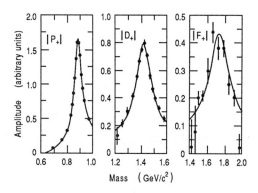

Fig. 28. Leading natural spin–parity amplitudes. A single Breit–Wigner resonance is assumed in each wave.

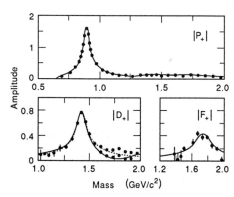

Fig. 29. Mass dependence of P, D, and F waves from threshold up to 2000 MeV.

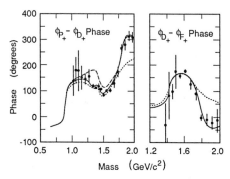

Fig. 30. Energy dependence of the phase difference between the P–D and D–F waves in the process $K^-p \to K^0\pi^-p$.

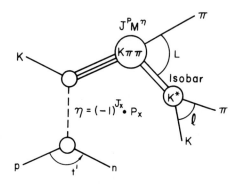

Fig. 31. The production of the three meson final state. The final state is produced via an intermediate isobar–bachelor system (e.g. $K^*\pi$ as shown in the figure).

the superposition of interfering resonant amplitudes with arbitrary relative production phases; the result is shown as the solid curve in Figs. 29 and 30.

The dashed curve in Fig. 30 shows the behavior of the P–D phase, if the 1400 MeV/c^2 P-wave state is excluded. Similarly, the dotted curve shows the expected phase behavior if the D-wave state at 1980 MeV/c^2 is excluded. The parameters of the underlying resonances from this analysis are given in Table 21. The parameters for the two underlying P-wave states, and the D-wave state, are in good agreement with those obtained in our previous analysis.[17,24] This is independent confirmation of the radial excitation of the $K_2^*(1420)$ at 1980 MeV/c^2, and of the P-wave states—the radial excitation of the $K^*(892)$ at ~ 1400 MeV/c^2 and the $^3D_1$ member of the $L = 2$ triplet associated with the $K_3^*(1780)$.

Table 21. The P and D-wave resonance parameters.

Wave	Mass (GeV/c^2)	Width (GeV/c^2)	Phase (degrees)
	0.8905 (fixed)	0.045 (fixed)	0 (fixed)
1^-	1.367 ± 0.054	0.114 ± 0.101	69 ± 7
	1.678 ± 0.064	0.454 ± 0.270	-92 ± 17
2^+	1.425 (fixed)	0.1 (fixed)	40 ± 4
	1.978 ± 0.040	0.398 ± 0.047	2 ± 5

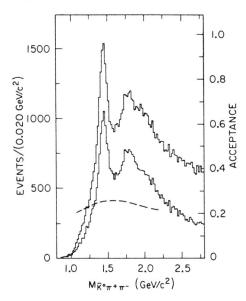

Fig. 32. The $\bar{K}^0\pi\pi$ invariant mass distribution. The outer histogram is the mass spectrum for the entire data sample. The inner histogram is the same distribution for events with momentum transfer, t', less than 0.3 (GeV/c)^2. The dashed line shows the acceptance of the experiment for this final data sample as a function of the $K\pi\pi$ mass.

e) **The Reaction $K^-p \to \bar{K}^0\pi^+\pi^-n$**

This reaction yielded a sample of 34,000 events with the $K\pi\pi$ system produced at small momentum transfers and analyzed in terms of quasi two-body decay of the meson systems with $K\rho$ or $K^*\pi$ isobars, using the SLAC-LBL partial wave analysis program, Ref. 6. See Fig. 31.

The mass spectrum for the $K^0\pi^+\pi^-$ system is given in Fig. 32; it is tempting to understand this picture in terms of production of the $\bar{K}_2^*(1420)$ and $\bar{K}_3^*(1780)$. However,

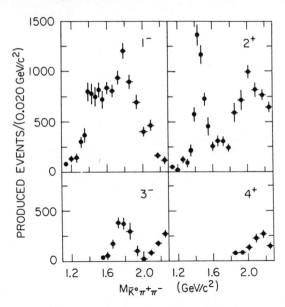

Fig. 33. The natural spin-parity amplitudes. The summed intensities of the natural spin-parity partial waves with the same J^P are plotted as a function of mass. The sums include the interference terms between coherent waves.

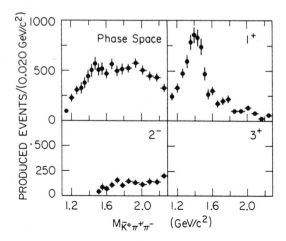

Fig. 34. The unnatural spin-parity amplitudes, and the phase space contribution.

more careful study indeed confirms that most of the cross section is two-body resonance production, but 30% of the first bump is $J^{PC} = 1^{--}$ (not 2^{++}!) and 70% of the second bump has $J^{PC} = 1^{--}$ (not $J^{PC} = 3^{--}$!!).

The scattering amplitudes for the natural spin-parity state are shown in Fig. 33, and for the unnatural J^P states in Fig. 34. A more detailed breakdown showing the individual isobar contributions—both in amplitude and in phase—for the main partial

238

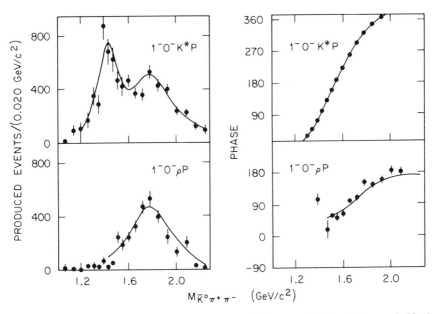

Fig. 35. The amplitude and phase of the 1^- partial wave for the $K^*\pi$ and $K\rho$ isobar decays, as a function of $K\pi\pi$ mass.

waves, is shown in Figs. 35, 36 and 37. The solid lines display Breit–Wigner fits to the scattering amplitudes and the resulting masses and widths for the resonances are given in Table 22.

It is interesting to note that the lower mass P-wave state at 1400 MeV, couples weakly to $K\rho$ while the state around 1700 MeV couples quite strongly to the $K^*\pi$ and $K\rho$. In Fig. 38 the production angular distribution (i.e., $d\sigma/dt$) for the P-wave scattering amplitude in $K^*\pi$ and $K\rho$ are shown for the two mass intervals—solid dots for the 1400 MeV region, and open circles for the 1700 MeV region. The higher mass cross section have the steep falloff characteristic of the π–exchange processes (i.e., $K\pi \to K^*\pi$, $K\pi \to K\rho$), whereas the low mass region shows little sign of π–exchange. This weak coupling to $K\pi$ is similar to what was learned from the analysis of $K\pi$ elastic scattering in section 3(b) above.

f) The Reaction $K^-p \to K^-p\,\pi^+\pi^-\pi^0$

The mass spectrum for the three pions from the reaction $K^-p \to K^-p\,\pi^+\pi^-\pi^0$ is shown in Fig. 39, where clear η and ω signals are observed.

The reaction $K^-p \to K^-\eta p$ has been measured with large statistics for the first time, in the LASS experiment. The mass spectrum for the $K^-\eta$ system is shown in Fig. 40. An amplitude analysis of the $K^-\eta$ angular distribution has established that the peak around 1800 MeV is mainly F-wave, resulting from the production of $K_3^*(1780)$—see Fig. 41. A comparison with the data from the reaction $K^-p \to \bar{K}^0\pi^-p$, described in section 3(c) above, allows a measurement of the $K\eta/K\pi$ branching ratio, which is found to be $7.7 \pm 1.0\%$—in good agreement with the SU(3) expectations. SU(3) also predicts that K^* states of even spin couple only weakly to $K\eta$, while K^* states of odd spin should couple strongly. This is confirmed in Fig. 41 where clear signs of the $J^P = 3^-$ state are observed, and no structure is seen around 1430 MeV in the $J^P = 2^+$ amplitude. The 95% confidence level upper limit on the branching ratio for $K_2^*(1430) \to K\eta$ is established at 0.45%. This value is an order of magnitude smaller limit than determined in previous studies.

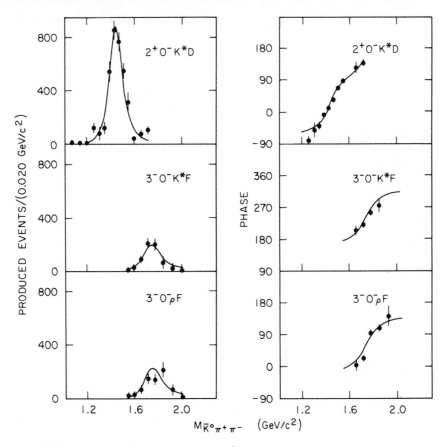

Fig. 36. The amplitude and phase of the 2^+ and 3^- partial waves for the $K^*\pi$ and $K\rho$ isobar decays, as a function of $K\pi\pi$ mass.

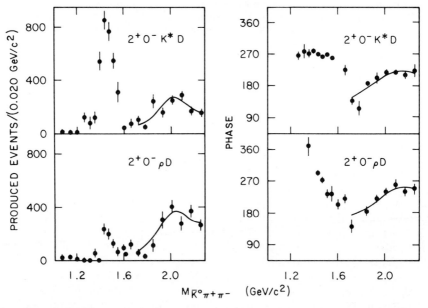

Fig. 37. The amplitude and phase for the higher mass region for the 2^+ partial waves, for the $K^*\pi$ and $K\rho$ decay isobars.

Table 22. Resonance parameters obtained from analysis of the $K^*\pi$ and $K\rho$ decays of the P, D and F wave K^* states.

J^{PC}	Mass (MeV)	Width (MeV)	Decay Mode
1^{--}	$1420 \pm 7 \pm 10$	$240 \pm 18 \pm 12$	$K^*\pi$
1^{--}	$1735 \pm 10 \pm 20$	$423 \pm 18 \pm 30$	$K^*\pi$, $K\rho$
2^{++}	$1434 \pm 4 \pm 6$	$124 \pm 15 \pm 15$	$K^*\pi$
	$1973 \pm 8 \pm 25$	$373 \pm 33 \pm 60$	$K^*\pi$, $K\rho$
3^{--}	$1740 \pm 14 \pm 15$	$171 \pm 42 \pm 20$	$K^*\pi$, $K\rho$

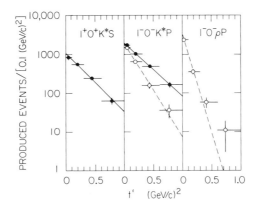

Fig. 38. The production differential cross-section, $\frac{d\sigma}{dt}$, for the P-wave ($J^P = 1^-$) amplitudes in the 1400 MeV (solid points and solid lines) and 1700 MeV (open points and dashed lines) mass regions.

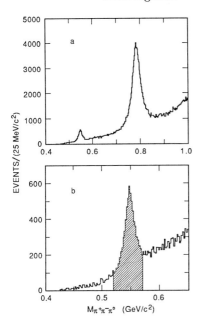

Fig. 39. (a) The $(\pi^+\pi^-\pi^0)$ mass distribution for the reaction $K^-p \to K^-p\,\pi^+\pi^-\pi^0$. (b) A detailed look at the region around 550 MeV.

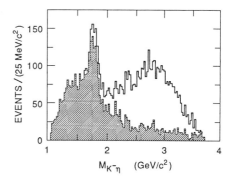

Fig. 40. The $K^-\eta$ mass distribution for the reaction $K^-p \to K^-p\,\pi^+\pi^-\pi^0$. The shaded region corresponds to $M(\eta p) > 2$ GeV and $M(K^-p) > 1.85$ GeV.

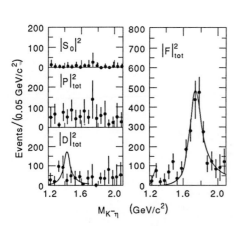
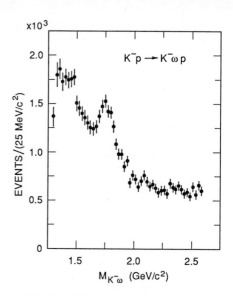

Fig. 41. The total F and D wave intensity distributions for the reaction $K^-p \to K^-\eta p$. The F wave curve corresponds to a $K_3^*(1780)$ Breit–Wigner. The D wave curve indicates the 95% confidence limit on $K_2^*(1420)$ production.

Fig. 42. The $K\omega$ mass spectrum for the reaction $K^-p \to K^-p\,\pi^+\pi^-\pi^0$.

The mass distribution for the $K\omega$ system is shown in Fig. 42. This process is just coming under study now. There are ~ 50 times more events for this reaction than previous experiments, and we hope to learn more of the unnatural spin-parity states around 1400 MeV and 1700 MeV.

g) Summary of K^* Spectroscopy

Let us briefly review what we have learned from the SLAC K^* studies.

In the reaction $K^-p \to K^-\pi^+n$ at 11 GeV/c, we were able to isolate a rather clean data sample to study elastic $K\pi \to K\pi$ scattering. In this study we found clear evidence for the leading K^* states from $J^{PC} = 1^{--}$ all the way through 2^{++}, 3^{--}, 4^{++} to 5^{--}. These states correspond to the quark spin parallel case and with orbital angular momentum $L = 0, 1, 2, 3, 4$ between the q and the $\bar{q}$. In the same analysis, we found some of the underlying structure:

- a 0^{++} state below the 2^{++} at ~ 1400 MeV,
- a 1^{--} state below the 3^{--} at ~ 1700 MeV,
- a 2^{++} state at ~ 1900 MeV, an excited K_2^*,
- a 0^{++} state at ~ 1900 MeV, an excited K_0^*,
- a 1^{--} state at ~ 1450 MeV, an excited K_1^*.

We also studied $K\pi$ inelastic scattering, leading to $K^*\pi$ and $K\rho$ final states, and charge exchange $K\pi$ scattering via w exchange. These studies confirmed some of the observations reported above; and demonstrated new structures. Particularly, the existence of radially excited states:

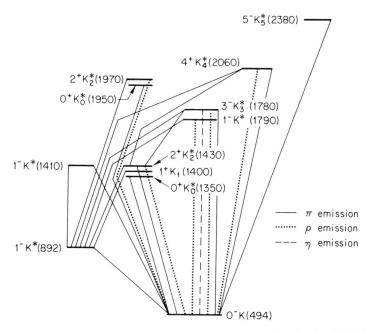

Fig. 43. The various transitions among the K^* states studied in the LASS program.

- a radially excited heavy K meson—$K_0(1400)$,
- a radially excited K_1^* at 1420 MeV and again at ~ 2100 MeV,
- a radially excited K_2^* at 1970 MeV and again at ~ 2200 MeV,
- a radially excited K_0^* at 1915 MeV,
- a radially excited K_3^* at 2100 MeV.

These states are summarized in Fig. 10. The various transitions studied are recounted in Fig. 43.

The spin-spin effects are very large for the ground states:

$M(1^1S_1) \sim 500$ MeV and $M(1^3S_1) \sim 900$ MeV

(i.e., $\Delta M \sim 400$ MeV),

but not for higher states:

$M(2\ ^1S_1) \sim M(2\ ^3S_1) \sim 1400$ MeV   and $M(1\ ^1P_1) \sim M(1\ ^3P_1) \sim 1450$ MeV.

(i.e., $\Delta M \sim 0\text{-}50$ MeV).

All of the states expected in the simple $q\bar{q}$ model are found, and no extra states are found in the K^* sector.

The $L \cdot S$ triplet structure —i.e., the $(L+S, L, L-S)$ states—is clearly observed, for $L = 1$ and $L = 2$.

The Godfrey-Isgur QCD relativistic potential model does a fair job in describing the meson mass spectrum, for light as well as for heavy quark mesons. We shall discuss this more in the concluding section. This model predicts that the splitting in the $s\bar{u}$ sector should be larger than for the $c\bar{c}$ and $b\bar{b}$ cases. It is not!

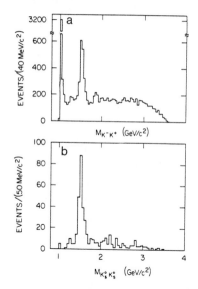

Fig. 44. The $(K\bar{K})$ mass spectrum for the reactions (a) $K^-p \to K^-K^+\Lambda$ and (b) $K^-p \to K^0_s K^0_s \Lambda$.

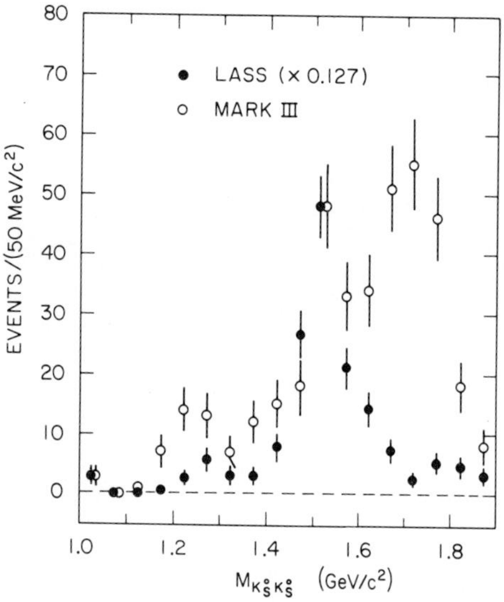

Fig. 45. The comparison of the $K^0_s K^0_s$ mass distribution for the reaction $K^-p \to K^0_s K^0_s \Lambda$ in the LASS experiment and the same spectrum from the MARK III experiment's study of radiative J/ψ decays.

4. THE $(s\bar{s})$ MESONS, OR STRANGEONIUM

a) Introduction

Strangeonium spectroscopy is of interest, both in terms of the complementary information it provides on the $q\bar{q}$ spectrum, and as a potential place to find exotic mesons. Several candidate exotic mesons—glueballs, hybrids and multiquark states—are observed to couple strongly to $s\bar{s}$ final states.[25]

It should be pointed out, that in peripheral hypercharge exchange dominated processes, one is usually studying different meson production depending on which reaction is observed. Specifically, in the process $\pi^-p \to K\bar{K}X$ the majority process is the production of $(u\bar{u})$ mesons and their subsequent decay into a (minority) $K\bar{K}$ channel. In the reaction $K^-p \to (K\bar{K})X$, the majority process is the production of an $s\bar{s}$ meson and the subsequent observation of its decay into a $K\bar{K}$ system.

We will consider mainly the data from the LASS experiment at 11 GeV/c, looking at the reactions:

$$K^-p \to K^+K^-\Lambda \qquad (a)$$

$$\to K^0_s K^0_s \Lambda \qquad (b)$$

$$\to K^0_s K^\mp \pi^\pm \Lambda \,, \qquad (c)$$

all dominated by K and K^* peripheral exchange. The data are very clean, with all the final state particles being observed in the spectrometer. Reaction (a) proceeds through natural spin-parity exchange (i.e., 0^+, 1^-, 2^+, 3^-, 4^+), as does reaction (b); but given the identical mesons in the final state it will only involve even J (i.e., 0^+, 2^+, 4^+). Reaction (c) can proceed through many exchange processes.

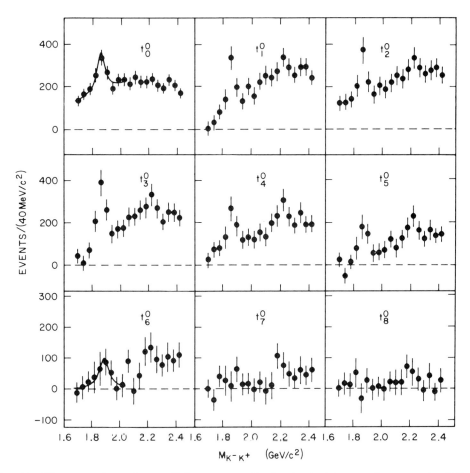

Fig. 46. The mass dependence of the unnormalized spherical harmonic moments of the K^+K^- system for the reaction $K^-p \to K^-K^+\Lambda$ in the mass region (1680-2440) MeV and for momentum transfers, t', less then 0.2 $(GeV/c)^2$.

b) The Natural Spin-Parity Resonances

The mass spectrum for the $K\bar{K}$ system in reactions (a) and (b) is displayed in Fig. (44). Clear structure is seen in the K^+K^- spectrum corresponding to the known $\phi(1020)$ and $f'_2(1525)$ leading orbital $(s\bar{s})$ states, as well as a small bump in the $\phi_J(1850)$ region. Only the $f'_2(1525)$ is observed in the $K^0_s K^0_s$ spectrum, since there the spin is restricted to even values by the identical particles.

Notice that there is no sign of the $\theta(1720)$ in either spectrum. The $K^0_s K^0_s$ mass spectrum is compared to the Mark III data on radiative J/ψ decay for the mass region below 1900 MeV, in Fig. 45. Both spectra show a small, but intriguing threshold rise, followed by activity in the $f_2(1270)$ —$A_2(1320)$ region and then the large $f'_2(1525)$ peak. At higher masses the Mark III spectrum is dominated by the $\theta(1720)$. There is no evidence for any such signal in the LASS data, and the upper limit on the production cross section is ≤ 94 nb. We will discuss this later in Section 5, when we discuss exotic mesons.

An amplitude analysis of the data from reactions (a) and (b) displays the expected P-wave structure for the $\phi(1020)$ and the D-wave structure for the $f'_2(1525)$—see Fig. 47. The S-wave intensity peaks around 1525 MeV and is not required below or

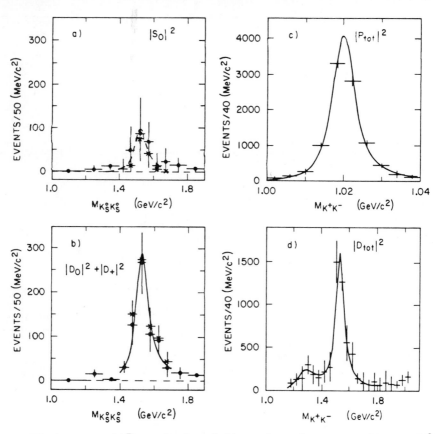

Fig. 47. The low mass $K\bar{K}$ amplitudes. (a,b) are from the reaction $K^-p \to K^0_s K^0_s \Lambda$ and (c, d) are from the reaction $K^-p \to K^+K^-\Lambda$.

above in mass. Although the errors on the individual points are large (and non-linear), the data require the existence of an S-wave amplitude in this region at about a 5σ level. This suggests the existence of a 0^+ resonance, which is most naturally interpreted as the triplet partner of the $f'_2(1525)$—(i.e., the $(L-S)$ member of the ^3P triplet, where the $f'_2(1525)$ is the $(L+S)$ member). This would lead to the suggestion that the $f_0(975)$ (which is usually assigned to this multiplet) is perhaps not a $q\bar{q}$ state.

A study of the K^+K^- angular distribution, evaluated in the helicity system for a peripherally produced $K\bar{K}$ system, results in the moments shown in Fig. 46 for mass greater than 1.68 GeV. It should be noted that amplitudes with spin J can contribute to moments with $L \leq 2J$. Clear structure is seen at around 1850 MeV.

The presence of structure around 1850 MeV in all the moments in Fig. 46, up through $L=6$, but not in $L=7$ or 8, is an indication that this structure is probably $J^{PC} = 3^{--}$. The F-wave intensity distribution from the amplitude analysis discussed above shows structure in the 1850 MeV region. A fit to this amplitude, yields Breit-Wigner resonance parameters of $M = 1855 \pm 22$ MeV, $\Gamma = 74 \pm 67$ MeV (see Fig. 46), while a fit to the bump in the mass distribution (i.e., the t^0_0 moment) yields $M = 1851 \pm 7$, $\Gamma = 66 \pm 29$ MeV. This is the first measurement of the spin-parity of this state, but fortunately we get another chance at it. For high $K\bar{K}$ mass, this reaction has serious background from the diffractive dissociation process where the target proton dissociates into a low mass $K^+\Lambda$ system. This is clearly seen in the Dalitz plot (Fig. 48)—as the dense horizontal band at the bottom of the plot. The two processes—meson exchange

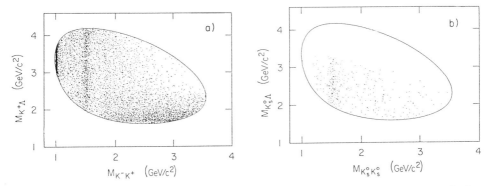

Fig. 48. The Dalitz plots for the reactions $K^-p \to K^+K^-\Lambda$ and $K^-p \to K_s^0 K_s^0 \Lambda$. Notice the dark band along the bottom of the plot (a); this corresponds to the diffractive dissociation of the proton into a ΛK^+ system.

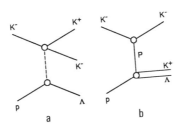

Fig. 49. Diagrammatic representations for (a) peripheral meson production and for (b) diffractive dissociation of the target proton.

and diffractive dissociation—are shown, diagrammatically, in Fig. 49. The diffractive dissociation process is almost purely imaginary and has a large S-wave $K^+\Lambda$ component, less P-wave, even less D-wave. Our second chance comes from trying to detect the small spin 3 ($s\bar{s}$) resonant amplitude *interfering* with this large, dominantly imaginary, diffraction dissociation amplitude.

Fig. 50 shows the projection of the interference terms involving the F-wave amplitude and the main contributions to the diffraction discrimination process. The details of how to project out the bi-linear products of production amplitudes, may be found in Ref. 26. The solid line is the result of a fit to the data using the imaginary part of a Breit-Wigner amplitude interfering with a linear background, and yields estimates of the mass and width of the $\phi_3(1850)$ of $M = 1858 \pm 12$, $\Gamma = 58 \pm 36$ MeV. The results agree well with the fits described above.

An extension of this interference method has been used to analyze the G-wave amplitude in the 2200 MeV mass region. The $K_s^0 K_s^0$ mass distribution for this high mass region is shown in Fig. 45, together with the data from the Mark III experiment studying radiative J/ψ decays. The LASS $K\bar{K}$ data favor spin 4 in this region, but the statistical weight of the data is not large. When the interference of the G-wave amplitude with the diffractive dissociation background is projected out using reaction (a), a clear bump in the 4^{++} amplitude is observed at 2200 MeV, see Fig. 51. This is a good candidate for the mainly $s\bar{s}$ member of the 4^{++} nonet, expected in the quark model as the leading $L = 3$ state. See Fig. 51.

To recap, we have seen evidence of the leading 1^{--}, 2^{++}, 3^{--}, and 4^{++} $s\bar{s}$ states in our study of $K\bar{K}$ scattering. We have also seen an indication of the 0^{++} partner to the 2^{++} state at 1525 MeV.

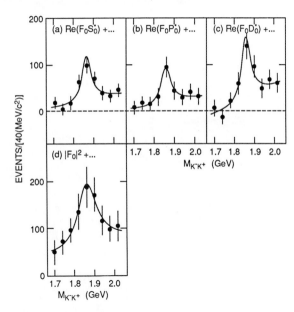

Fig. 50. The projection of the interference terms between the F- wave $K\bar{K}$ amplitude and the large, imaginary diffractive dissociation amplitude.

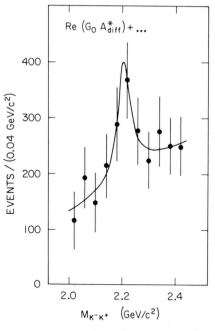

Fig. 51. The mass dependence of the interference between the G_o and diffractive background amplitudes from the reaction $K^-p \to K^-K^+\Lambda$.

c) **The Axial Mesons**

A complete understanding of the low mass axial-vector mesons has yet to be achieved. The quark model predicts two nonets below 1.6 GeV/c^2, one with $J^{PC} = 1^{++}$

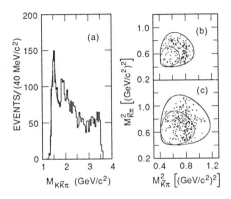

Fig. 52 The $K\bar{K}\pi$ mass distributions for the reactions $K^-p \to K_s^0 K^\pm \pi^\mp \Lambda$.

and the other with $J^{PC} = 1^{+-}$. Although there are clear candidates for the I=1 and I=1/2 members of these multiplets, the situation for the I=0 states remains confused. The singlet-octet mixing angle is unknown, and while the $1^{++} f_1(1285)$ and the $1^{+-} h_1(1190)$ are good candidates to be I=0 members of their respective nonets containing mostly first generation quarks, their mostly strangeonium partners are not yet established. In particular, there are no good candidates for the 1^{+-} state, while there are several possibilities for the 1^{++} state. The $f_1(1420)$ (formerly called the E meson) is usually assigned as the mainly $s\bar{s}$ member of the 1^{++} nonet,[12] but its strong production in πp interactions,[27] and possibly in hadronic J/ψ decay against an ω but not against a ϕ,[28] is inappropriate for a dominantly $s\bar{s}$ object. Moreover, in peripheral production via the hypercharge exchange in the reactions

$$K^-p \to K^+ K_S^0 \pi^- \Lambda \qquad (d)$$

$$K^-p \to K^- K_S^0 \pi^+ \Lambda \qquad (e)$$

where $s\bar{s}$ states are expected to dominate, the production of $f_1(1420)$ has not been clearly observed. An earlier study of reactions (d) and (e) provided evidence for another 1^{++} meson, called the $D'(1530)$,[29] which is a good candidate for the $s\bar{s}$ member of this nonet. The situation is confused, and recent evidence[30] for several additional objects in this vicinity decaying to $K\bar{K}\pi$, some of which are candidates for glueball states, has served to highlight the importance of understanding the $s\bar{s}$ spectrum in this mass region.

We now examine the results of a three-body partial wave analysis of the $K\bar{K}\pi$ system in the mass region below 2.0 GeV/c^2 using new data on reactions (d) and (e) at 11 GeV/c obtained from a 4.1 ev/nb study of K^-p interactions using the LASS spectrometer at SLAC.

After track reconstruction, candidates for reactions (d) and (e) are selected requiring a topology with two V^0's and two charged tracks associated with the primary vertex. The event sample is then restricted to those events with approximate energy-momentum balance, and with good K^0 and Λ candidates separated from the primary vertex. After assigning the charged particle masses in accordance with the available particle identification information, a fully constrained fit is performed. The identification of one of the two charged tracks is sufficient for unambiguous assignment to reaction (d) or (e) and creates no holes in the acceptance. Low mass baryon resonances reflect primarily into the $K\bar{K}\pi$ mass region above 2.0 GeV/c^2, and removal of the $\Sigma^{*+}(1385)$ region $(1.34 < M_{\Lambda\pi^+} < 1.42$ GeV/$c^2)$ from reaction (e) has no effect on the analysis presented here. Finally, events with $t' \leq 3.0$ GeV/c^2 ($t' \equiv |t| - |t|_{min}$), where t is the 4-momentum transfer between the incoming beam and the $K\bar{K}\pi$ system) are selected to enhance the dominant hypercharge-exchange production process. The resulting data sample, which contains 3900 events (1787 events for reaction (d) and 2113 for (e)), has nearly uniform acceptance, good resolution, and a negligibly small background.

Fig. 52 shows the corresponding $K^{\pm}K_s^0\pi^{\mp}$ invariant mass distribution. After a small signal at ~ 1.28 GeV/c^2 which is attributed to the production of $f_1(1285)$, the mass spectrum shows a sharp rise right at $K^*\bar{K}$ threshold, followed by peaks at ~ 1.5 and ~ 1.85 GeV/c^2, close to the positions of the leading $L=1$ and $L=2$ $s\bar{s}$ resonances (the $f_2'(1525)$ and the $\phi_3(1860)$, respectively). However, several states are expected in these mass regions, so these peaks cannot be attributed to specific states without an amplitude analysis. The threshold rise appears to be substantially sharper than would be expected for $K^*\bar{K}$ phase space, but a detailed understanding of this behavior requires the partial wave analysis (PWA) discussed below.

Both final states are dominated by the production of of K^* and $\bar{K}^*$ isobars at all masses, while there is no significant structure attributable to the $a_0(980)$ (formerly known as the δ). This can be seen in the Dalitz plots in Fig. 52(b) for the low mass region $1.34 \leq M_{K\pi\pi} \leq 1.46$ GeV/c^2, and in Fig. 52(c) for the region $1.46 < M_{K\pi\pi} \leq 1.58$ GeV/c^2. The amounts of K^* and $\bar{K}^*$ are substantially different. Moreover, the $K^*\bar{K}^*$ overlap region at lower mass is somewhat depleted, indicating a destructive interference effect, while at higher mass there is a hint of constructive interference. For an $I=0$ state, a pure $C=-1$ resonance would show destructive interference while for a $C=+1$ state the interference would be constructive. In any event, the unequal K^* and $\bar{K}^*$ production makes it clear that the region below ~ 1.6 GeV/c^2 is not dominated by the production of a single resonance.

The SLAC-LBL three-body PWA[31] is used to study the spin-parity structure of the $K\bar{K}\pi$ system. The data are fit using a set of partial wave amplitudes that are specified by the quantum numbers $J^P M^{\eta}$ (isobar)L : J^P is the spin parity of the $K\bar{K}\pi$ system, M is the z projection of J, η is the naturality of the t-channel exchange, isobar is the two-body isobar combination ($K^*(892), \bar{K}^*(892), a_0(980)$...), and L is the orbital angular momentum between the isobar and the bachelor K or π. The two reactions are analyzed separately and significant isobars are parametrized by relativistic Breit-Wigner line shapes with nominal resonance parameters and energy dependence widths.[32] Figs. 52(b) and 52(c) show no clear indication for $a_0(980)$ production; since states decaying to $a_0(980)\pi$ have been observed in this mass region in other experiments,[11] this isobar is included, and is described by a parametrization from a coupled channel analysis of the $\eta\pi$ and $K\bar{K}$ final states.[33] In addition to these isobars, the final wave set includes an incoherent three-body phase space term to absorb contributions from events that are not attributable to isobar production.

The wave set used to obtain the final amplitudes is chosen by an extensive iterative search procedure which uses many different wave sets and starts from different initialization values for the parameters. Only waves which significantly improve the likelihood are retained. Inconsistencies between wave sets and amplitudes obtained in adjacent bins are resolved by requiring continuity in mass. The number of events required to obtain stable results is rather large, so the fits are performed in overlapping 120 MeV/c^2 intervals. Given the amplitude structure and the limited number of events, the analysis produces little significant phase information.

The acceptance-corrected spin-parity intensities required to describe the $K\bar{K}\pi$ system below 2.0 GeV/c^2 are shown in Fig. 53. The corrected mass distribution, displayed by the points in Fig. 53(a), exhibits the same features as the raw mass spectrum of Fig. 52(a). This is shown explicitly by the histogram of Fig. 53(a), which is raw mass spectrum, in 120 MeV/c^2 bins, corrected by the average geometrical acceptance; clearly, the observed structure is unaffected by the correction procedures. As we can see in Figs. 53(b)-(h), the 1^+ amplitude is dominant below 1.6 GeV/c^2, contributing $\sim 70\%$

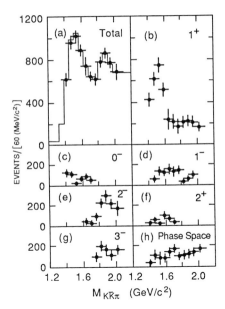

Fig. 53. The intensity distributions corresponding to the partial wave decomposition of the $K\bar{K}\pi$ system produced in the reactions $K^-p \to K_s^0 K^{\mp}\pi^{\pm}\Lambda$.

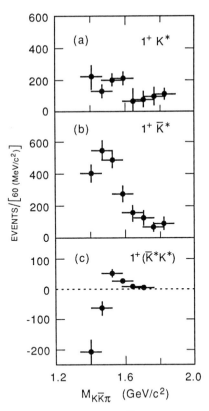

Fig. 54. The K^* and $\bar{K}^*$ intensities for the $J^P = 1^+$ wave, and the interference between the K^* and $\bar{K}^*$ amplitudes as a function of $K\bar{K}\pi$ mass.

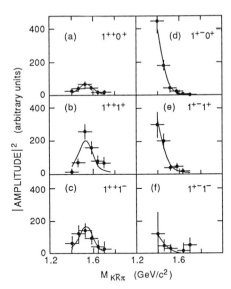

Fig. 55. The production amplitude intensity distributions from the reactions $K^-p \to K_s^0 K^{\mp}\pi^{\pm}\Lambda$ for the $K\bar{K}^*$ and $\bar{K}K^*$ G-parity eigenstate combinations; (a-f) are labelled by $J^{PG}M^\eta$, where M is the helicity and η the naturality of the t-channel exchange.

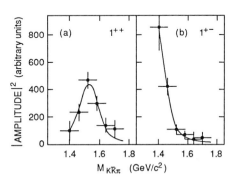

Fig. 56. The $J^P = 1^+$ G-parity eigenstate amplitudes squared. The curves are explained in the text.

of the total intensity at the peak and the $0^- a_0(980)$ amplitude is small. At higher mass, the structure near 1.9 GeV/c^2 results primarily from the 2^- and 3^- amplitudes. The other amplitudes are relatively small and featureless.

Only the low mass 1^+ amplitudes will be discussed in this paper.

The clear peak in the 1^+ intensity distribution, which is suggestive of resonance behavior, cannot be described as a single resonance. Figs. 54(a) and (b) show that the magnitude of the $1^+ \bar{K}^*$ intensity is about twice that of $1^+ K^*$ below 1.7 GeV/c^2; a single resonance should decay equally into both channels. Moreover, the interference term between the $1^+ K^*$ and $1^+ \bar{K}^*$ amplitudes, shown in Fig. 54(c), is negative (denoting destructive interference) at $K^* \bar{K}$ threshold but becomes positive (denoting constructive interference) at $\sim$ 1.52 GeV/c^2. These features of the 1^+ amplitude corroborate those suggested by the Dalitz plots of Fig. 52. Since, in the decay of a resonance with definite G-parity, the relative sign between between $|K^* \bar{K}\rangle$ and $|\bar{K}^* K\rangle$ is positive (negative) for G positive (negative) states, the observed interference structures can be qualitatively explained as resulting from two objects, one with G=$-$1 near threshold, and the other with G=$+$1 near 1.5 GeV/c^2. In such a model, the unequal production of K^* and $\bar{K}^*$ results from the interference between these two opposite G-parity objects with the appropriate production phase.

The amplitudes determined above can be combined to form G-parity eigenstates. The resulting intensity distributions are shown in Fig. 55 for the separate exchanges and in Fig. 56 for the total intensities. The $J^{PG} = 1^{++}$ amplitudes peak at $\sim$ 1.52 GeV/c^2 while the $J^{PG} = 1^{+-}$ amplitudes is strongly peaked toward $K^* \bar{K}$ threshold. In peripheral hypercharge-exchange production of the $K\bar{K}\pi$ final state as seen here, it is natural to expect that production of $I = 0$ $s\bar{s}$ mesons dominates; this implies that G-parity and the charge-conjugation quantum number C should coincide. These expectations are reinforced by the observation of strong $s\bar{s}$ meson production in the reactions $K^- p \to K\bar{K}\Lambda$.[29] It follows that the simplest explanation of the data are that two $I = 0$ $s\bar{s}$ resonances are being observed; one in the $J^{PC} = 1^{+-}$ amplitude near threshold and the other in the $J^{PC} = 1^{++}$ amplitude near 1.52 GeV/c^2. Fits to the total 1^{++} and 1^{+-} intensity distributions using S-wave Breit-Wigner forms yield the parameters shown in Table 23 and the curves of Fig. 56.

Given the wide bin size required to perform the partial wave analysis, it is useful to verify that a simplified model which includes these two resonances is also capable of producing the sharp structure seen in the raw mass distribution. Fig. 52(a) has been fitted by a model which attempts to incorporate the most important features of the observed amplitudes and contains the following ingredients; (1) an $f_1(1285)$ resonance with mass and width fixed at the PDG values;[11] (2) two resonances to represent the 1^{+-} and 1^{++} states observed in Fig. 56; and (3) a polynomial representing incoherent $K^* \bar{K}$ background. The resonances are represented by S-wave Breit-Wigner forms; a fit assuming no coherence between the resonances represents the data well, as shown by the curve in Fig. 52(a). Fits have also been performed with different coherence assumptions and with different background forms. These fits represent the data equally well, and the resulting parameter values are quite consistent with those given in Table 23. In general, two resonances are required to account for the rapid rise at $K^* \bar{K}$ threshold and for the structure at 1.52 GeV/c^2.

Table 23. Breit-Wigner Fits to the Axial ($s\bar{s}$) Mesons.

Amplitude	Mass (GeV/c^2)	Width (GeV/c^2)
1^{++}	1.53 ± 0.01	0.10 ± 0.04
1^{+-}	1.38 ± 0.02	0.08 ± 0.03

We have seen evidence for two axial-vector resonances decaying to $K^*\bar{K}(+$ c.c.): one has mass ~ 1530 MeV/c^2, and quantum numbers $J^{PC} = 1^{++}$, in good agreement with the parameters of the D',[28] while the other has mass ~ 1380 MeV/c^2, width ~ 80 MeV/c^2, and quantum numbers $J^{PC} 1^{+-}$.

These states are good candidates to be the mostly $s\bar{s}$ members of a 1^{++} nonet which also includes the $a_1(1270)$, the $f_1(1285)$, and the K_{1a}, and a 1^{+-} nonet which also includes the $b_1(1235)$, the $h_1(1190)$, and the K_{1b}. The precise values obtained for the mixing angles in these nonets depend on the masses assumed for K_{1a} and K_{1b} which may be a mixture of the physically observed $K_1(1400)$ and $K_1(1280)$ states. A recent model[34] suggests that K_{1a} and K_{1b} may be close to the physical $K_1(1400)$ and $K_1(1280)$ states, respectively. The quadratic mass formula then implies that the nonets are almost ideally mixed. In contrast, earlier analyses of the decay properties of the K_1 states[35] suggest that the physical states are mixed in such a way that that K_{1a} and K_{1b} both lie at about 1340 MeV/c^2. With this value, the quadratic mass formula gives a mixing angle of $\sim 55°$ for the 1^{++} nonet, implying that the $f'_1(1530)$ is mainly $s\bar{s}$ whereas the $f_1(1285)$ has only $\sim 10\% s\bar{s}$ content. For the 1^{+-} nonet, the quadratic mass formula yields a mixing angle $\sim 10°$, with the implication that the $h'_1(1380)$ is $\sim 80\% s\bar{s}$, whereas the $h_1(1190)$ is only $\sim 20\% s\bar{s}$. Even though this model yields nonets which are not ideal, the mixing angles are compatible with the observed production properties of these states.

It follows that the two axial vector states discussed above are good candidates to be the mostly $s\bar{s}$ members of the ground state 1^{++} and 1^{+-} nonets predicted by the quark model. The $f_1(1420)$ no longer has a place in such a scheme, thus it may not be a simple $q\bar{q}$ state, but rather an object of quite a different kind.

d) **Summary**

We have seen:

- The leading $s\bar{s}$ states with $L = 0, 1, 2$ and 3 units of orbital angular momentum, giving rise to the J^{PC} states 1^{--}, 2^{++}, 3^{--} and 4^{++}.

- The $f'_0(1530)$ a 0^{++} state, lying under the $f'_2(1535)$. This is a candidate for the $(L-S)$ triplet $q\bar{q}$ state. A similar degeneracy of states is observed in the K^* sector with K^*_2, Q_A, Q_B and K^*_0 all about the same mass, 1430 MeV.

- Two candidates for the axial mesons with $J^{PC} = 1^{++}$ and 1^{+-}. They are called the $h'_1(1380)$ and $f'_1(1530)$.

- No sign, whatsoever, of the $\theta(1720)$. This state is probably **not** a $q\bar{q}$ meson!

Table 24 and Fig. 57 summarizes the status for the $s\bar{s}$ mesons, while Fig. 58 represents the various level transitions studied.

Table 24. The $s\bar{s}$ States.

J^P	M (MeV)	Γ (MeV)
1^-	1019.6 ± 0.1	4.5 ± 0.3
0^+	~ 1530	~ 100
1^{++}	1530 ± 10	100 ± 40
1^{+-}	1380 ± 20	80 ± 30
2^+	1527 ± 4	90 ± 12
2^-	~ 1860	~ 180
3^-	1851 ± 9	66 ± 29
4^+	2209 ± 27	~ 60

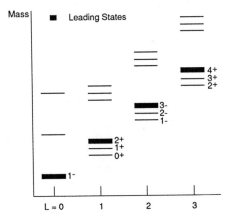

Fig. 57. Grotrian plot for $(s\bar{s})$ states.

5. SOME COMMENTS ON GLUONIUM

a) Introduction

The non-abelian character of QCD motivates the search for bound states of multigluons, called gluonium or *glueballs*. These states may exist as two gluon or three or more gluon states. See Ref. 25 for a recent review of the spectroscopy.

For two gluons bound in a glueball, the spin-space-color symmetry leads to the following expectations for the spin-parity of the state with a given angular momentum, L, between the two gluons, as given in Table 25. The quantum numbers outlined by the small squares in Table 25 are exotic. That is, they cannot be made by a normal $q\bar{q}$ meson system. Discovery of such a state would be a "smoking gun" indicator of the existence of non-$q\bar{q}$ mesonic matter. A candidate for such an exotic state has been reported, and will be discussed below.

We expect gluonium to:
- decay strongly,
- decay with the properties of an SU(3) singlet,
- decay in a flavor independent manner,
- be strongly favored in radiative ψ decays (since the decay should proceed dominantly through a three gluon intermediate state).
- *not* be strongly produced in $\gamma\gamma$ collisions.

Table 25. Quantum number for glueballs.

L	J^P
0	$0^{++}, 2^{++}$
1	$0^{-+}, \boxed{1^{-+}}, 2^{-+}$
2	$2^{++}, 0^{++}, 4^{++}$
3	$2^{-+}, \boxed{3^{-+}}, 4^{-+}$

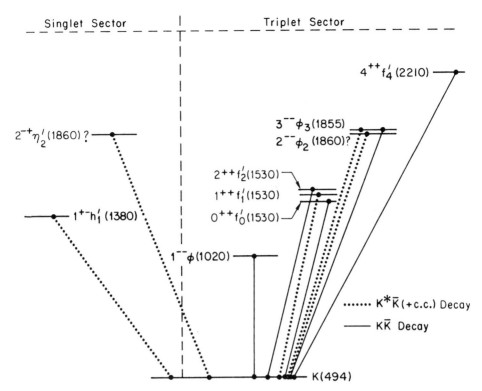

Fig. 58. The various transitions between $(s\bar{s})$ levels studied in the LASS experiment.

The gluon mass scale expected from the naive BAG model calculations[36] was in the range around 1 GeV for the ground state 0^{++} and 2^{++} mesons, and around 1300 MeV for the 0^{-+} and 2^{-+} states. Recent lattice calculations[37] have increased the expected mass values, such that the scalar is now expected around 1500 MeV and the tensor meson between 2000 and 2500 MeV.

The main candidates for the glueball states today are:
- $\iota/\eta(1440)$ seen in $K\bar{K}\pi$,
- $E/f_1(1420)$ seen in $K\bar{K}\pi$,
- $\theta/f_2(1720)$ seen in $K\bar{K}$,
- $G_T/f(2010)$ seen in $\phi\phi$,
- $G_T/f(2300)$ seen in $\phi\phi$,
- $G_T/f(2340)$ seen in $\phi\phi$.

Of these, the strongest claim must be that of the $\theta/f(1720)$.

Let us examine the experimental situation, in each of these areas.

b) **The E Story**

By E, we mean the $J^{PC} = 1^{++}$ state seen in $K\bar{K}\pi$ with mass around 1400 MeV.

The E is observed by the Mark II and TPC/2γ experiments in two photon collisions where one photon is slightly off the mass (i.e., $\gamma\gamma^* \to K\bar{K}\pi$), but not at all in real $\gamma\gamma$ collisions (i.e., $\gamma\gamma \to K\bar{K}\pi$) (see Ref. 38).

The E is observed by the Mark III experiment in hadronic ψ decays recoiling against an ω meson, but *not at all recoiling* against a ϕ meson.[39] No sign of the E is observed in the LASS experiment studying $K\bar{K}$ and $K\bar{K}\pi$ production in K^-p collisions at 11 GeV.[40]

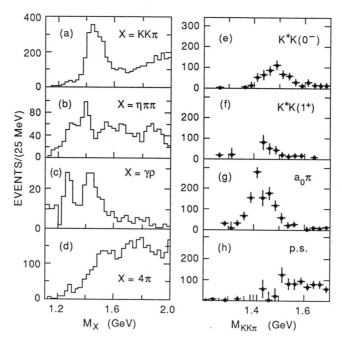

Fig. 59. The mass distribution for the hadronic final state X for the radiative decays $J/\psi \to \gamma X$ as observed in the MARK III experiment, (a)–(d). The partial wave decomposition of the $K\bar{K}\pi$ amplitudes from a Dalitz plot fit, from the same experiment, (e)–(h).

In the WA76 experiment at CERN, the E is observed in central collisions of $\pi^+ p$ and pp at 85 GeV/c.[41]

The lepton-F experiment studied $\phi\gamma$ production in the reaction $\pi^- p \to \phi\gamma n$ at 32.5 GeV at Serpukov. The $\phi\gamma$ decay mode should be a good probe of the $s\bar{s}$ content of mesons. Indeed strong production of the D (the $J^{PC} = 1^{++}$ meson at 1280 MeV, that is reckoned to be 10% $s\bar{s}$ quark) is observed, but no sign of the E !

In conclusion, there clearly is a meson state coupling strongly to $K\bar{K}\pi$ with a mass of about 1420 MeV and with $IJ^{PC} = 01^{++}$, which we call the E meson. The above observations argue that the E is unlikely to be an $s\bar{s}$ quark-antiquark state, but it could be a glueball, a four quark state or a $q\bar{q}$ molecule. If it is a gluonium state, it cannot be a $2g$ state—see the selection rules of Table 25 and it is not very likely to be a $3g$ state so close to threshold.

A stronger conclusion awaits further data and more serious analysis.

c) **The Iota Story**

By the iota we mean the broad bump in $K\bar{K}\pi$ and $\eta\pi\pi$ observed in the range (1350-1550) MeV and dominantly with $J^{PC} = 0^{-+}$.

This state was discovered in the CERN experiment studying stopping $\bar{p}p$ interactions:

$$\bar{p}p \to (K\bar{K}\pi)\pi\pi$$

in which they observed a peak in $K\bar{K}\pi$ at 1420 MeV with $J^{PC} = 0^{-+}$.[42]

The Mark III experiment observe the state in radiative ψ decays, and in hadronic ψ decays recoiling from vector mesons. See Fig. 59.

A partial wave analysis of the reaction $\bar{p}p \to (\bar{K}K\pi)X$ at 6 GeV, in a BNL experiment finds a peak in the $K\bar{K}\pi$ $J^{PC} = 0^{-+}$ amplitude around 1420 MeV.[43] A similar analysis of the reaction $\pi^-p \to \eta\pi^+\pi^-$n at 8 GeV in a KEK experiment finds clear signals of resonant behavior in the $\eta\pi^+\pi^-$ system in the 0^{-+} partial wave at energies of 1280 MeV and 1420 MeV.[44]

A BNL study of $\pi^-p \to K\bar{K}\pi$n at energies of 6 and 8 GeV sees a clear signal of $K^*\bar{K}$ and $\delta\pi$ resonance around 1420 MeV with $J^{PC} = 0^{-+}$.[45]

Finally the LASS experiment sees no sign of resonances in 0^{-+} for the $K\bar{K}\pi$ system.

The conclusion drawn from these data is that the 0^{-+} state at 1280 MeV should be thought of as the first radial excitation of the η meson, while the state at 1420 MeV may be a gluonium state, a hybrid meson, or the first radial excitation of the η'.

d) The Theta Story

The $\theta/f(1720)$ is the bump at 1720 MeV observed in $K\bar{K}$ with spin-parity 2^{++}.

The Mark III experiment has studied this state in radiative ψ decays, which is supposedly a gluonium rich environment.

$$\psi \to \gamma\theta$$

where θ decays into $\eta\eta$, $K\bar{K}$ and $\pi\pi$. See Fig. 60.[46] Clear signals have been identified for $K\bar{K}$ (B.R. = 5.10^{-4}), and $\pi\pi$—(B.R. = 1.10^{-4})—but no signal is observed for $\eta\eta$. The mass and width are reported as $M = 1720$ MeV and $\Gamma = 132$ MeV.

The LASS experiment, as described in section 4 above, sees no sign of the θ in $K\bar{K} \to K\bar{K}$ or in $K\bar{K}^* \to K\bar{K}$ in K^-p reactions at 11 GeV.

It is concluded that the θ does not strongly couple to $s\bar{s}$, and is probably not a $q\bar{q}$ meson. This state is the best candidate for a glueball that we have to date.

e) The $\phi\phi$ States from BNL

The MPS at BNL has studied the reaction $\pi^-p \to \phi\phi$n at 22 GeV, and isolated a clean sample of about 7000 such events. A partial wave analysis of the $\phi\phi$ system uncovers resonant behavior in the $J^{PC} = 2^{++}$ amplitude at three masses —$M = 2011$ MeV with $\Gamma = 202$ MeV, $M = 2297$ MeV with $\Gamma = 149$ MeV and $M = 2339$ MeV with $\Gamma = 315$ MeV.[47] See Fig. 61.

These states are candidates for glueballs.

f) The GAMS Experiment States.

The GAMS experiment at CERN and Serpukov has observed three separate states which may well be exotic mesons.

First, the $C(1590)$ with $J^{PC} = 0^{++}$ and decaying into $\eta\eta$ and $\eta\eta'$ [48], the $C(1755)$ which is a 0^{++} or 2^{++} state coupling to $\eta\eta$,[49] and the $M(1406)$ which couples to $\pi\eta$, and has explicitly exotic quantum numbers, $J^{PC} = 1^{-+}$![50]

The latter effect is observed in the reaction $\pi^-p \to \pi^0\eta$n at 100 GeV. The reaction gives rise to 4γ final states, which are detected in a lead–glass calorimeter array. The γ's are paired to find the π^0 and η combinations. A study of the forward to backward production of the η's around a $\pi\eta$ mass of 1400 MeV shows a dramatic effect—see Fig. 62—which is interpreted as the interference between the (very) large resonant A_2 amplitude and a new resonance in the 1^{-+} amplitude.

Chanowitz claims that this state may be a hybrid meson[51] while Close and Lipkin say it is a multi-quark state.[52]

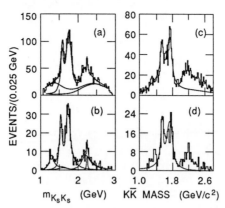

Fig. 60. (a) The K^+K^- mass spectrum measured in the decay $J/\psi \to \gamma K^+K^-$ by MARK III, with a coherent fit to five Breit–Wigner resonances. (b) The $K_s^0 K_s^0$ mass spectrum measured in the decay $J/\psi \to \gamma K_s^0 K_s^0$ by MARK III, with an incoherent fit to five Breit–Wigner resonances. (c)(d) The K^+K^- and $K_s^0 K_s^0$ mass spectra measured in the radiative J/ψ decay in the DM2 experiment, with a fit to two Breit–Wigners in the (1500 - 1800) MeV region.

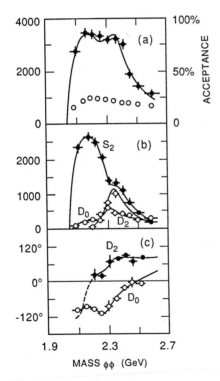

Fig. 61. (a) The acceptance corrected $\phi\phi$ mass spectrum, (b) intensity, and (c) phase difference for the three $J^{PC} = 2^{++}$ waves. The curves show the fit to three Breit–Wigner resonances.

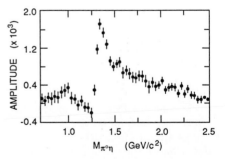

Fig. 62. The forward-backward ratio of the η relative to the incident π^- direction as evaluated by the GAMS experiment

(g) Lepton-F Experiments

The Lepton-F collaboration have observed a resonance the—$C(1480)$—in a study of the reaction $\pi^- p \to \pi^0 \phi n$ at 32 GeV.[53] The isovector meson has $J^{PC} = 1^{--}$ and $M = 1480$ MeV with $\Gamma = 130$ MeV. No sign of coupling to $\pi\omega$ is observed and a regular vector meson should have favored decay with $\omega\pi$ over $\phi\pi$ —for this reason it is considered an exotic meson candidate. It could be a hybrid state or a four quark state.

h) The CERN WA62 State

A heavy, narrow state with strangeness -1 has been observed in hyperon experiments at CERN—the WA62 experiment[54]—and in the Serpukov BIS-2 experiment.[55]

The reactions studied are:

$$\Sigma^- + Be \to (\Lambda \bar{p} + n\pi's) + X \text{ at 135 GeV}$$

and

$$n + Z \to (\Lambda \bar{p} + n\pi) + X \text{ at 40 GeV}$$

Very narrow peaks are observed in both experiments at a mass of 3100 MeV with charge states of $+1$, 0, and -1 being separately observed. They are candidates for multiquark states.

i) Conclusion

In Table 26, we draw together all of the candidates for exotic mesons, and try to categorize them.

6. A BRIEF VISIT TO HEAVY QUARK SPECTROSCOPY

a) Onia

Research in heavy quark meson spectroscopy, divides itself between the study of quarkonia—the states with hidden flavor ($s\bar{s}$ strangeonium, $c\bar{c}$ charmonium, $b\bar{b}$ bottomnium), and the study of states with open-flavor—the charm, charm-strange and bottom mesons.

Table 26. The Odd Balls.

J^{PC}	Name	Status	Probable Assignment
0^{-+}	$\eta/\iota(1420)$	good	gluon
1^{++}	$f_1/E(1440)$	good	multiquark
2^{++}	$f_2/\theta(1720)$	excellent	gluon
2^{++}	$f_2/G_T(2010)$	OK	gluon
2^{++}	$f_2/G_T(2300)$	OK	gluon
2^{++}	$f_2/G_T(2340)$	OK	gluon
0^{++}	$f_0/G(1590)$	needs confirmation	gluon or $f_0'(1535)$
1^{--}	Exotic/$C(1480)$	needs confirmation	hybrid or multiquark
$0^+/2^+$	$f_0, f_1/C(1755)$	needs confirmation	gluon (could be $\theta(1780)$)
1^{-+}	Exotic/(1300)	needs confirmation	hybrid
?	Exotic/$U(3100)$	OK	strange multiquark state

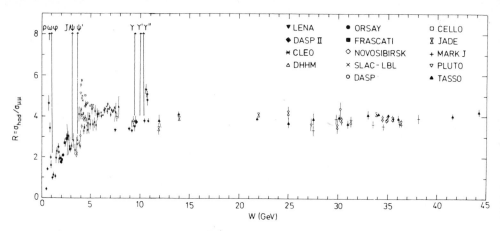

Fig. 63. The ratio of $R = \sigma(e^+e^- \to \text{hadrons})/\sigma_{\mu\mu}$ where $\sigma_{\mu\mu} = 4\pi\alpha^2/3s$.

The vector nature of the e^+e^- annihilation process provides a powerful tool for detailed study of the onia states. They are directly produced in the e^+e^- process with large cross section and with very little background. The narrow J/ψ state—the first heavy quark meson was discovered jointly at SPEAR and at BNL. The first radial excitation of the ψ—the $\psi'(3770)$ was observed at SPEAR shortly after the ψ discovery. Since then, the charm and upsilon spectroscopy has been a healthy cottage industry of the e^+e^- storage rings at Stanford, Hamburg and Cornell, and the spectroscopy has been rather well explored. A composite of the excitation function is shown in Fig. 63, where the ratio of the total hadronic annihilation cross section normalized to the point cross section for $e^+e^- \to \mu^+\mu^-$, is shown as a function of center of mass energy of the e^+e^- collision.[56]

The spectroscopy is summarized in Fig. 64, for the charm sector, and in Fig. 65 for the bottom sector.

Radiative transitions within the onia system probe the $q\bar{q}$ wave function at distances away from the origin. This part of the potential describes the confinement of the quarks within the meson. Light quark decays, on the other hand, give information on the value of the wave function at the origin. Measurement of the hadronic partial widths of meson decays also determines this part of the potential. Such studies probe the behavior of the $q\bar{q}$ wave function, and together with detailed studies of the onia energy levels (which determine the potential at intermediate distances) should allow the development of a coherent picture of the $q\bar{q}$ interaction. See Fig. 66.

In as much as the quarks making up the onia state are really heavy, then we may hope that relativistic effects are small, and that a description based on Schrödinger's equation may explain the quark binding, namely,

$$\left[-\frac{1}{M_q} \cdot \nabla^2 + V(r)\right]\psi(r) = E\psi(r)$$

There has been substantial activity in attempting to choose the best form of the potential, $V(r)$, and in determining the free parameters of that potential from fits to the data. A number of successful potentials are given in Table 27. The fits are set within the range of quark separations of 0.1 to 1 fermi (i.e., where the data exist), and

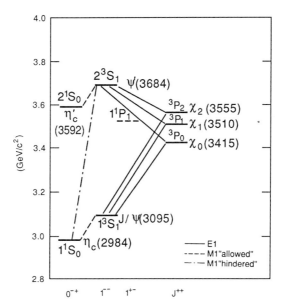

Fig. 64. Spectrum of the $c\bar{c}$ family.

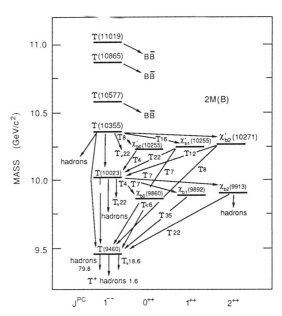

Fig. 65. Spectrum of the $b\bar{b}$ family.

Table 27. Potentials used to describe the heavy quarkonium systems.

Author	Reference	Potential
Otto and Stack	57	– – –
Quigg and Rosner	58	$V(r) = c \ell n \left(\frac{r}{r_0} \right)$
Martin	59	$V(r) = A + B r^\epsilon$
Eichten et al.	60	$V(r) = \frac{4}{3} \frac{\alpha_s}{r} + kr$
Appelquist et al.	61	– – –
Richardson	62	$V(q^2) = \frac{4}{3} \frac{12\pi}{33-2n_f} \frac{1}{q^2 \ell n \, [q^2/(\Lambda^2+1)]}$

all of the successful potentials, of course, look very similar within these limits, and then strongly diverge for smaller and larger quark separations. When discovered, toponium will provide a very rich laboratory to understand the quark potential.

Examples of the fits are shown in Fig. 67 and 68 together with an indication of which states constrain the potential over what distances.

The radiative widths of onia states may be quantitatively described by, for example,

$$\Gamma(\psi' \to \gamma \chi_c) = \frac{4}{9} \cdot \frac{2J_f + 1}{2J_i + 1} \cdot q^2 \, \alpha \, k^3 \, |E_i f|^2$$

where q is the charge of the quark, α is the fine structure constant, k is the energy of the radiated photon, J is the total spin of the initial/final system and E is the transition matrix element.

Figure 69 shows the data from the Crystal Ball experiment, with clear signals for almost all the lower $q\bar{q}$ transitions.[63]

b) Open–Flavor Mesons

To set the stage, I remind us of the Grotrian plot that we have used to described the K^* and $s\bar{s}$ mesons in sections 3 and 4 above. See Fig. 70.

The pseudo-scalar states (i.e., $J^{PC} = 0^{-+}$), for charmed and charm-strange mesons are shown in Fig. 71; in Fig. 71(a) the D^0 and D^+ signals from the Mark II experiment at SPEAR,[64] and in Fig. 71(b) the D_s^0 signal from the Mark III experiment at SPEAR.[65] These are the heavy quark ground state mesons corresponding to S-wave $q\bar{q}$ pairing for $c\bar{c}$ and $c\bar{s}$ systems, with anti-parallel quark spins.

The vector states (i.e., $J^{PC} = 1^{--}$), corresponding to the parallel quark spin pairing, for the charmed and charm-strange mesons are shown in Fig. 72 and 73. The D^* data are from the ARGUS experiment at DESY,[65] and the D_s^* data are from the Mark III experiment at SPEAR.[66]

The $L = 1$ $q\bar{q}$ states are examined next. Fig. 74 shows the $D^*\pi$ mass spectrum from the ARGUS experiments.[67] The bump around 2400 MeV is resolved into two peaks at 2420 and 2470 MeV, and a crude analysis of the angular correlations indicate that the higher mass state could be the $J^{PC} = 2^{++}$ state and the lower mass peak may be a 1^{++} meson.

In Fig. 75 an analogous study of the charm–strange state is shown from the ARGUS experiment.[68] The $D^*K_s^0$ mass spectrum shows a peak at 2535 MeV and is ascribed to the expected 2^{++} state.

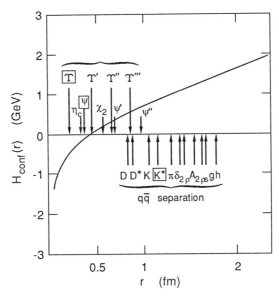

Fig. 66. The radial dependence of the $q\bar{q}$ potential. The part of the potential explored by different meson states is also indicated.

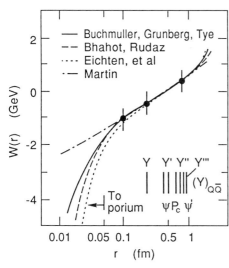

Fig. 67. The fit to the radial dependence of the $q\bar{q}$ potential for some specific parametrizations. See text for details.

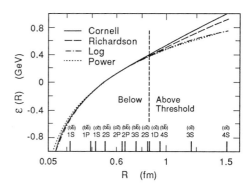

Fig. 68. Some more fits to the radial dependence of the $q\bar{q}$ potential.

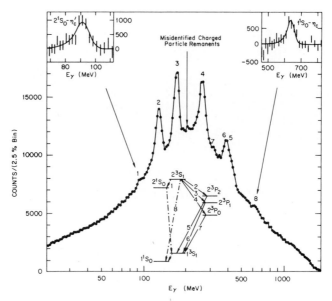

Fig. 69. Inclusive photon spectrum at the ψ' obtained by the Crystal Ball experiment. Note that the logarithmic energy scale yields bin sizes approximately proportional to photon energy resolution. The numbers over the spectrum correspond to the expected radiative transitions shown in the spectrum inset.

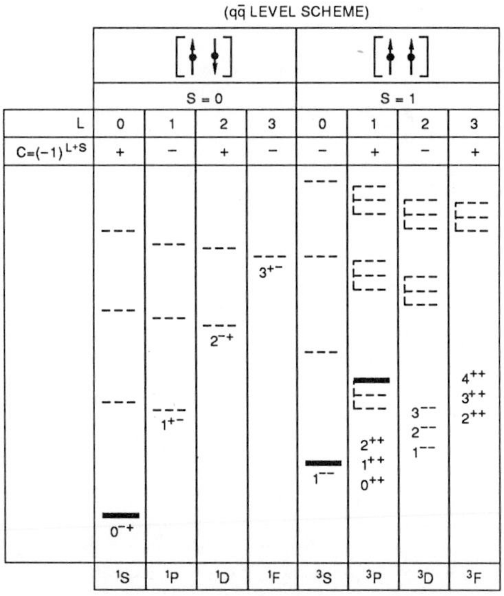

Fig. 70. The Grotrian plot, showing the levels for the charmed and charm–strange mesons that have been observed to date.

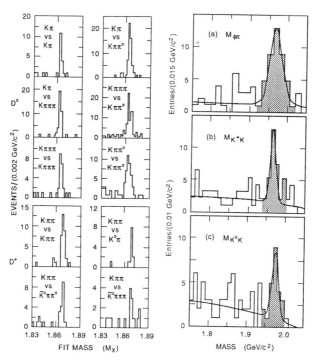

Fig. 71. (a) The mass spectra for $K\pi$, $K\pi\pi\pi$ and $K\pi\pi$ showing the D^+ and D^0 mesons from the Mark III experiment. (b) The mass spectra for $\phi\pi$, $K^*\bar{K}$ and $K^*\bar{K}$ showing the D_s meson, also from the Mark III experiment.

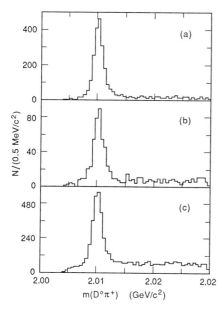

Fig. 72. The $D^0\pi^+$ mass spectra showing the vector D^* state, from the ARGUS experiment.

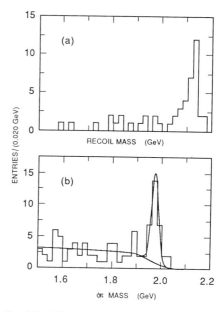

Fig. 73. The $\phi\pi$ mass spectrum for the Mark III experiment, showing the vector D_s^* state.

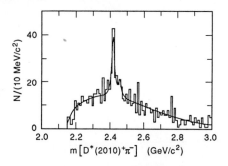

Fig. 74. The $D^{*+}\pi^-$ mass spectrum from the ARGUS experiment showing a peak at 2470 MeV which may be the tensor D^* state.

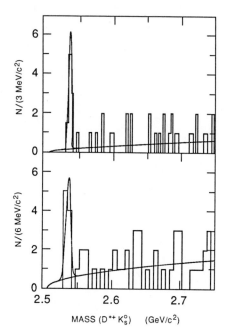

Fig. 75. The $D^{*+}K_s^0$ mass spectrum from the ARGUS experiment, showing a peak at 2535 MeV which may be the tensor D_s^* state.

c) **Summary**

Figs. 76 through 80 show the Grötrian plots for the $c\bar{c}$, $c\bar{u}$, $c\bar{s}$, $b\bar{b}$, $b\bar{u}$ meson states that have been observed so far.

The field is still very active, and we may expect to see experimental progress from several fronts. The new e^+e^- storage ring at Beijing is now operating at the ψ energy with substantially higher luminosity than SPEAR. We may expect more information on the charm and the charm–strange sectors from there. The studies on b meson spectroscopy will continue at DESY and CORNELL. The new CLEOII experiment and the improved luminosity at CESR promise new advances in the b sector. Beyond that, new machines are being proposed to push back the luminosity frontier—a τ/c Factory in Spain and a b Factory in the US, Japan, Soviet Union or Europe—where a systematic assault on the detailed study of the heavy quark spectroscopy could be performed.

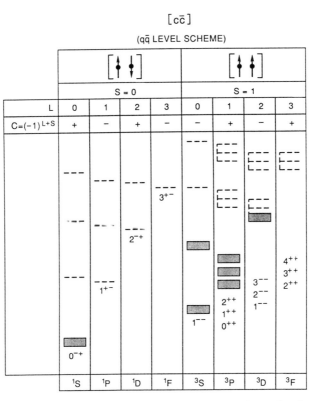

Fig. 76. The Grotrian plot for the $c\bar{c}$ sector mesons that have been clearly experimentally identified.

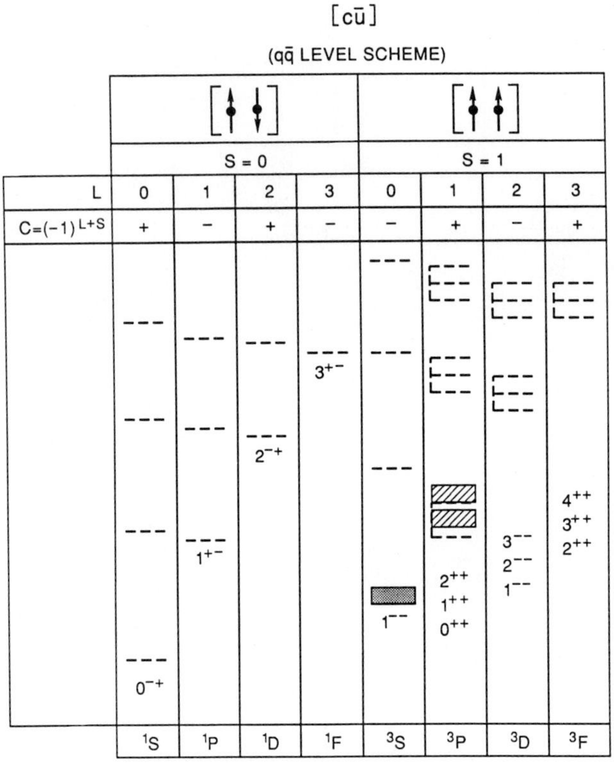

Fig. 77. The Grotrian plot for the $c\bar{u}$ sector mesons that have been clearly experimentally identified.

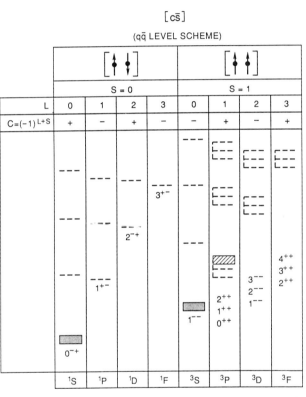

Fig. 78. The Grotrian plot for the $c\bar{s}$ sector mesons that have been clearly experimentally identified.

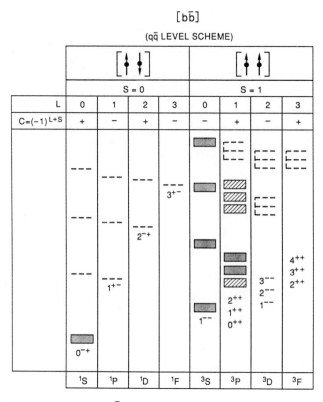

Fig. 79. The Grotrian plot for the $b\bar{b}$ sector mesons that have been clearly experimentally identified.

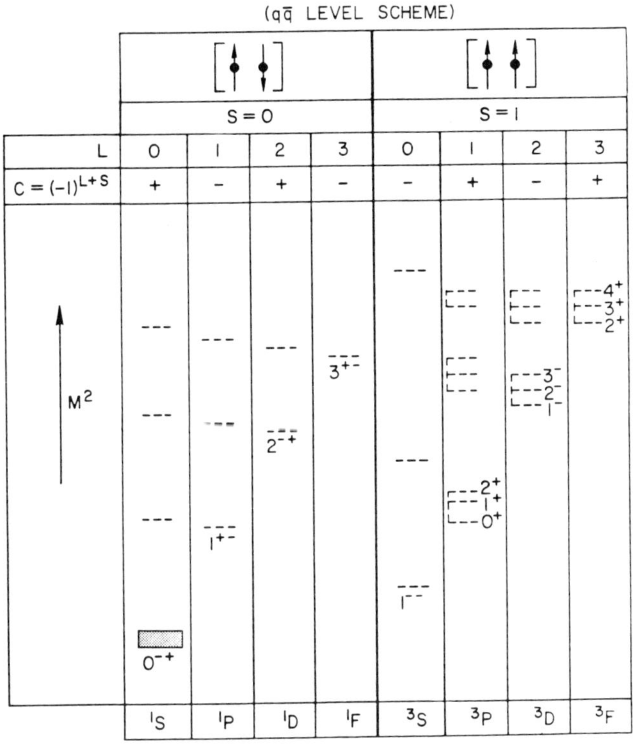

Fig. 80. The Grotrian plot for the $b\bar{u}$ sector mesons that have been clearly experimentally identified.

Lastly, and certainly not least, major short term breakthroughs can be expected from the fixed target experiments at FNAL. With the recently demonstrated power of silicon vertex detectors in the photoproduction of charm at FNAL,[69] two new experiments—one focussing on hadron-production, the other on photoproduction of heavy quark mesons, are gearing up in Chicago. Look for interesting heavy quark results from these experiments. Also, the collider experiments will begin to focus their efforts on the smaller $p_\perp$ physics of b and c spectroscopy now that some of the heat is off in the large $p_\perp$ sector.

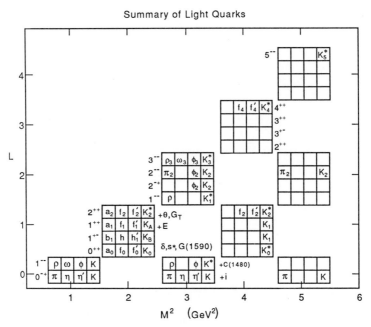

Fig. 81. The current experimental status for mass spectroscopy in the light quark sector.

7. SUMMARY AND CONCLUSIONS

Figure 81 summarizes the light quark meson states that we have been discussing in these lectures. The format in the figure follows that of Fig. 8 in our introduction, where the orbital angular momenta between the quark and the antiquark gives the vertical scale and the mass squared of the meson system, the horizontal scale. Radial excitations appear in this plot as a repeat of the ground state configuration at higher masses.

The $L = 0$ sector is in good shape, and many of the expected first radial excitations are also observed. For the $L = 1$ sector, everything is in good shape except for the scalars.

Prior to the finding of the $f_1(1530)$ and the $f_0'(1535)$, there were candidate states—E, S^*, δ—all of which are probably now to be understood as multi-quark states.[71] The $\theta(1720)$ and the $i(1420)$ are most likely not $q\bar{q}$ states, and are indeed probably the best candidates for glueballs. Some pieces of the radial states have been identified. For the $L = 2$ sector, much of it has been found, and pieces of the $L = 3$ and $L = 4$ states.

In general, the simple valence $q\bar{q}$ model works rather well for classification,

- the predicted regularities are observed in the data;
- no extra states have been found in the K^* sector up to 2400 MeV mass, and almost all of the expected $q\bar{q}$ states are found. This is quite a remarkable data set.

The spin-spin force is observed to be large in the ground state [e.g., $(K^*\ K)$ or (ρ, π) mass difference], but is rather weak for all other states [e.g., (Q_A, Q_B) or (f_1, f_1') mass differences].

The spin-orbit force is either unimportant, or has large internal cancellations for the light quark mesons (e.g., the $L = 1 : 2^+, 1^+, 0^+$ states are essentially degenerate, as are the $L = 2 : 3^-, 2^-, 1^-$ states.)

Beautiful measurements in the $c\bar{c}$ and $b\bar{b}$ systems are allowing a quantitative study of the heavy quarkonia systems.

It is becoming clear that the $q\bar{q}$ description of mesonic matter does not explain all of the data, and that other degrees of freedom are required—probably gluonic and multi-quark states will be shown to exist, although there is not unequivocal evidence to date. (See Table 27 in Section 5).

The relativized $q\bar{q}$ model of Isgur and Godfrey[19] does a fair job of predicting the masses of the meson states from the pion to the heaviest upsilon states. This give hope that some quantitative studies of $q\bar{q}$ effects can be confronted even in the light quark systems. The model employs a Hamiltonian with a potential that depends on both spatial and momentum coordinates. The spin independent part has both a coulombic-type interaction with a running coupling constant and a linear confining part. The spin dependent part has spin-spin, spin-orbit and tensor terms along with Thomas precession and annihilation terms. The relativistic effects are taken into account by (a) smearing the spatial coordinates out over a distance of order the inverse quark mass, and (b) making the coefficients of the various pieces of the potential be dependent on the momentum of the interacting quarks.

Their fit to the meson spectrum is surprisingly good, and is shown in Table 28.

Finally, we revisit the table format introduced in the Introduction (Table X), and fill in all the known meson states for each J^{PC} value, for all of the $q\bar{q}$ flavors that are known. This is given in Tables 29, 30, 31, 32, and 33.

The gaps in these tables will be addressed in the experiments currently underway or being planned, as listed in Table 34.

ACKNOWLEDGEMENTS

I would like to thank the participants—both students and lecturers—at the St.Croix school for a stimulating and enjoyable time, especially Tom Ferbel, for inviting me to take part. I gratefully acknowledge my colleagues in the LASS experiment, especially William Dunwoodie and Blair Ratcliff. Finally, I would like to thank the Publication Department at SLAC and Lilian Vassilian for their careful and patient work in helping put these lecture notes together.

Table 28. Comparison of the meson masses, as calculated by Godfrey and Isgur, and the experimentally measured quantities.

State		Experimental Mass (MeV)	Theory (MeV)	Difference Δ_E (MeV)
J^{PC}	Quark Constituent			
	$[\bar{s}u]$			
0^+		1430 ± 8	1230	200
		1950 ± 24	1890	60
1^-		$896 \pm .6$	904	-8
		1404 ± 28	1574	-170
		1695 ± 31	1775	-80
2^+		1431 ± 1.8	$--$	input
		1973 ± 25	1940	33
3^-		1781 ± 9	1791	-10
		~ 2100	2180	-80
4^+		2064 ± 51	2109	-45
5^-				
	$[s\bar{s}]$			
1^-		1019.6 ± 0.1	1020	0
0^+		~ 1530	1350	180
1^{++}		1530 ± 10	1480	50
1^{+-}		1380 ± 20	1470	-90
2^{++}		1527 ± 4	1520	-7
2^{-+}		~ 1860	1890	-30
3^{--}		1851 ± 9	1897	-46
4^{++}		2209	2210	1

(Continued)

Table 28 (Continued, page 2 of 3)

J^{PC}	State Quark Constituent	Experimental Mass (MeV)	Theory (MeV)	Difference Δ_E (MeV)
	$[c\bar{c}]$			
0^{-+}		2982	2970	11
1^{--}		3096.9	3100	input
1^{--}		3686.0	3680	+6.0
1^{--}		3770.0	3820	+6.0
0^{++}		3414.9	3440	−25
1^{++}		3510.7	34510	input
2^{++}		3556.3	3550	6.3
0^-		1869.3	1880	−10.7
		1864.6		−15.4
1^{--}		2010.1	2040	−29.9
		2007.2		−32.8
1^+		2420.0	2490	−70
2^+		2470.0	2500	−30
0^-		1970.5	1980	−9.5
1^-		2100.0	2130	−30

(Continued)

Table 28 *(Continued, page 3 of 3)*

State		Experimental	Theory	Difference
J^{PC}	Quark Constituent	Mass (MeV)	(MeV)	Δ_E (MeV)
	$[b\bar{b}]$			
1^{--}		9460.0	9460	input
1^{--}		10023.4	10000	23.4
1^{--}		10355.5	10350	5.5
1^{--}		10577.0	10630	-53
1^{--}		10865.0	10880	-15
1^{--}		11019.0	11100	-81
0^{++}		9858.9	9850	8.9
1^{++}		9891.9	9880	11.9
1^{++}		10255.0	10250	5.0
2^{++}		9913.3	9900	13.3
2^{++}		10271.0	10260	11.0
0^{-}		5271.0	5340	-38.8
		5275.2	5340	-34.8

Table 29. The full meson spectrum that has been experimentally identified for the $J^{PC} = 0^{-+}$ state.

J^{PC}	$u\bar{u}, u\bar{d}, d\bar{d}$ $I = 1$	$u\bar{u}, d\bar{d}, s\bar{s}$ $I = 0$	$\bar{s}u, \bar{s}d$ $I = 1/2$	$c\bar{c}$ $I = 0$	$c\bar{u}, c\bar{d}$ $I = 1/2$	$c\bar{s}$ $I = 0$	$b\bar{b}$ $I = 0$	$\bar{b}u, \bar{b}d$ $I = 1/2$	Other
0^{-+}	π	η, η'	K	η_c^{2980}	D	D_s		B	$\iota/\eta(1430)$ Glue or Multiquark
	π^{1300}	η^{1295}	K^{1430}	η_c'					
	π^{1770}		K^{1830}						

Table 30. The full meson spectrum that has been experimentally identified for the $J^{PC} = 1^{--}$ state.

J^{PC}	$u\bar{u}, u\bar{d}, d\bar{d}$ $I = 1$	$u\bar{u}, d\bar{d}, s\bar{s}$ $I = 0$	$\bar{s}u, \bar{s}d$ $I = 1/2$	$c\bar{c}$ $I = 0$	$c\bar{u}, c\bar{d}$ $I = 1/2$	$c\bar{s}$ $I = 0$	$b\bar{b}$ $I = 0$	$\bar{b}u, \bar{b}d$ $I = 1/2$	Other
1^{--}	ρ	ω, ϕ	K_1^{*890}	$J/\psi_{1s}^{3.097}$	D^*	D_s^*	Υ_{1s}^{9460}		$C(1480)$ Hybrid or Multiquark
	ρ^{1450}	ϕ^{1680}	K_1^{*1400}	ψ_{2s}^{3686}			Υ_{2s}^{10023}		
							Υ_{3s}^{10355}		
	ρ^{1700}		K_1^{1715}	ψ^{3770}			Υ_{4s}^{10580}		
							Υ^{10860}		
							Υ^{11020}		

Table 31. The full meson spectum that has been experimentally identified for the 2^{++}, 3^{--} 4^{++}, 5^{--} J^{PC} states.

J^{PC}	$u\bar{u}, u\bar{d}, d\bar{d}$ $I=1$	$u\bar{u}, d\bar{d}, s\bar{s}$ $I=0$	$\bar{s}u, \bar{s}d$ $I=1/2$	$c\bar{c}$ $I=0$	$c\bar{u}, c\bar{d}$ $I=1/2$	$c\bar{s}$ $I=0$	$b\bar{b}$ $I=0$	$\bar{b}u, \bar{b}d$ $I=1/2$	Other
2^{++}	A_2	f_2, f_2'	K_2^{*1430}	χ_2^{3536}	D^{**}	D_s^{**}	χ_{b1}^{9913}		$\theta/f_2(1700)$ $C(1755)$
		$f_2^{1810}, f_2^{\prime 2150}$	K_2^{*1960}				χ_{b2}^{10271}		$f_2(2010, 2300, 2340)$
3^{--}	ρ_3^{1690}	w_3, ϕ_3	K_3^{*1780}						
4^{++}		f_4, f_4' f_4^{2300}	K_4^{*2050}						
5^{--}			K_5^{*2380}						

Table 32. The full meson spectum that has been experimentally identified for the $J^{PC} = 1^{++}$, 1^{+-} 0^{++} states.

J^{PC}	$u\bar{u}, u\bar{d}, d\bar{d}$ $I=1$	$u\bar{u}, d\bar{d}, s\bar{s}$ $I=0$	$\bar{s}u, \bar{s}d$ $I=1/2$	$c\bar{c}$ $I=0$	$c\bar{u}, c\bar{d}$ $I=1/2$	$c\bar{s}$ $I=0$	$b\bar{b}$ $I=0$	$\bar{b}u, \bar{b}d$ $I=1/2$	Other
1^{++}	A_1^{1260}	$f_1^{1285}, f_1^{\prime 1530}$	K_A^{1400}	χ_1^{3511}	D_1^*		χ_{b1}^{9842}		$\epsilon/f_1(1420)$ Multiquark or Glue
							χ_{b2}^{10255}		
1^{+-}	B_1^{1235}	$h_1^{1170}, h_1^{\prime 1380}$	K_B^{1270}						
0^{++}		$f_0^{\prime 1525}$	K_0^{*1430}	χ_0^{3415}			χ_{b0}^{9859}		$S^*/f_0(975)$ $\delta/a_0(980)$ $\epsilon/f_0(1400)$ Multiquark States?
0^{++}									$G(1590)$ gluon

Table 33. The full meson spectum that has been experimentally identified with $J^{PC} = 2^{-+}$ states.

J^{PC}	$u\bar{u}, u\bar{d}, d\bar{d}$ $I = 1$	$u\bar{u}, u\bar{d}, s\bar{s}$ $I = 0$	$\bar{s}u, \bar{s}d$ $I = 1/2$	$c\bar{c}$ $I = 0$	$c\bar{u}, c\bar{d}$ $I = 1/2$	$c\bar{s}$ $I = 0$	$b\bar{b}$ $I = 0$	$\bar{b}u, \bar{b}d$ $I = 1/2$	Other
2^{-+}	π_2^{1670}		K_2^{1770}						Exotic 1^{-+} at 1300
	π_2^{2100}		K_2^{2250}						$U(3100)$

Table 34. Where Can We Look For Breakthroughs?

Experiments	Light Quarks and Glueballs	Charm Mesons	Bottom Mesons
Beijing e^+e^- on ψ, ψ', ψ''	X	X	
MPS with Neutral Spectrometer at BNL	X		
GAMS and Other Spectrometers at Sepukov	X		
KAON Facility in Canada (*if built*)	X		
FNAL Hadron and Photoproduction Experiments		X	X
CLEO II at Cornell	X	X	X
ARGUS at DESY	X	X	X
τ/c Factory in Europe (*if built*)	X	X	
B Factory (*if built*)	X	X	X

REFERENCES

1. N. Isgur, "Proceedings of Glueballs, Hybrid and Exotic Hadrons" Conf., Brookhaven Nat. Lab., 3 (1988).

    F. Close, 421, same proceedings.

    Walter Toki, 692, same proceedings.

    S. Godfrey and H. Willutzki, 703, same proceedings.

2. R. Huff and J. Kirz, Lawrence Rad. Lab. Phys. Note 474, taken from R. D. Tripp Ann. Rev. Nucl. Part. 15:325 (1965).

3. M. Gell-Mann, Phys. Lett 8:214 (1964).

    G. Zweig, CERN Report 8182 (1964).

4. See, for example D. Hitlin "Techniques and Concepts in High Energy Physics IV", T. Ferbel, ed., 275 (1987), and H. J. Schnitzer, Phys. Lett. 134B:253 (1984).

    M. Frank and P. O'Donnell, Phys. Lett. 157B:174 (1985).

5. D. Aston et al., SLAC-Report 298 (1986).

6. D. J. Herndon, P. Söding and R. J. Cashmore, Phys. Rev. D11:3165 (1975).

    D. Aston, T. Lasinski and P. K. Sinervo, SLAC-Report 287 (1985).

7. P. F. Kunz, Nucl. Inst. Meth. 135:435 (1976).

    P. K. Kunz et al. IEEE NS-27:582 (1980).

    P. M. Ferran et al., "Proceedings of the Amsterdam Computer" Conf., 322 (1985).

8. G. Chew and F. Low, Phys. Rev. 113:1640 (1959).

9. P. Estabrooks and A. D. Martin, Nucl. Phys. B95:322 (1975).

10. J. Blatt and V. Weisskopf, "Theoretical Nuclear Physics," Wiley, N.Y., 361 (1952).

11. Review of Particle Properties, Phys. Lett. B170:1-350 (1986).

12. D. Aston et al., Phys. Lett. B106:235 (1981).

13. N. A. Tornquist, Phys. Rev. Lett. 49:624 (1982).

14. P. Estabrooks et al., Part. Rev. D19:2678 (1979).

15. P. Estabrooks et al., Nucl. Phys. B133:490 (1978).

16. D. Aston et al., SLAC-PUB-3972 (1986), Nucl. Phys. B292:693 (1987).

17. D. Aston et al., Phys. Lett. B180:308 (1986).

18. D. Aston et al., SLAC-PUB-4260 (1987), Nucl. Phys. B296:493 (1988).

19. S. Godfrey and N. Isgur, Phys. Rev. D32:189 (1985).

20. A. Bradley, J. Phys. G4:1517 (1978).

    A. Bradley and D. Robson, Z. Phys. C4:67 (1980).

    C. Ayala et al., Z. Phys. C26:57 (1984).

21. M. Atkinson et al., Nucl. Phys. B243:269 (1984).

    S. I. Dolinsky et al., Phys. Lett. 174B:453 (1986).

22. A. Donnachie and H. Mirzaie Z. Phys. C33:407 (1987).

23. F. Bird, SLAC-Report-332, Stanford University Thesis (1988).

24. D. Aston et al., Nucl. Phys. B296:493 (1988).

    D. Aston et al., Nucl. Phys. B292:693 (1987).

    D. Aston et al., SLAC-PUB-4202 and SLAC-PUB-4652.

25. W. Toki, "Proceedings of Glueballs, Hybrid and Exotic Hadrons," Conf., BNL, 692 (1988).

26. D. Aston et al., Phys. Lett. 208B:324 (1988).

27. C. Dionisi et al., Nucl. Phys. B169:1 (1980).

28. J. J. Becker et al., Phys. Rev. Lett. 59:186 (1987).

29. Ph. Gavillet et al., Z. Phys. C16:119 (1982).

30. H. Aihara et al., Phys. Rev. Lett. 57:51 (1986) and Phys. Rev. Lett. 57:2000 (1986).

    S. U. Chung et al., Phys. Rev. Lett. 55:779 (1985).

    C. Edwards et al., Phys. Rev. Lett. 49:259 (1982).

31. D. Aston et al., SLAC-Report-287 (1985).

32. D. Aston et al., Nucl. Phys. B292:693 (1987).

33. J. B. Gay et al., Phys. Lett. B63:220 (1976).

    A. C. Irving, Phys. Lett. B70:217 (1977).

34. S. Oneda and A. Miyazaki, RIFP-710 (1987).

35. R. V. Carnegie et al., Phys. Lett. B68:287 (1977).

    C. Daum et al., Nucl. Phys. B187:1 (1981).

36. R. Jaffe and K. Johnson, Phys. Lett. 60B:201 (1976).

37. G. Shierholz, "Proceedings of Glueballs, Hybrid and Exotic Hadrons," Conf., BNL, 281 (1988).

    S. Sharpe, "Proceedings of Glueballs, Hybrid and Exotic Hadrons," Conf., BNL, 55 (1988).

38. D. Caldwell, "Proceedings of Glueballs, Hybrid and Exotic Hadrons," Conf., BNL, 465 (1988).

    H. Aihara et al., Phys. Rev. Lett. 57:2500 (1986).

    G. Gidal, "Proceedings of Glueballs, Hybrid and Exotic Hadrons," Conf., BNL, 171 (1988).

39. T. Burnett for the Mark III Collaboration, "Proceedings of Glueballs, Hybrid and Exotic Hadrons," Conf., BNL, 182 (1988).

40. D. Aston et al., "Proceedings of Glueballs, Hybrid and Exotic Hadrons," Conf., BNL, 350 (1988).

41. A. Kirk for the WA76 Collaboration, "Proceedings of Glueballs, Hybrid and Exotic Hadrons," Conf., BNL, 340 (1988).

42. P. Baillon et al., Nuov. Cim., 50A:393 (1987).

43. D. Reeves et al., Phys. Rev. D B4:1960 (1986).

44. A. Ando et al., Phys. Rev. Lett. 57:1296 (1986).

45. S. U. Chung et al., Phys. Rev. Lett. 55:779 (1985).

46. U. Mallik, SLAC-PUB-4238.
    R. M. Baltrusaitis et al., Phys. Rev. D 35:2077 (1987).
47. A. Etkin et al., Phys. Lett. B165:217 (1985).
48. F. Binon et al., Nuov. Cim. 78A:313 (1983); Nuov. Cim. 80A:363 (1984).
49. D. Alde et al., Phys. Lett. B182:105 (1987).
50. M. Boutemeur, "Recontre de Moriond Les Arcs," France (1987).
    M. Boutemeur, "Proceedings of Glueballs, Hybrid and Exotic Hadrons," Conf., BNL, 373 (1988).
    D. Alde et al., Phys. Lett. 205B:397 (1988).
51. M. Chanowitz, Phys. Lett. 187B:409 (1987).
52. F. Close and H. Lipkin, Phys. Lett. 196B:245 (1987).
53. S. I. Bityukov et al., Phys. Lett. 188B:383 (1987).
54. M. Bourquin et al., Phys. Lett. 172B:113 (1986).
55. A. N. Allen et al., JINR 19 (1986).
56. S. L. Wu, Phys. Rept. 107:59 (1984)
57. S. Otto and J. Stack, Phys. Rev. Lett. 52:2328 (1984).
58. C. Quigg and J. L. Rosner, Phys. Rev. D 23:2625 (1981).
    W. Kwong, J. L. Rosner, and C. Quigg, Ann. Rev. Nucl. Part. 37:325 (1987).
59. A. Martin, Phys. Lett. 933:338 (1980); Phys. Lett. 100B:511 (1981).
60. E. Eichten et al., Phys. Rev. D 17:3090 (1978); Phys. Rev. D 21:203 (1981).
61. T. Appelquist et al., Ann. Rev. Nucl. Sc. 28:387 (1978).
62. J. L. Richardson, Phys. Lett. 82B:277 (1979).
63. E. D. Bloom and C. W. Peck, Ann. Rev. Nucl. Part. 33:143 (1983).
64. K. Riles et al., Phys. Rev. D35:2914 (1987).
    G. Blaylock et al., Phys. Rev. Lett. 58:2171 (1987).
65. H. Albrecht et al., Phys. Lett. 150B:235 (1985).
66. G. Blaylock et al., Phys. Rev. Lett. 58:2171 (1987).
67. H. Albrecht et al., Phys. Lett. B221:422 (1989).
68. H. Albrecht et al., Phys. Lett. B230:162 (1989).
69. J. C. Anjos, Phys. Rev. Lett. 62:1587 (1989).
70. J. Weinstein and N. Isgur, Phys. Rev. D 41:2236 (1990).

INTRODUCTION TO THE PHYSICS OF PARTICLE ACCELERATORS

Robert Siemann

Newman Laboratory of Nuclear Studies
Cornell University
Ithaca, N. Y. 14853

INTRODUCTION

Progress in science is closely connected to the capabilities of instruments. Accelerators and detectors are the instruments of particle physics; these lectures are devoted to the former. Examples of relationship between accelerator science and particle physics are the discovery of the W and Z bosons which was a result of the development of stochastic cooling and the extensive studies of c- and b-quarks have been critically dependent on understanding and improvement of e^+e^- storage rings. In the future work on superconducting magnets could lead to the discovery of the t-quark and the Higgs.

In an accelerator charged particles move in electromagnetic fields. The motion can be classical or quantum mechanical, and the fields can be static or time dependent, externally applied or beam generated, and linear (proportional to displacement) or nonlinear. Combinations of all these possibilities occur and can be important. For example:
1. The SSC and LHC apertures are determined by classical motion in static, externally applied, nonlinear fields.
2. The luminosity limit of heavy quark factories depends in part on classical motion in time dependent, beam generated, nonlinear fields.
3. Quantum mechanical effects due to time dependent, beam generated fields at the collision point influence linear collider parameters strongly.

The usual approach is to understand the dominant contributions to the motion first and then treat other aspects as perturbations. This dominant motion is that of single particles undergoing classical motion in externally applied, linear fields. Both static and time dependent fields are important.

SINGLE PARTICLE MOTION

Figure 1 is a sketch of a linear accelerator consisting of accelerating RF cavities with quadrupole lenses interspersed. There is a fictitious *Ideal* particle that is on-axis and at the nominal energy (which increases linearly with s, the distance along the accelerator). Real particles can deviate from the *Ideal* by being off-axis and/or at other than the nominal energy. Concentrate on the energy motion first.

Energy Motion

The RF wave in the accelerating cavities can be thought of as propagating in the +s direction with phase velocity equal to c, the speed of light. At a fixed s the accelerating electric field is

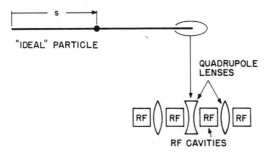

Fig. 1. A sketch of a linear accelerator.

$$E_{RF} = G \cos(\omega_{RF} t) \tag{1}$$

where G is the accelerating gradient (with units of V/m) and ω_{RF} is the RF (angular) frequency. Consider a particle that passes at a time τ later than the *Ideal* which is at the crest of the wave (Figure 2a). Assuming the particle and the *Ideal* are relativistic, the equations of motion are

$$\frac{d\gamma}{ds} = \frac{e}{mc^2} G \cos(\omega_{RF}\tau) \quad \text{and} \quad \frac{d\tau}{ds} = 0 \tag{2}$$

where e is the electric charge of the particle and γ is the energy of the particle in units of the rest energy mc^2. A bunch of such particles has a stable shape - particles at the head of the bunch remain there over the entire length of the accelerator, and there is a correlation between energy and τ - the fractional energy spread σ_γ/γ is proportional to σ_τ^2 where σ_τ is the rms time spread of the bunch.

Figure 3 is a sketch of a storage ring. It consists of dipole magnets that bend particles in a circle, quadrupoles for focusing, and an RF cavity. Again, there is an *Ideal* particle that is on-axis and at the nominal energy, γ_0, which is constant for a storage ring. The *Ideal* is located at a different place on the RF wave because the nominal energy is constant (Figure 2b). The equations of motion of a particle passing through the RF cavity are given by eq. (2) with the argument of the cosine being $\omega_{RF}\tau + \pi/2$ instead of $\omega_{RF}\tau$. For one passage through the cavity

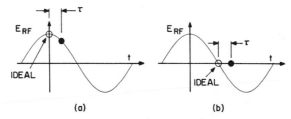

Fig. 2. The RF wave at a fixed position. The *Ideal* is located as shown for a linear accelerator (a) and a storage ring (b).

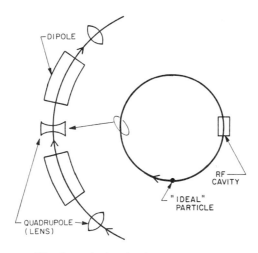

Fig. 3. A sketch of a storage ring.

$$\Delta\gamma \cong \frac{-eG\,L}{mc^2}\omega_{RF}\tau \quad \text{and} \quad \Delta\tau = 0 \tag{3}$$

where L is the RF cavity length.

The particle is bent in a circle between passages through the RF cavity. Consider only the dipoles for a moment. In bending through an angle $d\theta$ the *Ideal* travels a distance $ds = Rd\theta$ where R is the bending radius of the *Ideal*, and a particle with energy γ travels $ds' = R(1 + (\gamma-\gamma_0)/\gamma_0)d\theta$. This particle takes a different amount of time to travel $d\theta$ because $ds' \neq ds$, and τ varies as $d\tau/ds = (\gamma-\gamma_0)/(c\gamma_0)$. The quadrupoles are important because they provide additional bending. When this is accounted for $d\tau/ds = \alpha(\gamma-\gamma_0)/(c\gamma_0)$ where α is called the momentum compaction and is in the range 0.001 to 0.05 depending on the quadrupole configuration. For a passage around the arc

$$\Delta\gamma = 0 \quad \text{and} \quad \Delta\tau = \alpha T_0 \frac{\gamma-\gamma_0}{\gamma_0} \tag{4}$$

where T_0 is the revolution period.

The motion described by eqs. (3) and (4) is an oscillation about the *Ideal*. This can be seen by looking at particles at different positions in the longitudinal phase space shown in Figure 4. This oscillation is called a synchrotron oscillation and corresponds to a particle interchanging energy and time displacements. The underlying principle was discovered by McMillan in 1945.[1] The frequency of oscillation can be determined by making the approximations $d\tau/dt \cong \Delta\tau/T_0$ and $d\gamma/dt \cong \Delta\gamma/T_0$ where $\Delta\tau$ and $\Delta\gamma$ come from eqs. (4) and (3), respectively. The resultant expressions can be combined to give

$$\frac{d^2\tau}{dt^2} = \frac{\alpha}{\gamma_0}\frac{d\gamma}{dt} = \frac{-eG\,L\alpha\omega_{RF}}{mc^2\gamma_0 T_0}\tau. \tag{5}$$

This is the equation for simple harmonic motion with angular frequency, the synchrotron frequency,

$$\omega_s = \left(\frac{eG\,L\alpha\omega_{RF}}{mc^2\gamma_0 T_0}\right)^{1/2}. \qquad (6)$$

The "synchrotron tune" is the number of synchrotron oscillations per revolution. It is $Q_s = \omega_s T_0/2\pi$.

Particles oscillate about the position of the *Ideal* which is the point where the RF voltage is zero and has a negative slope. The RF phase advances by an integer multiple of 2π each time the *Ideal* makes a turn - a passage around the ring. This is the only way it can have the same position on successive turns. This means that the orbital period must be a multiple of the RF period, $T_0 = h\,(2\pi/\omega_{RF})$ where h is an integer called the harmonic number. The RF frequency, not the dipole field strength, determines the accelerator circumference! This simple argument has another consequence. There are h positions equally spaced around the ring where the *Ideal* could be located. Therefore, h is the number of bunches the ring can hold.

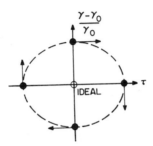

Fig. 4.   Longitudinal phase space with four particular points shown. The arrows indicate the direction of motion given by eqs. (3) and (4). If these points are connected together, oscillatory motion results.

The analysis above that led to eqs. (4) and (5) assumed relativistic particles. There are two effects that are important for non-relativistic particles traveling in a circle between passages through the RF cavity. If the energy is greater than that of the *Ideal* the path length is longer and the velocity is greater. The latter is more important below the "transition energy". Non-relativistic particles undergo synchrotron oscillations also; the difference is that the position of stable equilibrium is located at the point where the RF voltage is zero but has a positive slope.

Transverse (Betatron) Motion

Particles have transverse displacements and angles with respect to the trajectory of the *Ideal*. The quadrupole magnets provide the focusing needed to contain the beam.

In a current free region the magnetic field is the gradient of a scalar potential, $\mathbf{B} = -\nabla\phi$. For the quadrupole magnet shown in Figure 5, $\phi = gxy$; g is called the quadrupole gradient and has units of T/m. The equation of motion, $d\mathbf{p}/dt = e\mathbf{v}\times\mathbf{B}$, can be broken-up into x and y components. For the x component $dp_x/dt = -ecB_y = ecgx$ becomes

$$\frac{d^2x}{ds^2} = \frac{eg}{\gamma m c} x \qquad (7)$$

when p_x is approximated as $p_x \cong \gamma mc\, dx/ds$. This is the equation for simple harmonic motion if $eg/\gamma mc < 0$, and the x motion is bounded in this case. When $eg/\gamma mc > 0$ there is exponential growth and unbounded motion.

For the y component the equation of motion is

$$\frac{d^2y}{ds^2} = -\frac{eg}{\gamma m c} y \,. \qquad (8)$$

If the motion is bounded in x, it grows exponentially in y and vice versa. This is a consequence of the magnetic field being derived from a scalar potential which in turn is a consequence of the beam traveling in a current free region.

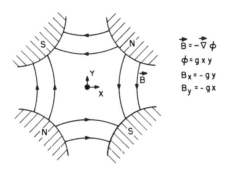

Fig. 5. A quadrupole magnet.

A single quadrupole can focus in only one transverse dimension at a time, and to have focusing in both x and y multiple quadrupoles arranged as a compound optical system are needed. A simplified arrangement is shown in Figure 6. The fact that such compound systems can be constructed with bounded motion in both x and y is the principle of "strong focusing".[2] The equation of motion in one transverse dimension is

$$\frac{d^2x}{ds^2} = \frac{eg(s)}{\gamma m c} x = k(s) x \qquad (9)$$

where $k(s) < 0$ in focusing quadrupoles, $k(s) = 0$ in field free regions, and $k(s) > 0$ in defocusing quadrupoles. The motion looks like simple harmonic motion, but the variation of k with s makes it more complicated. However, it is natural to look for an oscillatory solution of the form

$$x = A\, w(s)\, e^{i \psi(s)} \qquad (10)$$

Such a solution exists if $w(s)$ and $\psi(s)$ satisfy[2]

$$\frac{d^2w}{ds^2} + k(s) w - \frac{1}{w^3} = 0 \qquad (11)$$

and

$$\frac{d\psi}{ds} = \frac{1}{w^2}. \tag{12}$$

Equation (11) is called Hill's equation.

There are a number of important comments that can be based on eqs. (11) and (12).
1. w(s) depends on the "lattice" - the configuration of quadrupoles - through the second term on the left-hand-side of eq. (11).
2. However, w(s) is not uniquely determined by the lattice; eq. (11) is a second order differential equation and two constants of integration are needed for a unique solution.
3. w(s) has an absolute scale; i.e. if $w_1(s)$ is a solution, $2 w_1(s)$ is not.
4. The phase advance, $d\psi/ds$, is not constant but is given by eq. (12).

Accelerator physicists often call transverse motion "betatron motion" and talk about the "β-function" of a storage ring, a beam transport, or a linear accelerator. It is related to w(s) by $\beta(s) = \{w(s)\}^2$. In terms of β eq. (12) is

$$\frac{d\psi}{ds} = \frac{1}{\beta(s)}, \tag{13}$$

and the motion of a particle is

$$x = A \{\beta(s)\}^{1/2} e^{i\psi(s)}. \tag{14}$$

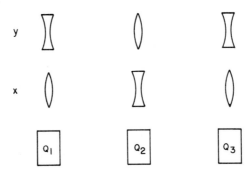

Fig. 6. A sketch showing alternating gradient focusing - an array of alternating focusing and defocusing quadrupoles. This arrangement is used in the simplified SSC discussed below. In that example the quadrupoles are 5.05 m long, there is 180 m from the center of Q_1 to the center of Q_3, and each quadrupole has |g| = 211 T/m.

Some initial conditions must be specified to determine β for a beam transport or linear accelerators, but there is a unique definition for a circular accelerator. The lattice of a circular accelerator is periodic, $k(s + C) = k(s)$ where C is the circumference, and the β-function of a circular accelerator is the periodic solution to Hill's equation, $\beta(s + C) = \beta(s)$.

Storage ring lattices usually consist of arcs where the β-functions are repetitious and interaction regions where the beams are focused for collisions. As a first example consider a simplified SSC without interaction regions and consisting of 500 cells one of which is illustrated in Figure 6. The β-functions are shown in Figure 7; they are large where the quadrupoles are focusing and small where they are defocusing. The β-function is periodic with period C, but the motion given by eq. (14) is not. Figure 8a shows the trajectory of a particle on two successive turns, and Figure 8b shows the trajectories on 100 successive turns.

Figure 9 shows the phases; they advance rapidly when β is small and slowly when β is large. The number of betatron oscillations per revolution is

$$Q_\beta = \frac{1}{2\pi} \oint_0 \frac{ds}{\beta(s)} \; ; \tag{15}$$

Q_β is called the betatron tune. For the simplified SSC $Q_\beta = 127.80$.

The β-function measures the sensitivity to errors in addition to giving the beam envelope and the rate of phase advance. If there is a gradient error δg in a quadrupole of length L located at a position s_e, there will be a change in tune

$$\delta Q_\beta = -\frac{1}{4\pi}\beta(s_e)\frac{e\delta g L}{\gamma m c} \; . \tag{16}$$

Large β leads to a large change of tune and high sensitivity to field errors.

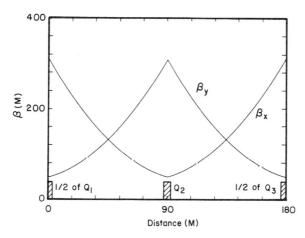

Fig. 7. The β-functions for one cell of the simplified SSC lattice.

Collider interaction regions are designed to have low β's for two reasons. First, the luminosity is given by

$$L = \frac{N^2 f_c}{4\pi\sigma_x\sigma_y} \tag{17}$$

where N is the number of particles per bunch, f_c is the collision frequency and $4\pi\sigma_x\sigma_y$ is the effective collision area. Since σ_x and σ_y are proportional to $\sqrt{\beta_x}$ and $\sqrt{\beta_y}$, respectively, low β reduces the effective area and increases the luminosity. Second, particles in one beam interact with the electromagnetic fields produced by the opposing beam. These fields are roughly equivalent to an error, and small β reduces the sensitivity to this error. This is discussed in more detail in the section about beam-beam effects.

The β-function at the collision point is usually denoted by β^*. At a distance S from the collision point before the first quadrupole is encountered $\beta(S) = \beta^* + S^2/\beta^*$. (This can be shown from eq. (11) using the fact that β is a minimum at the collision point and $k(s) = 0$

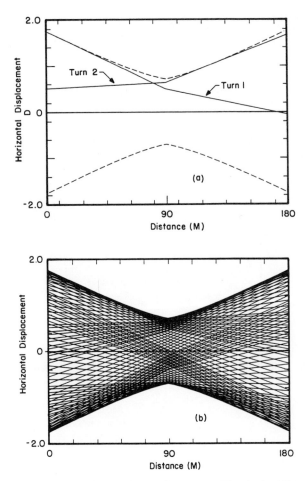

Fig. 8. Horizontal (x) trajectories on successive turns. The dashed lines are proportional to $\pm\sqrt{\beta_x}$. (a) shows two turns and (b) shows 100 turns.

in the region before the first quadrupole.) With a small β^* the beam size increases rapidly as it leaves the collision point. Figure 10 shows a low β^* interaction region for the SSC.[3] There are three strong quadrupoles arranged in an alternating gradient configuration called a quadrupole triplet that gives $\beta^* = 0.5$ m in both dimensions. The first quadrupole begins 20 m from the collision point; $\beta \sim 800$ m at that point. This quadrupole can focus in only one dimension, and the beam is defocused in the other. The β-function in that dimension grows rapidly. It reaches a peak value of about 8000 m before being turned around by the next quadrupole in the triplet. The beam size has grown by about a factor of 125, and because of the large β's in these quadrupoles, their field quality in important. The combination of demands on the interaction region quadrupoles - high gradient, large aperture and good field quality - make these quadrupoles one of the technological marvels of modern colliders. After the triplet the lattice is arranged to restore the smooth β-functions such as shown in Figure 7.

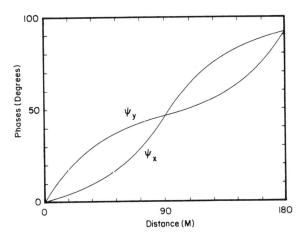

Fig. 9. The betatron phases for the simplified *SSC* lattice.

Emittances

The β-function gives the variation of the beam size with position s, but it doesn't give the actual beam size. For that one needs to know the typical amplitude of oscillation, A, in eq. (14). The "transverse emittance" of a beam, ε, is defined as

$$< A^2 > \equiv 2\varepsilon . \qquad (18)$$

(A word of caution is necessary; different laboratories use different proportionality constants between ε and $< A^2 >$.) The rms beam size at position s is

$$\sigma_z(s) = [\beta_z(s)\varepsilon]^{1/2} \qquad (19)$$

where z is either x or y. A longitudinal emittance associated with the synchrotron oscillations can be defined in the same manner. The discussion below is restricted to one transverse dimension, but it holds equally well for the other transverse dimension, and a similar one holds for the longitudinal emittance.

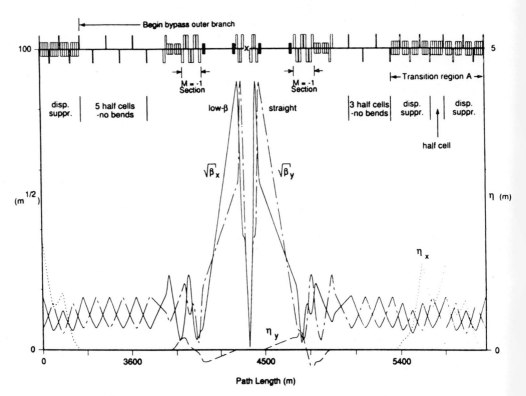

Fig. 10. A low β^* interaction region at the SSC. The η-functions will be discussed later.

The amplitude of a particle can be written in terms of its position and angle

$$A^2 = \frac{x^2}{\beta} + \beta\left(x' - \frac{\beta'}{2\beta}x\right)^2 \qquad (20)$$

where $\beta' = d\beta/ds$. Equation (20) is the equation of an ellipse in *transverse phase space*, a space with axes x and x'. The amplitude A is independent of position, s, even though x, x', β and β' all depend on position; i.e. a particle with amplitude A remains on this ellipse as it moves around the accelerator. The area of the ellipse given by eq. (20) is πA^2, and the emittance measures the phase space area occupied by the beam.

Liouville's Theorem[4] puts constraints on the emittance; it says that the area occupied in a phase space with axes x and $\gamma\beta_v x'$ (the canonical momentum of x) cannot change in time; β_v is the usual relativistic velocity, $\beta_v = v/c$. Therefore, $\gamma\beta_v\varepsilon$ is constant as a beam is accelerated, transported down a beam line, etc. whenever Liouville's Theorem is valid. It is in proton accelerators and linear electron accelerators, but it isn't in electron storage rings and anti-proton sources. For specificity consider a proton accelerator first. The beam originates in an ion source, and the emittance is determined by the properties of ion sources. The beam is accelerated first in a linac and then in a number of synchrotrons before reaching collision energy. During all these process the emittance shrinks as $1/\gamma\beta_v$. Except for this decrease with energy the beam size at collision energies is determined by the physics of ion sources! It isn't surprising that particle sources are an active area of research.

The above discussion is not complete because it is impractical. A beam may develop a complicated, filament-like structure enclosing empty regions of phase space as it is accelerated. Theoretically, these filaments can be untangled, but that is impractical, and the effective phase space area of a beam tends to increase. The emittance behavior implied by Liouville's Theorem is the best possible, and in most cases emittances increase from the minimum given by Liouville's Theorem.

Liouville's Theorem isn't valid if there is damping or energy loss. There is energy loss in the form of synchrotron radiation in electron storage rings and anti-proton sources have stochastic cooling systems for damping. Therefore, $\gamma\beta_v\varepsilon$ is not constant, but it can be reduced. The possibilities depend on the physics and technology of the energy loss and damping mechanisms. These play the same role as the physics and technology of ion sources do for the proton accelerator discussed above.[5] This is too complex a subject for these notes, and it will have to suffice to say that within limits these accelerators can be designed to give a desired emittance.

WAKEFIELDS AND THEIR EFFECTS

The discussion has concentrated on single particles so far. Many interesting phenomena and performance limits come from interactions between particles. This section deals with one class of these interactions - beam current limits due to beam generated electromagnetic fields.

A beam passing a change in vacuum chamber profile radiates electromagnetic fields that propagate down the beam pipe in waveguide modes and, if a resonant structure is involved, excite normal modes. This is illustrated graphically in Figure 11.[6] At t = 0 the beam is entering the cavity, and the electric field is predominantly the space charge field of the beam. As time passes fields penetrate into the cavity, and cavity modes are excited. When the beam leaves at t = 2.0 nsec, it has lost energy to electromagnetic fields in the cavity. These beam induced fields, called wakefields, have many consequences:
1. The beam losses energy. This energy must be replaced by the RF system, and it can cause localized heating and damage to elements of the vacuum chamber.
2. Energy can be transferred from the head to the tail of a bunch.
3. The head of a bunch can deflect particles in the tail of the bunch.
4. Energy and deflections can be transferred between bunches if there are normal modes with large enough quality factors.
The last three can lead to emittance growth and instabilities.

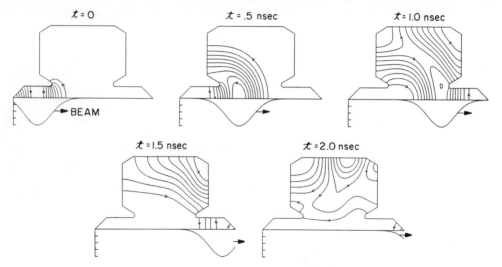

Fig. 11. The electric field lines produced by a Gaussian bunch passing through a LEP RF cavity.[6]

Wakefields

Consider the situation shown in Figure 12. Two ultrarelativistic particles Q and q are travelling in the z direction through a rotationally symmetric structure; Q has unit charge. The particles are a distance ct_0 apart; q trails Q when $t_0 > 0$. The particle positions with respect to the symmetry axis are given by vectors $\vec{R}_T$ and $\vec{r}_T$; ϕ is the angle between $\vec{R}_T$ and $\vec{r}_T$.

First, let Q travel on-axis, $\vec{R}_T = 0$. The "longitudinal wake potential" is the beam induced voltage seen by q

$$V_{\delta 0}(t_0) = \int_{-\infty}^{\infty} dz \int_{-\infty}^{\infty} dt' \, E_z(z, t') \, \delta(t' - (t_0 + z/c)) \quad . \tag{21}$$

The quantity E_z is the longitudinal component of the electric field produced by Q. The wake potential $V_{\delta 0}$ has some simple properties illustrated by Figure 13. First, it is causal; $V_{\delta 0} = 0$ for $t_0 < 0$. Second, for positive t_0 as $t_0 \to 0$ $V_{\delta 0}$ must be negative because Q loses energy into the structure. At larger values of t_0 the wake potential depends on the structure geometry.

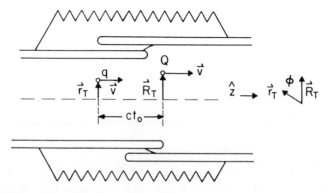

Fig. 12. Definition of the variables used to describe wakefields. The structure is assumed to be rotationally symmetric about the z-axis.

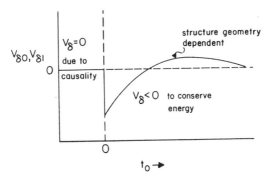

Fig. 13. Time dependence of the longitudinal wake potential.

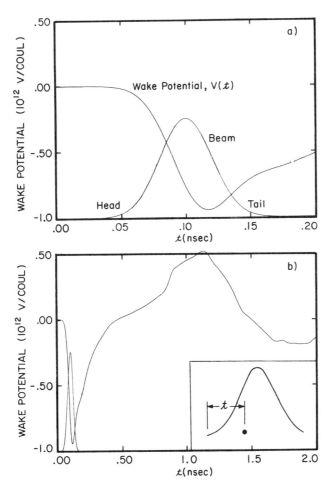

Fig. 14. The wake potential for a 6 mm long, 1 C bunch traveling through a CESR cavity shown on two different time scales. The inset defines the time t.

295

The wake potential for a bunch is given by the convolution of $V_{\delta 0}$ with the bunch charge density. It is proportional to N, the number of particles in the bunch, and depends on the rms bunch length and the structure geometry. A wake potential for a CESR RF cavity is plotted in Figure 14. Several of the consequences of beam induced fields mentioned above can be seen in this figure. First, the beam loses energy as is expected from a comparison of the electromagnetic field energies at t = 0 and t = 2 nsec in Figure 11. The energy loss is

$$\text{Energy change} = \int_{-\infty}^{\infty} dt\, I(t)\, V(t) < 0 \tag{22}$$

where I(t) is the bunch current. The wake potential and I(t) are each proportional to N, so the energy loss is proportional to N^2. This "higher order mode" power is an important design consideration in heavy quark factories.[7] Second, Figure 14a shows that V(t) varies along the bunch; the wake potential is different at the head and tail. This introduces additional τ dependence into eq. (3) that can lead to the single bunch instabilities. Figure 14b shows that the wake potential can be positive at longer times, and a particle passing through the cavity at those times gains energy. Therefore, it is possible to transfer energy between bunches; this leads to coupled bunch instabilities.

If Q is not on-axis[8]

$$V_\delta(t_0) = \int_{-\infty}^{\infty} dz \int_{-\infty}^{\infty} dt'\, E_z(z, t')\, \delta(t' - (t_0 + z/c))$$

$$\cong V_{\delta 0}(t_0) + r_T R_T\, V_{\delta 1}(t_0) \cos\phi . \tag{23}$$

The new term comes from modes that have $E_z = 0$ on-axis and are excited by Q being off-axis. The arguments about the qualitative behavior of $V_{\delta 0}$ hold for $V_{\delta 1}$, so it has the general shape shown in Figure 13. When $V_{\delta 1} \neq 0$, $\partial E_z/\partial r \neq 0$ and it follows from Maxwell's equations that **E** and **B** have radial and azimuthal components that cause deflections. The deflection produced by Q is the $\vec{R}_T$ direction, and the magnitude, the "transverse wake potential", is

$$W_\delta(t_0) = \frac{1}{R_T} \int_{-\infty}^{\infty} dz \int_{-\infty}^{\infty} dt'\, (\mathbf{E}(z,t') + \mathbf{v}\times\mathbf{B}(z,t'))_T\, \delta(t' - (t_0+z/c)) \tag{24}$$

where **E** and **B** are the fields produced by Q and the subscript T indicates the component in the $\vec{R}_T$ direction. The wake potentials $V_{\delta 1}$ and W_δ are not independent. They are related by the Panofsky-Wentzel Theorem[9]

$$W_\delta(t_0) = -\int_{-\infty}^{t_0} dt\, V_{\delta 1}(t) . \tag{25}$$

Given the form of $V_{\delta 1}$ in Figure 13 and the Panofsky-Wentzel Theorem, the transverse wake potential is shown in Figure 15. For small t_0 q is displaced in the same direction as Q.

Impedances. Beam stability problems are often solved in the frequency domain. There one uses impedances rather than wake potentials. The longitudinal and transverse impedances, Z_L and Z_T, are related to the wake potentials by

$$Z_L(\omega) = -\int_{-\infty}^{\infty} dt\, V_{\delta 0}(t)\, e^{-i\omega t} \quad \text{and} \quad Z_T(\omega) = i\int_{-\infty}^{\infty} dt\, W_\delta(t)\, e^{-i\omega t} . \tag{26}$$

Emittance Growth in Linear Colliders

This is the first example of the effects of wakefields. It is discussed with the simplifying assumption that there are only two particles in the beam. Such two-particle models are valuable for understanding the underlying physics. In addition, they can be generalized to include more particles. This is usually the way computer simulations of beam stability are performed.

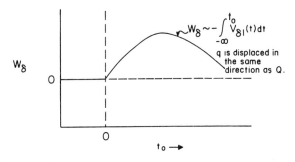

Fig. 15. Time dependence of the transverse wake potential.

If a particle is injected off-axis, the quadrupoles focus it and produce an oscillation about the axis. If a second particle, the *tail*, is injected following the first, the *head*, it experiences the wakefields produced by the *head*. For realistic bunch lengths the *tail* is in the short-time region of the transverse wake potential, and it is deflected in the direction of the *head*. Therefore, the *tail* is always deflected away from the axis by the *head*; see Figure 16. Its amplitude of oscillation grows thereby increasing the effective emittance of the beam. Figure 17a shows the results of a multiparticle simulation of the SLC with a 30 μm injection error.[10] Particles in the tail have experienced large deflections and can lower the luminosity and cause backgrounds while those at the head of the beam are near the axis.

Possible ways to control this emittance growth include removing misalignments and injection errors to the extent feasible, using strong quadrupoles to maximize focusing, choosing an accelerator with small wakefields, and using Balakin, Novokhatsky, and Smirnov (BNS) damping.[11] Concentrate on the last two of these. The bunch length, $c\sigma_\tau$, should be much less than the wavelength of the accelerating RF to keep the energy spread small. If $\sigma_\tau \omega_{RF}$ and the proportions of the accelerator are held fixed as ω_{RF} is changed, the transverse wake potential at the tail varies as ω_{RF}^3.[12] This argues for low frequencies and long wavelengths. On the other hand, the efficiency for transferring energy from the accelerating fields to the beam depends on ω_{RF} as ω_{RF}^2,[12] and there is a trade-off between reducing wakefields and efficient operation. Frequencies that are currently favored by linear collider designers are in the 10 GHz to 30 GHz range. Those are to be compared with the SLAC frequency of 2856 MHz. The need for good operating efficiency has driven people to higher frequencies, and BNS damping is critical for controlling emittance growth.

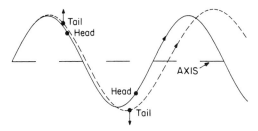

Fig. 16. The *head* is injected off-axis, and the quadrupoles produce an oscillation. The *tail* is always deflected in the direction of the *head* as indicated by the arrows.

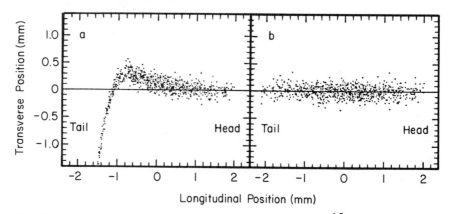

Fig. 17. The results of multiparticle simulations of the SLC.[10] (a) shows the large deflection of particles in the tail due to transverse wakefields. (b) shows the same situation when BNS damping has been applied.

BNS damping is analogous to an oscillator with an external drive. The driven response is greatest when the drive frequency equals the natural frequency, and there are beats between the driven response and oscillations (at the natural frequency) caused by non-zero initial conditions. In the linear collider the *head* drives the *tail*, and the emittance growth shown in Figure 17a can be reduced if $\omega_{head} \neq \omega_{tail}$, i.e. $\omega_{drive} \neq \omega_{nat}$. BNS damping is produced at the SLC by making the energies of the head and tail different. That gives the quadrupoles different focusing strengths (eq. (7)) and the desired frequency inequality results. Figure 17b shows the power of BNS damping in the SLC. RF quadrupoles are being considered as an alternative approach to get different focusing at the head and tail.[13]

Instabilities in Circular Accelerators

There are a number of instabilities caused by wakefields in circular accelerators. The next two subsections are examples of them. They are chosen to illustrate the nature of instabilities and the interplay between various aspects of accelerator physics discussed to this point. Simple one and two particle models are used.

The Fast Head-Tail Instability. This important instability was discovered and interpreted at PETRA.[14] It affected the performances of both PETRA and PEP. In addition, the single bunch current in LEP is expected to be limited to about 0.75 mA by the fast head-tail instability, and, because of this, luminosity upgrades of LEP are concentrating on storing more bunches rather than increasing the single bunch current.

Consider two particles in a single bunch in a circular accelerator. The *head* produces a transverse wakefield that acts on the *tail*. The difference from the linear accelerator is that the *head* and *tail* interchange due to synchrotron oscillations (see Figure 18). This introduces new physics. As a simplification take the transverse wake potential to be $W(t_0) = W\Theta(t_0)$ where Θ is a step function.[15] Particle 1 is the *head*, and particle 2 is the *tail* for the first half of the synchrotron oscillation period. The *head* moves in simple harmonic motion

$$\frac{d^2 y_1}{dt^2} + \omega_\beta^2 y_1 = 0 \qquad (27)$$

where the betatron motion has been approximated as smooth. The *tail* is driven by the *head*. Its equation of motion is

$$\frac{d^2 y_2}{dt^2} + \omega_\beta^2 y_2 = \frac{N e^2 W}{\gamma m} y_1 \qquad (28)$$

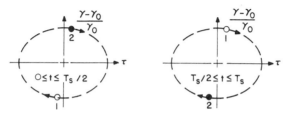

Fig. 18. The *head* and *tail* interchange every one-half synchrotron oscillation period, T_s.

where Ne is the charge of the *head*. Particle 2 is the *head* for the second half of the period, and the equations of motion are given by eqs. (27) and (28) with the indices interchanged, $1 \leftrightarrow 2$. Look for solutions of the form $y = Y(t)\exp(i\omega_\beta t)$ where $dY/dt \ll Y\omega_\beta$. Substituting into eqs. (27) and (28) Y_1 and Y_2 change by the following amounts during the first half period

$$\Delta Y_1 = 0 \text{ and } \Delta Y_2 = -\frac{i W N e^2 T_s}{4\omega_\beta \gamma m} Y_1 = \Xi Y_1 \qquad (29)$$

where T_s is the synchrotron oscillation period and this equation defines Ξ. The solution for a complete synchrotron oscillation period can be written in matrix form

$$\begin{pmatrix} Y_1 \\ Y_2 \end{pmatrix}_{T_s} = \begin{pmatrix} 1 & \Xi \\ 0 & 1 \end{pmatrix}\begin{pmatrix} 1 & 0 \\ \Xi & 1 \end{pmatrix}\begin{pmatrix} Y_1 \\ Y_2 \end{pmatrix}_0 = M \begin{pmatrix} Y_1 \\ Y_2 \end{pmatrix}_0 . \qquad (30)$$

For n synchrotron oscillations

$$\begin{pmatrix} Y_1 \\ Y_2 \end{pmatrix}_{nT_s} = M^n \begin{pmatrix} Y_1 \\ Y_2 \end{pmatrix}_0 . \qquad (31)$$

The motion is stable if all elements of M^n remain bounded as $n \to \infty$. Stability can be determined by looking at the eigenvalues of M. They are $\exp(\pm i\mu)$ where

$$\cos \mu = \tfrac{1}{2} \times \text{Tr}(M) = \tfrac{1}{2}\left(2 - \left(\frac{W N e^2 T_s}{4\omega_\beta \gamma m}\right)^2\right) . \qquad (32)$$

The motion is a stable oscillation as long as μ is real, i.e. $|\cos \mu| \leq 1$. This gives the condition for stability

$$N \leq \frac{8\omega_\beta \gamma m}{W e^2 T_s} . \qquad (33)$$

Equation (33) gives the instability threshold. Below that value this two particle system is stable. The amplitudes Y_1 and Y_2 oscillate but do not grow. Synchrotron motion has stabilized the beam and removed the potentially harmful effects of wakefields. This is in contrast to linear accelerators where there is no threshold for emittance growth. Above the threshold the amplitudes grow exponentially. The beam is unstable and doesn't fit into the vacuum chamber after a short amount of time.

Better calculations can be performed using realistic wakefields and many particles. These calculations verify the parametric dependences of eq. (33). Figure 19 shows results when the current is just above threshold. These computer simulations can start with the

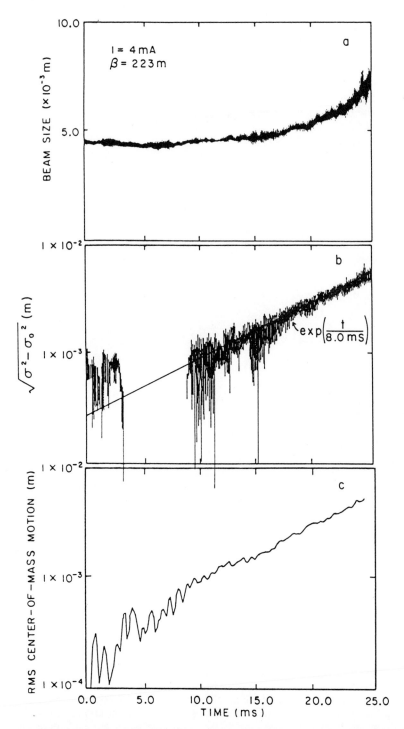

Fig. 19. The results of a computer simulation of beam stability in PEP. The current is just above threshold. The beam size ((a) and (b) where the natural size has been subtracted in (b)) and centroid motion (c) grow exponentially with an 8 ms time constant.[8]

geometry of components in the vacuum system, calculate wakefields from these geometries, and predict instability thresholds within a factor of two with NO free parameters! They have become important accelerator design tools.

The "Robinson Instability".[16] If beam induced fields last long enough, it is possible for energy and deflections to be transferred between bunches. This leads to multibunch and multiturn instabilities. A multiturn instability called the "Robinson instability" can result from the interaction of a beam with the fundamental (accelerating) mode of the RF cavity. It is modelled below with a single particle[15] and illustrates many aspects of multiple bunch instabilities.

When longitudinal wake potential of the RF cavity is included, eq. (3) becomes

$$\Delta\gamma \cong \frac{-eGL}{mc^2}\omega_{RF}\tau + \frac{Ne^2}{mc^2}\sum_n V_{\delta 0}(t_n) \quad \text{and} \quad \Delta\tau = 0. \quad (34)$$

The sum is over all previous turns. The equation of motion is that of a simple harmonic oscillator with a driving force from the wakefield

$$\frac{d^2\tau}{dt^2} + \omega_s^2\tau = \frac{Ne^2}{\gamma_0 T_0 mc^2}\sum_n V_{\delta 0}(t_n). \quad (35)$$

The wake potential has contributions from all the cavity resonant modes. The Robinson instability is due to the interaction with one of them, the fundamental, and this discussion concentrates on the fundamental mode. (Other modes lead to similar instabilities, and these are sometimes called Robinson instabilities also.) The fundamental mode has a quality factor Q and a natural frequency ω_n that is approximately equal to ω_{RF}. It is best to have $\omega_n \neq \omega_{RF}$ for the reasons that follow.

The wake potential due to the fundamental mode is shown in Figure 20a. Initially it is negative because of energy conservation (the same reason used when discussing Figure 13). It oscillates with frequency ω_n and decays as $\exp(-\omega_n t/2Q)$. The time elapsed between successive passages is $T_0 + \Delta\tau$ where $\Delta\tau$ is given by eq. (4). Only one term in the summation is kept for simplicity, and it can be approximated as

$$V_{\delta 0}(T_0+\Delta\tau) \cong V_{\delta 0}(T_0) + \Delta\tau \left.\frac{dV_{\delta 0}}{dt}\right|_{T_0} + \ldots$$

$$\cong V_{\delta 0}(T_0) + T_0\frac{d\tau}{dt}\left.\frac{dV_{\delta 0}}{dt}\right|_{T_0}. \quad (36)$$

Substituting eq. (36) into eq. (35) and ignoring the constant term because it doesn't have any effect on the dynamics gives

$$\frac{d^2\tau}{dt^2} - \left(\frac{Ne^2}{\gamma_0 mc^2}\left.\frac{dV_{\delta 0}}{dt}\right|_{T_0}\right)\frac{d\tau}{dt} + \omega_s^2\tau = 0. \quad (37)$$

If the overall coefficient of $d\tau/dt$ is positive, the motion is damped. However, if it is negative, τ grows exponentially. That is the Robinson instability. The criterion for stability can be restated as $dV_{\delta 0}/dt < 0$. Figure 20b shows the wakefield for three different values of ω_n. When $\omega_n = \omega_{RF}$, $\omega_n T_0$ is a multiple of 2π; the coefficient multiplying $d\tau/dt$ is zero and the cavity is "Robinson neutral". When $\omega_n > \omega_{RF}$, $dV_{\delta 0}/dt$ is positive, and the beam is unstable, but when $\omega_n < \omega_{RF}$, $dV_{\delta 0}/dt$ is negative, and there is no instability.

This conclusion is verified by more detailed calculations, and the Robinson instability is observed in storage rings when the RF cavities are improperly tuned. This and all other multibunch/multiturn instabilities are sensitive to the frequencies of resonant modes. When only the fundamental mode is of concern, the cavities can be tuned to avoid instability, but when additional resonant modes are important, it can be difficult to tune the cavities to avoid these instabilities. Often quality factors must be lowered and/or feedback systems used to control multibunch instabilities.

Instability Categories. The physics of two instabilities has been illustrated above. There are other instabilities; these can be classified as:
1. Single bunch or multibunch/multiturn - the fast head-tail instability was single bunch and the Robinson instability was multiturn.
2. Transverse or longitudinal - the fast head-tail was transverse and Robinson instability longitudinal.
3. With or without centroid motion - this determines whether or not feedback can be used.
4. Self-limiting or having unbounded growth - longitudinal instabilities can be self-limiting because the equation of motion (eq. (2)) becomes nonlinear as τ gets large.
All possible combinations occur and have different importance depending on the particular accelerator. Most modern colliders are pushing one or more instability limits, and a general strategy is to reduce wakefields whenever possible by careful design of the vacuum chamber. Particular attention must be paid to minimizing the number and degree of discontinuities because, as Figure 11 shows, these are the sources of wakefields.

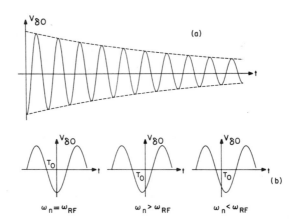

Fig. 20. The longitudinal wake potential for the fundamental mode. (a) shows the general behavior and (b) shows $V_{\delta 0}(t)$ near $t = T_0$ for three values of ω_n.

NONLINEAR FIELDS AND THEIR EFFECTS

The magnetic field of a quadrupole depends linearly on position (Figure 5), and a lattice of quadrupoles leads to betatron motion that is more complicated than simple harmonic motion but is linear - the oscillation frequency is independent of amplitude and the motion is stable for any amplitude. The size of the vacuum chamber determines the maximum amplitude. There are nonlinear fields in accelerators that arise from a number of sources. Some are intentional. For example, sextupoles are used to correct aberrations caused by quadrupoles. This is discussed in the next section about linear collider final focus systems. Other nonlinear fields are unavoidable consequences of methods that are used. Persistent currents that lead to nonlinear magnetic fields are such an unavoidable consequence of superconducting magnets, essential elements of high energy hadron colliders. Nonlinear fields often determine accelerator parameters and performance. This section discusses the causes and effects of these fields.

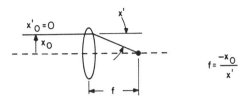

Fig. 21. The focal length of a thin lens.

Linear Collider Final Focus Systems

An ideal final focus system would have the same focusing properties independent of the energy of a particle, but a quadrupole focusing system has "chromatic aberrations" - energy dependent focusing properties. (A final focus system must focus in both transverse dimensions, and, therefore, must be a quadrupole triplet. This is ignored in the discussion that follows; only one transverse dimension and one quadrupole is analyzed.) Consider a focusing quadrupole of length L and gradient g. Equation (7) is the equation of motion. The position of a particle that enters the quadrupole with position and angle x_0 and x_0', respectively, is

$$x = x_0 \cos(\kappa s) + x_0' \sin(\kappa s) \qquad (38)$$

where s is the distance from the beginning of the quadrupole and

$$\kappa = \left(\frac{e|g|}{\gamma m c}\right)^{1/2}. \qquad (39)$$

By analogy with geometrical optics the focal length of the quadrupole can be found by looking at the trajectory with $x_0' = 0$ (see Figure 21). That trajectory leaves the quadrupole with a slope

$$x' = -x_0 \kappa \sin(\kappa L) \cong -x_0 \kappa^2 L . \qquad (40)$$

It is assumed that the quadrupole is thin ($\kappa L \ll 1$) in the last step. The focal length,

$$f = \frac{1}{\kappa^2 L} = \frac{\gamma m c}{e|g|L}, \qquad (41)$$

gets longer, the lens gets weaker, as the energy increases. This is a chromatic aberration.

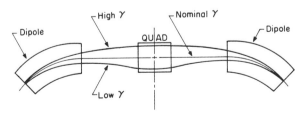

Fig. 22. A non-dispersive deflecting system.[17]

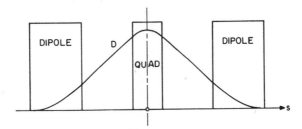

Fig. 23. The dispersion of the system in Figure 22.

This aberration can be corrected by a lens that gets stronger as the energy increases. That lens is a combination of dipoles, quadrupoles, and sextupoles. The dipoles and quadrupoles are arranged to form a "non-dispersive deflecting system".[17] A simple example is shown in Figure 22. A particle follows a path

$$x(s) = x_\beta(s) + D(s)\frac{\gamma-\gamma_0}{\gamma_0} \tag{42}$$

where x_β is the betatron motion that depends on the initial position and angle and $D(s)$ is the dispersion which is the trajectory of a particle with $x_0 = 0$, $x_0' = 0$, and energy γ. The first dipole produces dispersion. The quadrupole gives additional bending, and its strength is such that all trajectories with $x_0 = 0$ and $x_0' = 0$ are parallel at the symmetry point. The second dipole undoes the effect of the first one, and $D = 0$ at the exit of the system. The dispersion $D(s)$ is plotted in Figure 23.

A sextupole, a magnet with $\phi = Sx^2y$ (Figure 24), placed in a region where $D \neq 0$ completes the construction of the desired lens. The magnetic field,

$$\begin{aligned} B_y &= -Sx^2 = -S\left(x_\beta + D\frac{\gamma-\gamma_0}{\gamma_0}\right)^2 \\ &= -S\left(D\frac{\gamma-\gamma_0}{\gamma_0}\right)^2 - 2SD\left(\frac{\gamma-\gamma_0}{\gamma_0}\right)x_\beta - Sx_\beta^2 , \end{aligned} \tag{43}$$

has a quadrupole-like term (the second term) that increases with energy if S has the proper sign. The first term doesn't affect the betatron motion, so the price paid in constructing this

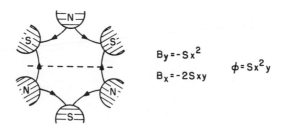

Fig. 24. A sextupole magnet.

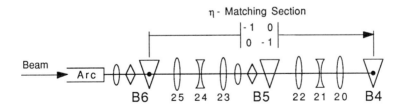
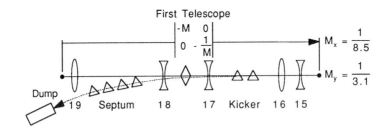
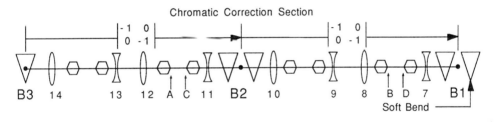
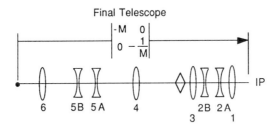

Fig. 25. The optical design of the SLC final focus system.[19] The discussion in the text is about a simplified "Chromatic Correction Section". Sextupoles are indicated by hexagons.

lens is the third term. The betatron equation of motion now has a nonlinear term! The chromatic aberration can be corrected by introducing a higher order "geometric aberration" - focusing that is not linear in the displacement.

This example points out: i) why nonlinear elements are used - with proper choice of magnet strengths a first order chromatic aberration can be eliminated, and ii) the problems they cause - a second order geometric aberration was introduced. The example was simplified because only one dimension was considered. Two dimensions and real **B** fields lead to more complicated systems. In addition, the example doesn't show how clever placement of the magnets can eliminate some of the geometric aberrations.[18] The design and construction of a final focus system with good optical properties is an art; Figure 25 illustrates this.[19]

Nonlinear Effects in Storage Rings

Chromaticity. Quadrupoles confine the beams in storage rings, and, therefore, the focusing, β-functions, phase advances, and tunes are momentum dependent. The chromaticity, defined as

$$\xi = \frac{\gamma}{Q_\beta} \frac{dQ_\beta}{d\gamma} , \qquad (44)$$

is negative, $\xi \sim -1$, because the quadrupoles get weaker as the energy increases. There are several reasons to want $\xi \geq 0$. First, the tune spread caused by chromaticity should be small to allow a tune spread from the beam-beam interaction. This is discussed in the section on beam-beam effects. Second, avoiding the "head-tail" instability (related to but different from the fast head-tail instability) requires $\xi > 0$.[20]

The orbit length in a storage ring is energy dependent (eq. (4)), and, therefore, the orbit itself must be energy dependent. This energy dependence is described by off-energy functions, called η-functions, which are solutions to the same equations as the dispersion, D, but with the boundary condition that η is periodic, $\eta(s + C) = \eta(s)$. The displacement of a particle with energy γ is given by eq. (42) with $\eta(s)$ replacing $D(s)$. The η-functions in the SSC interaction region are shown in Figure 10. In the arcs $\eta_y = 0$ and η_x has a smooth, periodic behavior similar to the β-functions. Sextupoles installed in the arcs are used to correct the chromaticity just as they are used to correct chromatic aberrations in final focus systems. Figure 26 shows a section of CESR as an example. The price for using sextupoles is the introduction of nonlinear fields and geometric aberrations that lead to an effective aperture - the "dynamic aperture".

A Single Octupole. The dynamic aperture is the dominant accelerator physics issue in the SSC, LHC, and synchrotron light rings. The chromaticity correcting sextupoles are strong in the latter, and they determine the dynamic aperture. The dominant nonlinear fields in high energy hadron colliders are due to persistent currents in superconducting magnets and not intentional sextupoles. Put aside the causes of nonlinear fields in this section and concentrate on their effects.

The easiest problem to analyze is a single octupole in a storage ring (Figure 27).[21] The octupole is treated as a perturbation of the betatron motion, and an octupole is chosen because all of the important effects show up in the first order perturbation expansion. The expansion must be done to second order to see all of the effects of a sextupole. The octupole is treated as a thin element, so in passing through half of it the position and angle change as

$$\Delta x = 0, \quad \Delta x' = B_3 x^3 . \qquad (45)$$

A particle entering the arc with initial coordinate x_0 and angle x_0' leaves with

$$x = x_0 \cos(2\pi Q_0) + \beta x_0' \sin(2\pi Q_0)$$
$$x' = -\frac{x_0}{\beta} \sin(2\pi Q_0) + x_0' \cos(2\pi Q_0) . \qquad (46)$$

Fig. 26. The Wilson Laboratory tunnel. The 10-GeV synchrotron is on the left and CESR on the right. The small CESR magnet in the foreground is a chromaticity correcting sextupole. It is followed by a quadrupole and dipoles. Prof. B. D. McDaniel, director of the laboratory at the time of CESR construction, is in the background.

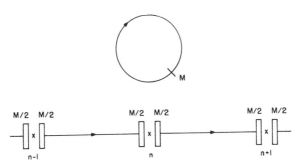

Fig. 27. A single thin octupole, M, in a storage ring. The arc has tune Q_0, and β is the value of the β-function at M. Turn numbers are defined in the middle of M.

The position on turn n, x_n, is

$$x_n = x_{n-1}\cos(2\pi Q_0) + (x'_{n-1} + B_3 x^3_{n-1})\beta\sin(2\pi Q_0). \quad (47)$$

Similar expressions can be written for $x'_n + \Delta x'_n$ and x_{n+1}, and these can be combined with eq. (47) to give a nonlinear equation

$$x_{n+1} - 2\cos(2\pi Q)x_n + x_{n-1} = 2\beta\sin(2\pi Q_0)B_3 x_n^3 \\ + 2x_n\{\cos(2\pi Q_0) - \cos(2\pi Q)\} \quad (48)$$

where Q is the actual tune that differs from Q_0 because of the octupole.

Solve this equation by Fourier analysis and perturbation. Write x_n as a Fourier series

$$x_n = \sum_{r=1}^{\infty} a_r \cos(2\pi r Q n) \quad (49)$$

where $a_1 \gg a_2, a_3, \ldots$. Writing the right-hand-side (RHS) of eq. (48) as a Fourier series also,

$$\text{RHS} = \sum_{r=1}^{\infty} c_r \cos(2\pi r Q n), \quad (50)$$

gives a relationship between the Fourier coefficients

$$a_r = \frac{c_r}{2\{\cos(2\pi r Q) - \cos(2\pi Q)\}}. \quad (51)$$

The first step in the perturbation treatment is to substitute the leading term in eq. (49), $x_n \cong a_1\cos(2\pi Qn)$ into the right-hand-side of eq. (48). Then

$$c_1 = 2a_1\left(\cos(2\pi Q_0) - \cos(2\pi Q) + \frac{3}{4}\beta B_3 a_1^2 \sin(2\pi Q_0)\right),$$

$$c_3 = \frac{1}{2}\beta B_3 a_1^3 \sin(2\pi Q_0), \quad (52)$$

$$c_r = 0 \text{ if } r \neq 1, 3.$$

Equation (51) has many of the features of nonlinear motion. First, the tune Q is a function of amplitude

$$\cos(2\pi Q) = \cos(2\pi Q_0) + \frac{3}{4}\beta B_3 a_1^2 \sin(2\pi Q_0) \quad (53)$$

because a_1 would be infinite unless $c_1 = 0$. There is no tune shift with amplitude for a sextupole if analyzed with the same first order perturbation treatment, but it does appear in second order and is proportional to the square of the sextupole strength. Second, there are resonances caused by the nonlinearity which have infinite response in this order of perturbation. In the octupole example $c_3 \neq 0$, and $a_3 = \infty$ when $\cos(6\pi Q) = \cos(2\pi Q)$, i.e. when $4Q = m$ where m is any integer. The response won't be infinite in a more complete analysis, but it is enhanced at these particular tunes. The particular values of Q are consequences of having chosen an octupole. Other multipoles have different resonant tunes; for a sextupole they are $3Q = m$.

The problem of a single octupole can be simulated also. Figure 28 shows the results. Particles are started at an initial amplitude A_0 and are followed for 1000 turns. When A_0 is small, the amplitude at the end of 1000 turns (calculated using eq. (20)) is approximately equal to A_0, $A_{1000} \sim A_0$. The initial amplitude is increased until a value is reached where $A_{1000} > 10\ A_0$. That value of A_0 is called the maximum initial amplitude and is plotted versus Q_0 in Figure 28. The amplitude of a particle that has reached $10\ A_0$ in 1000 turns is increasing rapidly, and the results are not sensitive to the definition of the maximum amplitude or the number of turns.

The most striking feature of these results is that there are such maximum amplitudes. A particle starting at $A_0(max)$ is unstable. Its amplitude grows rapidly, and it will hit the vacuum chamber wall no matter how far away. Figure 28 is a plot of the dynamic aperture of a linear storage ring with a single octupole! The dynamic aperture is tune dependent. There is an overall trend with some fine structure due to resonances. The $4Q = m$ resonances predicted by the first order perturbation expansion are seen clearly. They are much stronger when m is an even integer. When m is odd the resonances are weaker, and they seem to be displaced from the expected value; however, that is not the case because the resonant conditions depend on Q and not Q_0. Resonances are seen at $Q_0 \sim 3/8, 7/8$ also. These are higher order resonances, and they would appear in a higher order perturbation expansion.

This single octupole example has shown the important consequences of of nonlinear fields. They lead to tunes that depend on amplitude, resonances at particular tunes, and a dynamic aperture. Equations (51) and (52) show that effects depend on the strength of the nonlinear field and the β-function at the location of the nonlinearity.

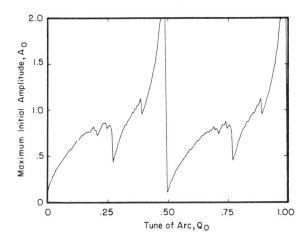

Fig. 28. The results of a computer simulation of a single thin octupole, M, in a storage ring. The maximum initial amplitude is plotted versus the tune of the arc, Q_0; This example used $\beta = 1$ m and $B_3 = 0.5$ T/m^3.

<u>The Dynamic Aperture in High Energy Hadron Colliders.</u> Superconducting magnets are current dominated; the **B** field is determined by the location of the current carrying conductors and not the location of the surrounding iron. A current density

$$J_z(\theta, r) = I\, \delta(r - r_c) \cos\theta \qquad (54)$$

produces a dipole field (Figure 29). The conductors in a real dipole are arranged in an approximation of a cos θ distribution as illustrated in Figure 30. Nonlinear fields can arise from errors in conductor placement and persistent currents.[21a] Persistent currents are currents that are induced in type II superconductors as the superconductor attempts to exclude flux (the Meisner effect). They flow in single filaments of superconductor. Each filament forms a magnetic dipole, $\mu = I_p D$, where D is the diameter of the filament and the persistent current I_p depends on many factors including the critical current density (which in turn depends on the local magnetic field and the operating temperature), the amount of transport current in the superconductor, the previous magnetic history of the superconductor, and time. The time dependence was unexpected and was discovered at the Tevatron;[22] the cause is still uncertain. The magnet multipoles due to conductor placement errors depend on the coil radius, r_c, as $B_n \sim 1/r_c^{n+1}$ where n = 1 is the quadrupole moment, n = 2 the sextupole moment, etc. The multipoles due to persistent currents vary as $B_n \sim I_p D/r_c^{n+2}$.

High energy hadron colliders, the SSC and LHC, will have a limited dynamic aperture due to the nonlinear fields in the dipoles. The aperture at the injection energy is the most critical because i) the persistent currents are large there, ii) the beam energy is low and the beam is less rigid, and iii) the beam is large - the size is proportional to $1/\sqrt{\gamma}$ because the emittance falls as $1/\gamma$. Having an adequate dynamic aperture is a clear and obvious design requirement, but going from that requirement to accelerator specifications requires a great deal of judgement for the following reasons:

1. Phenomena occur on different time scales ranging from the decay time constants of the persistent currents to the injection and storage times.
2. Corrections are possible. The beam itself can be used as a diagnostic to understand and improve the dynamic aperture. How much should one rely on this?
3. The specifications are sensitive to assumptions made about the parameters of subsystems and components. Superconductor properties, operating temperature spread, and even the power supply regulation are important.
4. The physics of the dynamic aperture is nonlinear dynamics, and that is an active field of research with articles appearing each month in <u>Physical Review Letters</u>.

Despite all these uncertainties the accelerator designers must make choices, and that is where the judgement enters.

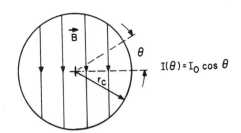

Fig. 29. An idealized superconducting dipole.

The discussion above is sufficient to understand the considerations that go into the choices. First, the superconducting filament size should be minimized. Research into the properties, metallurgy, and manufacturing of superconducting wires has been successful in increasing the critical current and reducing the filament size. This research was motivated in part by the SSC, and these advances are incorporated into the SSC and LHC designs. Second, raising the injection energy increases the effective dynamic aperture for the reasons enumerated above. Third, all of the nonlinear multipoles decrease as strong powers of the coil radius. For example, the persistent current sextupole varies as r_c^{-4}, and this sextupole moment can be reduced by a factor of two with less than 25% increase in r_c. Fourth, the sensitivity to nonlinear multipoles can be reduced by reducing the β-function. This was shown in the section on the single octupole. All of these ways of increasing the dynamic aperture were used in the recent changes of the SSC design.

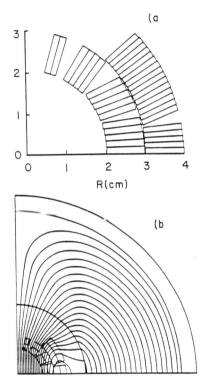

Fig. 30. A quadrant of a dipole cross section showing a) the distribution of superconducting cable and b) the magnetic field lines.[23]

Table 1. Storage Ring - Linear Collider Comparison

STORAGE RINGS	LINEAR COLLIDERS
The collision area is large, and high luminosity is made possible by a high collision frequency.	The collision frequency is small, and high luminosity is made possible by a small effective area.
The beams are reused.	The beams are used once.
Single collisions are "gentle", and the physics of the beam-beam interaction is that of nonlinear fields and nonlinear resonances.	Single collisions are violent, and the physics of the beam-beam interaction is plasma physics and quantum electrodynamics.
There is a wealth of experimental experience, and the understanding is guided by this experience.	There is limited experimental experience.

THE BEAM-BEAM INTERACTION

The center-of-mass energy, the types of beams, and the luminosity are the accelerator properties of primary importance to experimenters. For them

$$\text{Rate}(\text{sec}^{-1}) = L(\text{cm}^{-2}\text{sec}^{-1}) \times \sigma(\text{cm}^2) \tag{55}$$

where σ is the cross section of interest. The accelerator physicist thinks about the luminosity in terms of beam parameters

$$L = \frac{N^2 f_c}{4\pi \sigma_x \sigma_y} \tag{17'}$$

where N is the number of particles per bunch, f_c is the collision frequency and $4\pi\sigma_x\sigma_y$ is the effective collision area for Gaussian beams. The hard collisions that produce elementary particles are relatively rare, and the dominant interaction is between the particles of one beam and the electromagnetic fields of the other. This is the beam-beam interaction.

The beam-beam interaction is dramatically different in linear colliders and storage rings. They are compared in Table 1.

<u>The Electromagnetic Fields</u>

Begin by considering the electromagnetic fields of a flat, uniform charge distribution of width W, height H, and length D (Figure 31). The electric field at $x = 0$ and $y < H/2$ is

$$E_y = \frac{1}{\varepsilon_0} \frac{Ne}{WD} \frac{y}{H} \tag{56}$$

where ε_0 is the permittivity of free space. There is a magnetic field also. If the beams are relativistic, it is $B_x = E_y/c$ (see the coordinate system in Figure 31). The effects of the electric and magnetic fields add, and

$$\frac{dp_y}{dt} = -\frac{2Ne^2}{\varepsilon_0 WD} \frac{y}{H} . \tag{57}$$

where the minus sign comes from assuming oppositely charged beams. This equation of motion is that of a particle in a focusing quadrupole. When the collisions are gentle as in a storage ring this uniform charge distribution is a lens with focal length

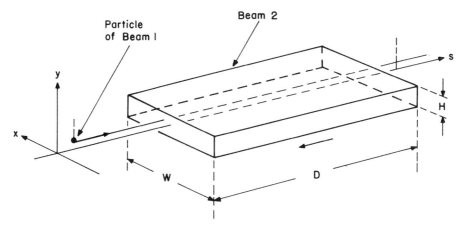

Fig. 31. A flat, uniform charge distribution. W >> H is assumed in the discussion.

$$f = \frac{\gamma W H}{8 \pi r_e N} \; ; \qquad (58)$$

$r_e = e^2/4\pi\varepsilon_0 mc^2$ is the classical electron radius.

The close connection between the luminosity and the beam-beam interaction can be seen by comparing eqs. (17') and (56). Both are proportional to the number of particles divided by the effective area ($4\pi\sigma_x\sigma_y$ for the Gaussian beam assumed in eq. (17') and WH for the uniform beams used for eq. (56)). Limitations on the beam generated fields have consequences for the luminosity.

Beams are not uniform. Gaussian distributions are a much better approximation. The electromagnetic fields for a Gaussian distribution are linear near the center of the beam where the equations of motion are[24]

$$\frac{d^2 z}{ds^2} = \frac{1}{\gamma m c^2} \frac{dp_z}{dt} = -\frac{2N r_e}{\gamma} \frac{1}{\sigma_x + \sigma_y} \frac{z}{\sigma_z} S(s) \qquad (59)$$

where $z = x$ or y and $S(s)$ is longitudinal density distribution. The quantities $N/[\gamma(\sigma_x+\sigma_y)\sigma_x]$ and $N/[\gamma(\sigma_x+\sigma_y)\sigma_y]$ set the scale of the beam-beam interaction near the center of the beam. The disruption parameters in linear colliders and the beam-beam tune spreads in storage rings are proportional to them. The fields fall as $1/(x^2 + y^2)^{1/2}$ when x, y >> σ_x, σ_y, and in the intermediate region they make the transition between the central and large distance behaviors. This is illustrated in Figure 32. The nonlinearity of the fields at y ~ σ_y is important for the beam-beam interaction in storage rings.

Beam-Beam Effects in Linear Colliders

A particle can be focused by the fields of the opposing beam - this is called "disruption", and it can radiate and produce pairs in those fields - the radiation is called "beamstrahlung". Figure 33 shows all of this. Consider disruption first.

Disruption. When the fields are weak, particles in the center of the beam are focused by a lens with focal lengths $f_{x,y} = \sigma_s/D_{x,y}$ where σ_s is the bunch length in meters and $D_{x,y}$ are the "disruption parameters" given by

$$D_z = \frac{2 r_e N \sigma_s}{\gamma(\sigma_x+\sigma_y) \sigma_z} \quad (z = x \text{ or } y) \; . \qquad (60)$$

This follows from eq. (59). As the disruption increases the focal length decreases, and when f ~ σ_s the beam-beam interaction cannot be described in the terms of geometrical

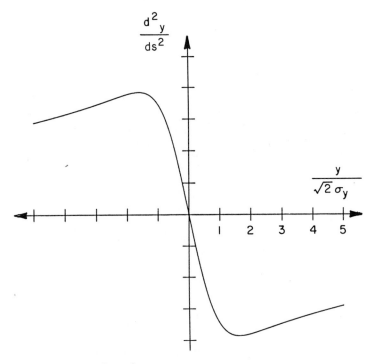

Fig. 32. d^2y/ds^2 near $x = 0$ for a flat beam ($\sigma_x \gg \sigma_y$).

optics. Particles in both beams are strongly focused, and the charge distributions evolve during the collision. This problem has to be studied with computer simulations. The result is that particles undergo a number of plasma oscillations given by $n \cong (D/10.4)^{1/2}$.[25] Figure 34 shows the evolution of the distribution of one of the beams during a collision with moderate disruption.

The luminosity is enhanced by disruption - the focusing reduces the effective collision area. Simulations have been used to estimate the enhancement factor.[25,26] These simulations make different assumptions and approximations and use different numerical methods. The results are sensitive to the differences (Figure 35). Some experimental results would clarify the situation, but it is unlikely that this will happen soon - in 1991 ~10% enhancement is expected at the SLC.[27] Future linear colliders are being designed with flat rather than round beams. The reasons are associated with beamstrahlung and pair production and are discussed below. The luminosity enhancement is not as great with flat beams; for $D_y \sim 5$ (and $D_x \ll D_y$) the enhancement factor is about two.

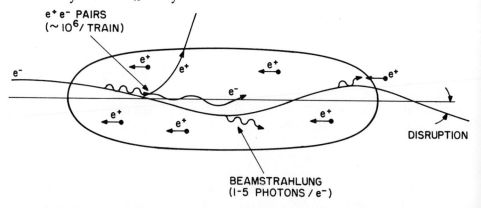

Fig. 33. The processes in the beam-beam interaction in linear colliders.

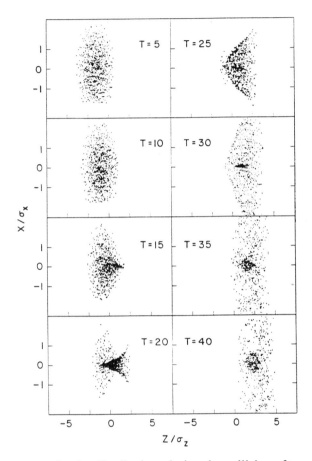

Fig. 34. The transverse density distributions during the collision of two round Gaussian beams with D = 3. The locations of test particles in the beam moving to the right are plotted as time advances during the collision. The particles in the beam moving to the left are not plotted.[25]

A consequence of disruption is that the beams have large angular divergences when they leave the interaction point. Simulation results for flat beams can be fit by[28]

$$\theta_y(rms) \sim \frac{0.55}{[1+(D_y/2)^5]^{1/6}} \frac{D_y \sigma_y}{\sigma_s}, \quad (61)$$

and $\theta_y(max) \sim 2.5\, \theta_y(rms)$. These angles are a fraction of a milliradian for typical parameters. The beam transport must be designed to accommodate the disrupted beams as they leave the interaction point to avoid backgrounds. The present thinking is to have the beam collide at a small angle and to have large aperture exit paths in the final focus quadrupoles. If that is done the disrupted beams probably will not be the critical source of backgrounds.

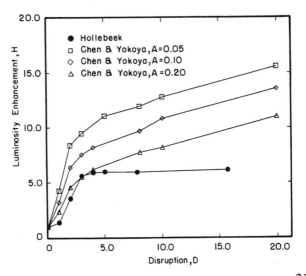

Fig. 35. Luminosity enhancement due to disruption from Hollebeek[25] and Chen and Yokoya.[26] These results are for round beams, and the parameter A is defined by Chen and Yokoya as $A = \varepsilon \gamma D / r_e N$. $A = 0$ in the Hollebeek calculation.

Beamstrahlung. The electromagnetic fields at the interaction point are so large that beamstrahlung (radiation) and pair production dominate linear collider designs. The beamstrahlung spectrum is characterized by a scaling parameter[29]

$$\Upsilon \sim \frac{r_e^2 \gamma N}{\alpha (\sigma_x + \sigma_y) \sigma_s} \quad (62)$$

where $\alpha = 1/137$ is the fine structure constant. When $\Upsilon \ll 1$ the beamstrahlung spectrum is a classical synchrotron radiation spectrum that peaks at an energy $E \sim E_c/3$ where E_c is called the critical energy. It is related to Υ by $\Upsilon = 2/3 \times (E_c/E_{beam})$. A collider with $\Upsilon \ll 1$ is said to operate in the "classical" beamstrahlung regime. The SLC and most likely the next linear collider are in the classical regime. When $\Upsilon \gg 1$ the spectrum must be calculated quantum mechanically,[30] and the collider is in the "quantum" beamstrahlung regime.

A second parameter characterizing beamstrahlung, δ, is the average fractional energy loss of a particle during the collision. The dependences of δ on beam sizes, N and γ are different in the classical and quantum regimes[12]

$$\delta_{cl} = 0.88 \frac{r_e^3 N^2 \gamma}{\sigma_s (\sigma_x + \sigma_y)^2}, \quad \delta_q = 0.56 \frac{\delta_{cl}}{\Upsilon^{4/3}} \quad (\Upsilon \gtrsim 10). \qquad (63)$$

The fractional energy loss is important for experimenters because it determines the average value of and the spread in the center-of-mass energy. Expressions for the luminosity, δ, and the beam power ($P_b = \gamma mc^2 N f_c$) can be combined to give scaling relationships for linear colliders[12]

$$\frac{L^2 \gamma^3}{\delta_{cl}} = 7.2 \times 10^{-3} \frac{(\sigma_x + \sigma_y)^2 \sigma_s H P_b^2}{r_e^3 (mc^2)^2 \sigma_x^2 \sigma_y^2}, \qquad (64)$$

and

$$\frac{L^2 \gamma}{\delta_q^3} = 5.0 \times 10^{-2} \frac{(\sigma_x + \sigma_y)^2 H P_b^2}{\alpha^4 r_e (mc^2)^2 \sigma_x^2 \sigma_y^2 \sigma_s} \qquad (65)$$

(H is the luminosity enhancement from disruption). The quantities on the left-hand-sides are determined by particle physics, and they give a relationship between the beam sizes and the beam power. The latter is the dominant contribution to the operating cost because the beam power must come from the AC mains with some efficiency for converting mains power to beam power.

Equations (64) and (65) show the constraints that beam-beam effects put on collider parameters. In particular, they show that the luminosity can be enhanced for a given δ with a flat beam, and this is one of the reasons that linear collider designers are concentrating on flat beams. The underlying reason comes from eqs. (17') and (56). They show that the typical electromagnetic field of the beam scales as

$$E_y(typ.) \propto \left(L \frac{H}{W}\right)^{1/2}, \qquad (66)$$

and this field determines the deflection of a particle and the amount it radiates.

Pair Production. It was realized about two years ago that there would be copious pair production in the beam-beam interaction in linear colliders. There are two processes. One is incoherent pair production - beamstrahlung photons interact with individual particles in the oncoming beam to produce pairs, for example $\gamma e^- \to e^+ e^- e^-$. The cross section is large, and $\sim 10^6$ pairs are produced per beam burst.[28] The second process is coherent pair production where beamstrahlung photons interact with the fields of the bunch rather than with an individual particle. The relative importance is determined by the scaling parameter Υ defined in eq. (62).[28] When $\Upsilon \ll 1$ (the classical beamstrahlung regime) coherent pair production is not important. When $\Upsilon \gg 1$ (the quantum regime) it is the dominant process, and the number of pairs per burst can be 10^7 or more.[28]

Linear colliders must be designed to minimize the impact of this potential background. Flat beams help by keeping Υ small and thereby avoiding coherent pair production. Equation (66) has the underlying reason again; the fields are reduced for a flat beam. The second way to control this background is having the beams collide at an angle. One particle in the pair is focused by the oncoming beam (see Figure 33) and exits along the axis with the disrupted beam. The other particle in the pair is expelled from the collision region by the high fields of the oncoming beam, but it is captured in a spiral trajectory by the solenoidal field of the detector. Therefore, both particles in the pair travel in the forward direction. A crossing angle separates the paths of the incoming and exiting beams and allows a large aperture exit for the disrupted beam and the shielding appropriate for dumping the large halo of pairs. A disadvantage of colliding at an angle is that the effective overlap area is increased and luminosity is lost, but this can be solved by tilting the beams with an RF deflector just before the interaction region.[31] This "crab crossing" idea appears again in the discussion about beam-beam effects in storage rings.

Beam-Beam Effects in Storage Rings

This is a problem in nonlinear dynamics. The force is not a linear function of the displacement (Figure 32), and this leads to amplitude dependent tunes, resonances, and the possibility of a dynamic aperture. The dynamics of the beam-beam interaction differs from that produced by nonlinear magnetic fields in several ways:
1. There are many resonances in the first order perturbation expansion. The single octupole problem had resonances at $4Q = m$, but the simplest beam-beam problem - a round beam, a linear transport, and a single interaction point - has resonances at $2nQ = m$ where m and n are integers.
2. The beam-beam fields fall-off rather than increase at large displacements. Large amplitude particles experience a small beam-beam force, and their motion is almost linear. There isn't a dynamic aperture except near resonant tunes.
3. The force depends on the beam distribution, and that can evolve in time. This opens up additional possibilities including collective beam-beam effects similar to the instabilities driven by wakefields.

Begin by estimating the amplitude dependent tunes. The electromagnetic fields are linear with displacement for small distances from the center of the beam (eq. (59)). They act like a focusing quadrupole and have all of the effects of such a quadrupole on the betatron motion. Particles with small oscillation amplitudes (eq. (14)) have betatron tunes, Q_x and Q_y, that are greater than the values they would have with no beam-beam interaction, $Q_z \cong Q_{z0} + \xi_z$ where

$$\xi_z = \frac{r_e}{2\pi} \frac{N \beta_z^*}{\gamma(\sigma_x + \sigma_y)\sigma_z} \quad (z = x \text{ or } y). \tag{67}$$

Equation (67) follows from the expressions for the fields (eq. (59)) and the effects of a quadrupole error (eq. (16)). Large amplitude particles experience a small beam-beam force that is not quadrupole-like. Therefore, their motion is almost linear, and they have tunes $Q_z \cong Q_{z0}$. The parameters ξ_x and ξ_y are the spreads in tunes. They are often called the "beam-beam tune shifts" because they are the tune shifts of the small amplitude particles. This name is misleading; it implies that the problem is a linear problem and it isn't - all particles do not have the same tune shifts. It is better to call the ξ's the "beam-beam tune spreads".

The tune spreads characterize the strength of the beam-beam interaction. The empirical limit in hadron colliders is based on experience in the Sp$\bar{\text{p}}$S and Tevatron; it is[32]

$$N_{ip}\xi \leq 0.024 \tag{68}$$

where N_{ip} is the number of interaction points and ξ ($\xi_x \cong \xi_y$) is the tune spread from one interaction point. There is a simple explanation for this limit.[33] The beam-beam interaction and nonlinear magnets produce resonances. If the discussion of a single octupole was extended to two dimensions and other nonlinearities, it would be found that the resonance condition is $mQ_x + nQ_y = p$ (m, n & p are integers). There are infinitely many such resonances, but their strength gets weaker as the resonance order, $|m| + |n|$, gets higher. Particles that fall on any low order resonances - have tunes that satisfy the resonance condition for a low order resonance - are lost from the storage ring. Low order resonances must be avoided, and eq. (68) is a measure of the amount of tune spread that can be accommodated without particles falling on low order resonances.

The situation is more complicated in e^+e^- storage rings. There are data from many more colliders, but these data do not show consistent patterns or trends.[34] A great deal of

Table 2. Beam-Beam Rules and Ideas

RULE	Ref.	EXPLANATION
$\beta_x^*, \beta_y^* > \sigma_s$	37	The betatron phase advance on successive turns is modulated by synchrotron oscillations. Recent work suggests that this rule may not be as stringent as once thought.[38,39]
$\eta^* = 0$	37	η^* is the η-function at the interaction point. If $\eta^* \neq 0$, the horizontal position is energy dependent; this modulates the beam-beam deflection at Q_s.
Head-on collisions	35	With non-zero crossing angle the beam-beam deflection is modulated by synchrotron oscillations.
IDEA	Ref.	EXPLANATION
Round beams	40	The modulation of the vertical deflection by horizontal betatron oscillations is removed.[41]
Crab crossing	36	Tilting the beam removes the modulation introduced by non-zero crossing angle.

attention has been focused on the effects of synchrotron radiation because it was thought that it would reduce further the strength of high order resonances and allow a larger tune spread than in hadron colliders. The data indicate that there are other, more important effects. A rough summary of the data is that in different e^+e^- colliders the maximum tune spreads (both ξ_x and ξ_y) are in the range

$$0.01 \leq \xi (\text{max}) \leq 0.07 \ . \tag{69}$$

Large tune spreads are important! Suppose that the number of particles per bunch or the total number of particles were limited by instabilities, the number of $\bar{p}$'s one can make, etc. Then

$$L = \xi_y \frac{\gamma N f_c (1+\sigma_y/\sigma_x)}{2 r_e \beta_y^*} \tag{70}$$

where $\xi_x = \xi_y$ has been assumed for simplicity. The tune spread measures the efficiency for using the particles to make luminosity. If there are no limits on the number of particles $L \propto \xi^2$, and a large tune spread is even more valuable.

There are a number of related questions. What limits the tune spreads? How can the tune spreads be increased? How can the data from the e^+e^- colliders be understood better? In my opinion the answers to all these questions are related to the nonlinearity of the beam-beam interaction and the nature of the resulting resonances. When the synchrotron oscillations are included along with the two transverse dimensions, the resonance condition becomes $mQ_x + nQ_y + pQ_s = r$ (m, n, p & r are integers and Q_s is the synchrotron tune), and the resonance order is $|m| + |n| + |p|$. Synchrotron motion is not important for hadron colliders because Q_s is low and resonances involving the synchrotron motion would be high order. However, it is important for e^+e^- colliders because the synchrotron tune can be large ($Q_s \sim 0.05$ is not unusual), and low order resonances involving synchrotron motion are possible.

The years of experience with e^+e^- storage rings have led to some empirical rules for getting large tune spreads. Most of these rules are ways to avoid resonances. Table 2 gives the rules, brief statements about the physics, and the original references. Understanding the

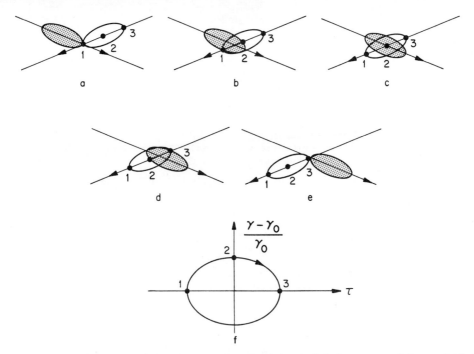

Fig. 36. Two bunches colliding at an angle. Particles 1, 2 & 3 are part of the unshaded bunch. a) through e) show the passage of time during the collision, and f) shows the positions of the particles in longitudinal phase space.

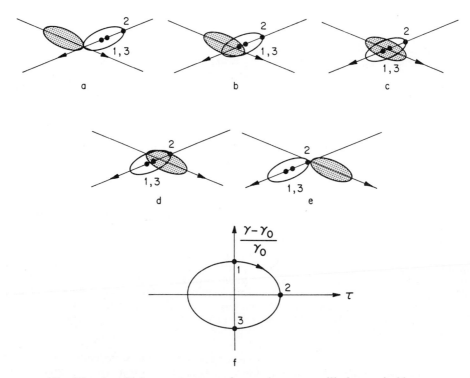

Fig. 37. A collision one-quarter of a synchrotron oscillation period later.

physics of these empirical rules leads to ideas about new modes of operation that could have higher tune spreads. Two such ideas are included in Table 2. Crab crossing and head-on collisions are related closely, and they are discussed as an example.

Figure 36 shows two bunches colliding at an angle. Particle #1 at the head of its bunch passes through the head of the oncoming bunch while particle #2 passes through the center and particle #3 passes through the tail. Particle #2 is deflected more than #1 and #3 because the charge density of the opposing bunch is greater at the center than at the head or tail. One-quarter of a synchrotron oscillation period later the situation is different (Figure 37). Particles #1 and #3 are in the center of the bunch, and #2 is at the tail. Therefore, #1 and #3 experience larger deflections than #2. The beam-beam deflection is modulated by synchrotron oscillations, and this leads to synchrobetatron resonances - resonances involving Q_s. The collisions must be head-on to avoid this modulation. The beams in DORIS I crossed at an angle, and DORIS I had a low maximum tune spread.[35] That experience led to the rule that collisions must be head-on.

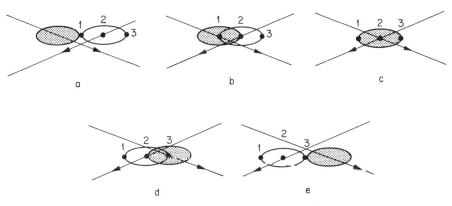

Fig. 38 A collision with crab crossing.

Colliding at an angle could be attractive for high luminosity because a high collision frequency would be possible. Crab crossing which was discussed above can be adapted to storage rings to permit crossing at an angle.[36] It is illustrated in Figure 38. The bunches cross at an angle, but they are tilted with respect to their directions of propagation. Each of the three particles pass through the head, center, and tail of the opposing bunch, and they have the same beam-beam deflection. The source of the modulation with synchrotron motion and the resonance are removed while still having a crossing angle. The cost is that RF cavities with deflecting fields are needed at each side of the interaction point to tilt the bunches before and remove the tilt after the collision.

Considerations such as those above are central to the design of heavy quark factories and the planned upgrades of the Tevatron collider because these accelerators are pushing other limits on N or $N \times f_c$. Efficient utilization of the current is essential for high luminosity.

CONCLUDING REMARK

Accelerators are key instruments for particle physics. They are interesting physical systems in their own right also. I hope these lecture notes show that and give experimenters some appreciation for the work of their colleagues in the accelerator part of our science.

REFERENCES

1. Edwin M. McMillan, Phys. Rev. **68**, 143(1945).
2. E. Courant and H. Snyder, Annals of Physics **3**, 27 (1958).
3. Site-Specific Conceptual Design for the SSC, ed. D. Matthews (1990).
4. Herbert Goldstein, Classical Mechanics, Addison-Wesley Publishing Company, Reading, Mass. (1959).
5. An excellent reference for electron storage rings is M. Sands, The Physics of Electron Storage Rings - An Introduction, SLAC-121 (SLAC, 1970).
6. T. Weiland, CERN/ISR-TH/80-07 (CERN, 1980).
7. R. H. Siemann, Proc. of the 17th SLAC Summer Institute, ed. E. C. Brennan, p. 263 (1990).
8. R. H. Siemann, AIP Conf. Proc. **127**, 368 (1985).
9. W. K. H. Panofsky and W. A. Wentzel, Rev. Sci. Inst. **27**, 967 (1956).
10. K. L. F. Bane, IEEE Trans. Nucl. Sci. **NS-32**, 3565 (1985).
11. V. E. Balakin, A. V. Novokhatsky, and V. P. Smirnov, Proc. of 12th Int. Conf. on High Energy Accel., ed. F. Cole, R. Donaldson, p. 119 (1983).
12. P. B. Wilson, SLAC-PUB-3674, (SLAC, 1985).
13. W. Schnell, CERN-LEP-RF/87-24, (CERN, 1987).
14. R. D. Kohaupt, Proc. of the 11th Int. Conf. on High Energy Accel., ed W. S. Newman, p. 562 (1980).
15. This analysis comes from A. Chao, AIP Conf. Proc. **105**, 353 (1983).
16. K. Robinson, CEAL Report TM-183 (CEA, 1969).
17. Klaus G. Steffen, High Energy Beam Optics, Interscience Publishers, New York (1965).
18. K. L. Brown and R. V. Servranckx, AIP Conf. Proc. **127**, 62 (1985).
19. R. A. Erickson, AIP Conf. Proc. **184**, 1554 (1989).
20. C. Pellegrini, Nuovo Cimento **LXIV**, 447 (1969).
21. This analysis is from R. Talman, AIP Conf. Proc. **153**, 835 (1987).
21a. H. E. Fisk et al, Proc. of the 1984 Summer Study on the Design and Utilization of the SSC, edited by R. Donalson and J. G. Morfin, p. 329 (1984).
22. R. P. Johnson, Proc. of the 1987 IEEE Accel. Conf., ed. E. R. Lindstrom and L. S. Taylor, p. 8 (1987).
23. SSC Conceptual Design, ed. J. D. Jackson (1986).
24. M. Bassetti and G. Erskine, CERN-ISR-TH/80-06, (CERN, 1980).
25. R. Hollebeek, Nucl. Inst. & Methods **184**, 333 (1981).
26. P. Chen and K. Yokoya, Phys. Rev. **D38**, 987 (1988).
27. M. Breidenbach et al, SLC Performance in 1991, (SLAC, 1990).
28. P. Chen, AIP Conf. Proc. **184**, 633 (1989).
29. T. Himel and J. Siegrist, AIP Conf. Proc. **130**, 602 (1985).
30. R. Blankenbecler and S. D. Drell, Phys. Rev. **D36**, 277 (1987).
31. R. Palmer, SLAC-PUB-4707 (SLAC, 1989).
32. R. Schmidt and M. Harrison, to be published in Proc. of 1990 European Part. Accel. Conf.
33. L. Evans, AIP Conf. Proc. **127**, 243 (1985).

34. J. Seeman, <u>Nonlinear Dynamics Aspects of Particle Accelerators</u>, Springer-Verlag, Berlin, edited by J. M. Jowett, S. Turner and M. Month, p. 121 (1986).

35. A. Piwinski, IEEE Trans. Nucl. Sci. **NS-24**, 1408 (1977).

36. K. Oide and K. Yokoya, Phys. Rev. **A40**, 315 (1989).

37. F. M. Izrailev and I. B. Vasserman, <u>7th All Union Conf. on Charged Part. Accel.</u>, p. 288 (1981).

38. D. Rice, Part. Accel. **31**, 1315 (1990).

39. S. Krishnagopal and R. Siemann, Phys. Rev. **D41**, 2312 (1990).

40. S. Krishnagopal and R. Siemann, <u>Proc. of the 1989 IEEE Part. Accel. Conf.</u>, edited by F. Bennett and J. Kopta, p. 836 (1989).

41. S. Peggs and R. Talman, Phys. Rev. **D24**, 2379 (1983).

CALORIMETRY IN HIGH ENERGY PHYSICS

Richard Wigmans

CERN
Geneva, Switzerland

1. INTRODUCTION AND SCOPE OF THE LECTURES

Experimental particle physicists study the fundamental structure of matter with a variety of approaches, which may be subdivided in two classes: accelerator and non-accelerator experiments. Accelerator experiments have the advantage of well-controlled experimental circumstances, non-accelerator experiments offer the possibility of studying processes that are not accessible to the available accelerator technology.

Accurate energy measurements are a prime tool for increasing the knowledge of the constituents of matter, and the forces by which they interact. Three examples may illustrate this. The fundamental nature of the cosmic background radiation emerged when it was realized that the energy spectrum corresponded to that of a black body radiating at a temperature of 3K. On the high-energy side of the scale, the elementary particles are unstable; without exception, the known ones were discovered by a kinematical reconstruction from their decay products, the quality of which is closely linked to the accuracy of the particle energy measurement. And finally, the crucial question whether the neutrino rest mass is different from zero needs an extremely accurate comparison between energy and momentum in order to be answered.

The energy of elementary particles is measured with instruments that are generally called *calorimeters*. There is a wide variety of them. The principle is simple. Basically, a calorimeter is a block of matter in which the particle to be measured interacts and transforms (part of) its energy into a measurable quantity. The resulting signal may be electrical, optical, thermal or acoustical. It is of course important that the signal be proportional to the energy that one wants to measure, which is not always easy to achieve.

In these lectures, I have chosen to concentrate on the calorimeters that are used in high-energy physics, i.e. on the detectors used for measuring particle energies in the GeV range and above. I will therefore ignore emerging (cryogenic) techniques that might in the future be employed in searching for low-energy fundamental particles, such as neutrinos and hypothetical species.

In high-energy physics, we have seen in the last 15 years a clear trend by which experiments changed from *electronic bubble chambers*, aiming for a precise measurement of the 4-vectors of all individual reaction products with spectrometric methods, to configurations with increasing emphasis on calorimetric particle detection. Some major discoveries, e.g. the existence of the intermediate vector bosons W and Z, became possible as a result of this development.

The reasons why calorimeters have emerged as the key detectors in almost any experiment in particle physics, can be divided into two classes. Firstly, there are reasons related to the *calorimeter properties*:

a) Calorimeters are sensitive to both charged and neutral particles.

b) Owing to differences in the characteristic shower patterns, some crucial particle identification is possible.

c) Since calorimetry is based on statistical processes, the measurement accuracy *improves* with increasing energy, in contrast to what happens for other detectors.

d) The calorimeter dimensions needed to contain showers increase only logarithmically with the energy, so that even at the highest energies envisaged it is possible to work with rather compact instruments (cost!).

e) Calorimeters do not need a magnetic field for energy measurements.

f) They can be segmented to a high degree, which allows accurate measurements of the direction of the incoming particles.

g) They can be fast -response times better than 100 ns are achievable- which is important in a high-rate environment.

h) The energy information can be used to trigger on *interesting* events with very high selectivity.

Secondly, there are reasons related to the *physics* to be studied. Here, the emphasis has clearly shifted from track spectroscopy to measuring more global event characteristics, indicative for interesting processes at the constituent level. These characteristics include missing (transverse) energy, total transverse energy, jet production, multijet spectroscopy, etc. Calorimeters are extremely well suited for this purpose.

In these lectures, I will mainly concentrate on the *fundamental aspects* of calorimetry. In Section 2 I will, therefore, rather extensively describe the various processes by which particles lose their energy when traversing dense matter and by which they eventually get absorbed. I will discuss shower-development phenomena, the effects of the electromagnetic and strong interactions, and the consequences of differences between these interactions for the calorimetric energy measurement of electrons and hadrons, respectively.

In Section 3, the performance of calorimeter systems is described. The section starts with a discussion of the so-called compensation mechanism, its relevance for the performance of hadron calorimeters, and the methods to achieve compensation in calorimeters. In a following subsection, the factors that determine and limit the energy resolution of different calorimetric detectors are discussed. The rest of the section describes in some detail the performance of existing and planned devices in terms of energy and position resolution and particle identification.

In section 4, I will discuss the project that I'm most familiar with, which aims to develop a calorimeter that might successfully operate under the extremely difficult experimental circumstances at future multi-TeV *pp* colliders, like the SSC and LHC.

Section 5 gives conclusions, and an outlook to further developments of this important experimental technique.

2. ABSORPTION MECHANISMS OF PARTICLES

When a particle traverses matter, it will generally interact and lose (a fraction of) its energy in doing so. The medium is excited in this process, or heated up, hence the word calorimeter. The interaction processes that play a role depend on the energy and the nature of the particle. They are the result of the electromagnetic, the strong and, more rarely, the weak forces reigning between the particle and the medium constituents. In this section, we will describe the various mechanisms by which particles may lose their energy and eventually be absorbed.

2.1 Electromagnetic absorption

The best known energy-loss mechanism contributing to the absorption process is the electromagnetic (e.m.) interaction experienced by charged particles that traverse matter. The charged particles ionize the medium, if their energy is at least sufficient for releasing the atomic electrons from the e.m. nuclear field. This process forms also the principle on which many detectors are based, since the liberated electrons may be collected by means of an electric field, and yield an electric signal.

The e.m. interaction may manifest itself, however, in many other ways. Charged particles may excite atoms or molecules without ionizing them. The de-excitation from these metastable states may yield (scintillation) light, which is also fruitfully used as a source of calorimeter signals. Charged particles travelling faster than the speed of light characteristic for the traversed medium lose energy by emitting Čerenkov light. At high energies, knock-on electrons (δ-rays) and bremsstrahlung are produced, and even nuclear reactions induced by the e.m. interaction may occur.

The e.m. field quantum, the photon, is affected by three different processes. First, there is the photoelectric effect, in which the photon transfers all its energy to an atomic electron. In the Compton process, only part of the energy is transferred in this way. At energies larger than twice the electron rest mass, the photon may convert into an electron-positron pair. The relative importance of these three processes depends strongly on the photon energy and the electron density ($\sim Z$) of the medium.

Except at the lowest energies, the *absorption* of electrons and photons is a multistep process, in which particle multiplication may occur (*shower development*). This phenomenon, which leads to the absorption of high-energy particles in relatively small volumes, is extensively discussed in the next subsections. The other particles subject to only the e.m. interaction, the muons, do not show such behaviour up to very high energies (100 GeV). They lose their energy primarily through ionization and δ-rays. These mechanisms account for an energy loss of typically $1 - 2$ MeV/g.cm^{-2} and, therefore, it takes very substantial amounts of material to absorb high-energy muons (1 TeV muons may penetrate several kilometres of the Earth's crust).

2.1.1 The 0 to 10 MeV range

Already at fairly low energies, relatively simple showers may develop. Let us consider as an example the γ's of a few MeV characteristic of nuclear de-excitation. The sequence of processes through which γ's of a particular energy are absorbed may be very different from event to event. An example of such a sequence is given in fig. 1. A 3370 keV γ enters the detector and converts into an electron-positron pair. Both particles get a kinetic energy of 1174 keV (point A), the remaining energy is needed for the mass of the e^+ and e^-. The electron loses its kinetic energy through ionization of the detector material and so does the positron. When the positron is stopped, it annihilates with an electron, thus releasing the energy $E = M_{e^+e^-}$ in the form of two γ's of 511 keV each (B). These γ's undergo Compton scattering (C,D), in which part of the energy is transferred to an electron and part to a new γ. The electrons lose their energy as described; the γ's may undergo either another Compton scattering (F), or photoelectric effect in which their full energy is transferred to an electron (E,G).

In this example, which is only one out of an infinite number of different possibilities, the energy of the original γ is absorbed through ionization of the detector medium by one positron and six different electrons. Events for which the whole sequence has taken place inside the sensitive volume of the detector will yield a signal peak at 3370 keV. In small detectors, leakage phenomena may occur (fig. 1). Either one or both 511 keV γ's may escape from the detector. This leads to peaks at energies lower by 511 and 1022 keV, respectively. And if *they* do not escape, some of the tertiary or higher-order γ's might, leading to a continuous background.

At these low energies, a modest role may also be played by photonuclear reactions, e.g. γn, γp or photo-induced nuclear fission. However, the cross-sections for these processes usually do not exceed 1% of the cross-sections for the processes mentioned before and may therefore, in general, be neglected.

2.1.2 From 10 MeV to 100 GeV

Most of the energy-loss mechanisms relevant to high-energy shower development were already mentioned in the previous subsection: Ionization for electrons

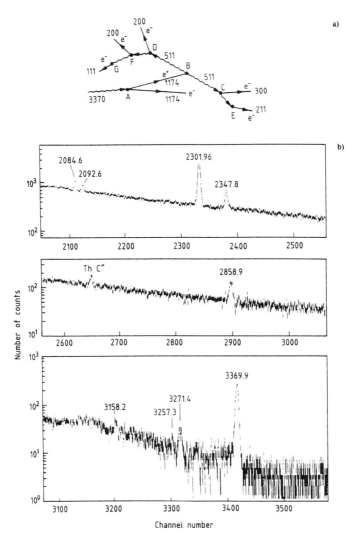

Fig. 1 Shower development induced by few-MeV nuclear γ's. In diagramme a), one possible sequence of absorption processes is shown, with the energies of the e^+, the e^-'s, and the γ's given in keV. The γ-spectrum measured with a Ge(Li) crystal that only partly contains the showers initiated by (among others) the 3370 keV γ is shown in diagrammes b). The total-containment peak (3369.9 keV), the single- (2858.9 keV) and double-escape peak (2347.8 keV), and the continuum background reflect the different degrees of absorption that may occur in this crystal (see text).

and positrons, pair production, Compton scattering, and the photoelectric effect for photons. There is one more, be it crucial mechanism that contributes at higher energies: *bremsstrahlung*.

In their passage through matter, electrons and positrons may radiate photons as a result of the Coulomb interaction with the nuclear electric field. The energy spectrum of these photons falls off like $1/E$. It extends, in principle, to the electron energy, but in general the emitted photon carries only a small fraction of this energy. In this process, the electron itself undergoes a (usually small) change in direction (multiple or Coulomb scattering). The deviation depends on the angle and the energy of the emitted photon, which in turn depend on the strength of the Coulomb field, i.e. on the Z of the absorber medium.

Bremsstrahlung is by far the principal source of energy loss by electrons and positrons at high energies. As a consequence, high-energy e.m. showers are quite different from the ones discussed in the previous subsection, since an important multiplication of shower particles occurs. A primary GeV-type electron may radiate on its way through the detector thousands of photons. The ones faster than 5-10 MeV will create e^+e^- pairs. The fast electrons and positrons from this process may in turn lose their energy by radiation as well, etc, etc. The result is a shower that may consist of thousands of different particles, electrons, positrons, and photons. The overwhelming majority of these particles is very soft. The average energy of the shower particles is obviously a function of the age of the shower, or the depth inside the detector: the further the shower has developed, the softer the spectrum of its constituents becomes.

The energy-loss mechanisms are governed by the laws of quantum electrodynamics (QED)[1]. They primarily depend on the electron density of the medium in which the shower develops. This density is roughly proportional to the (average) Z of the medium, since the number of atoms per unit volume is within a factor of ~ 2 the same for all materials in the solid state.

The results of calculations on the energy-loss mechanisms for photons and electrons are shown in fig. 2, as a function of energy, in three materials with very different Z-values: carbon ($Z = 6$), iron ($Z = 26$), and uranium ($Z = 92$)[2,3]. At high energies, above ~ 100 MeV, pair production by photons and energy loss by radiation dominate in all cases, but at low energies the differences between the various materials are considerable. Both the energy at which Compton scattering starts dominating pair production, and the energy at which ionization losses become more important than bremsstrahlung, are strongly material-dependent and are roughly inversely proportional to Z.

These conditions determine the so-called *critical energy* (ϵ_c), i.e. the point where no further particle multiplication occurs in the shower. Above this energy, γ's produce on average more than one charged particle (pair production), and electrons lose their energy predominantly by creating *new* γ's. Below ϵ_c, γ's produce only *one* electron each, and these electrons do not produce new γ's themselves.

Figure 2 also shows that the contribution of the photoelectric effect is extremely Z-dependent ($\sigma \sim Z^5$). In carbon, it plays a role only at energies below a few keV, while in uranium it is the dominating process below 0.7 MeV.

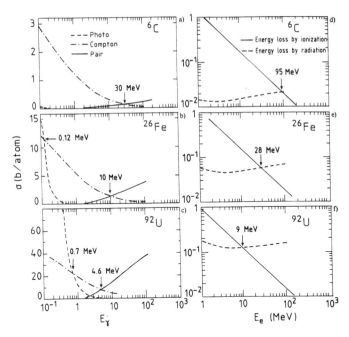

Fig. 2 The cross-sections for pair production, Compton scattering and photo-electric effect, as a function of the photon energy in carbon (a), iron (b), and uranium (c). The fractional energy loss by radiation and ionization, as a function of the electron energy in carbon (d), iron (e), and uranium (f).

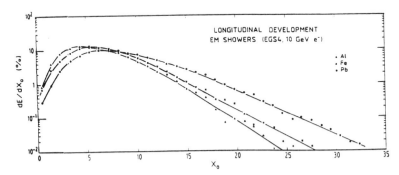

Fig. 3 The energy deposit as a function of depth, for a 10 GeV electron shower developing in aluminium, iron, and lead, showing approximate scaling of the longitudinal shower profile, when expressed in units of radiation length, X_0. Results of EGS4 calculations.

The approximate shape of the longitudinal shower profile can be deduced from figure 2. If the number of e^+ and e^- were to be measured as a function of depth in the detector, one would first find a rather steep rise due to the multiplication. This continues up to the depth at which the average particle energy equals ϵ_c. Beyond that point no further multiplication will take place and, since more and more electrons are stopped, the total number of remaining particles slowly decreases.

The positrons will predominantly be found in the early shower part, i.e. before the maximum is reached. Showers in high-Z materials will contain more positrons than in low-Z materials, because positron production continues until lower energies. The average energy of the shower particles is also lower in high-Z materials, since radiation losses dominate until lower energies. These effects will turn out to have interesting consequences.

Owing to the fact that the underlying physics is well understood and simple, e.m. shower development can be simulated in great detail by Monte Carlo techniques. One program, EGS4[4] has emerged as the world-wide standard for this purpose. It is extremely reliable, and in the following sections several results of it will be shown.

2.1.3 Above 100 GeV

At very high energies new effects will influence the absorption of electrons and photons in a block of matter. In the TeV region, the cross-sections for the e.m. and weak interactions become comparable and, therefore, processes involving hadron and/or neutrino production are no longer negligible.

Another, purely e.m. effect was first pointed out by Landau and Pomeranchuk[5] and treated quantitatively by Migdal[6]. They showed that at energies beyond 10 TeV, multiple scattering of the participating particles may lead to a significant decrease of the cross-sections for bremsstrahlung and pair production. Such effects will obviously change the gross behaviour of the shower development. The experimental information on these phenomena is scarce and, because of their limited practical relevance at present, we will not discuss them further.

2.1.4 Electromagnetic shower characteristics

Since the e.m. shower development is primarily determined by the electron density in the absorber medium, it is to some extent possible, and in any case convenient to describe the shower characteristics in a material-independent way. The units that are frequently used to describe the characteristic shower dimensions are the *radiation length* (X_0) for the longitudinal development and the *Molière radius* (ρ_M) for the transverse development.

The radiation length is defined as the distance over which a high-energy (> 1 GeV) electron loses on average 63.2% ($1 - 1/e$) of its energy to Bremsstrahlung. The average distance that very high energy photons travel before converting into an e^+e^- pair equals $9/7\ X_0$. The Molière radius is defined by the ratio of X_0 and ϵ_c, where ϵ_c is the electron energy at which the losses through radiation and ionization are the same. For approximate calculations, the following relations hold:

$$X_0 \approx 180 A/Z^2 \text{ (g/cm}^2\text{)} \text{ and } \rho_M \approx 7A/Z \text{ (g/cm}^2\text{)} .$$

Expressed in these quantities, the shower development is approximately material-independent. Figure 3 shows the longitudinal development of a 10 GeV electron shower in Al, Fe, and Pb, as obtained with EGS4 simulations. The profile is as expected from the discussion in section 2.1.2. Globally, it scales indeed with X_0. The differences between the various materials can be understood as well. The radiation length is defined for GeV-type particles and, therefore, does not take into account the peculiarities occurring in the MeV region. The shift of the shower maximum to greater depth for high-Z absorbers is a consequence of the fact that particle multiplication continues until lower energies; the slower decay beyond this maximum is due to the fact that lower-energy electrons still radiate.

The figure shows that it takes $\sim 25 X_0$ to absorb at least 99% of the shower energy. This corresponds to 14 cm Pb, 44 cm Fe, or 220 cm Al. If the energy is increased, only very little *extra* material is needed to achieve the same containment. A 20 GeV photon will travel on average 9/7 radiation length before converting into an $e^+ e^-$ pair of 10 GeV each. It therefore takes only an extra $1.3 X_0$ to contain twice as much energy.

The radiation length is, strictly speaking, defined for infinite energy and has no meaning in the MeV energy range. We just showed that ~ 15 cm lead absorb 20 GeV photon showers for more than 99%, whereas it takes *more* than that to make a proper shielding for a strong ^{60}Co source that emits 1.3 MeV γ's. The reason for this is clear from fig. 2 . The total cross-section around the region where Compton scattering takes over from pair production is considerably lower than at very high energies, particularly in high-Z materials. As a consequence, the mean free path in lead of photons of a few MeV is ~ 3 cm, or $\sim 5 X_0$!

The *lateral spread* of an e.m. shower is caused by two effects:

a) Electrons move away from the axis by multiple scattering.

b) In the energy region where the total cross section is minimal, bremsstrahlung photons may travel quite far from the shower axis, in particular if they are emitted by electrons that themselves travel under a considerable angle with this axis.

The first process dominates in the early stages of the shower development, while the second process is predominant beyond the shower maximum, particularly in high-Z media. Figure 4 shows the lateral distribution of the energy deposited by an e.m. shower in lead, at various depths[7] . The two components can be clearly distinguished (note the logarithmic ordinate). The radial profile shows a pronounced central core surrounded by a *halo*. The central core disappears beyond the shower maximum. Similar calculations in aluminium showed that the radial profile, expressed in ρ_M units, is indeed narrower than in lead. Like the radiation length, also the Molière radius does not take into account the peculiarities occurring in the MeV region.

Figure 4 shows that e.m. showers are very narrow, especially in the first few radiation lengths. The Molière radius of lead is ~ 1.7 cm. With a sufficiently fine-grained calorimeter, the showering particle can therefore be localized with a precision of ~ 1 mm (see section 3.3).

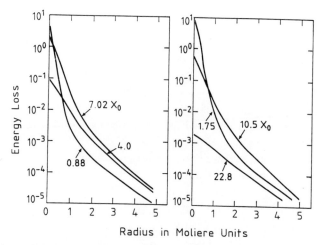

Fig. 4 The lateral distribution of the energy deposited by a 1 GeV e.m. shower in lead, at various depths. Results of EGS4 calculations.

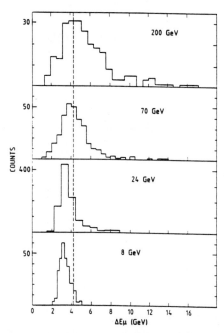

Fig. 5 The energy loss distributions, ΔE_μ, for 8 - 200 GeV muons, measured in a 8.5 nuclear interaction lengths deep uranium/plastic-scintillator calorimeter. The dashed line, drawn to guide the eye, corresponds to the most probable energy loss, measured for 200 GeV muons. Data from ref. 10.

2.1.5 Energy loss by muons

Muons passing through matter lose their energy also through e.m. processes. Compared to electrons, however, the cross-sections for higher-order QED processes, such as bremsstrahlung or e^+e^- pair production are suppressed by a factor of $(m_\mu/m_e)^2 \approx 40000$. The critical energy is, for example, at least 200 GeV. At energies below 100 GeV, the energy loss by muons will, therefore, be dominated by ionization processes.

The mean energy loss per unit path length for these processes, $<dE/dx>$, is given by the well-known Bethe-Bloch formula[8]. For relativistic muons, $<dE/dx>$ falls rapidly with increasing β, reaches a minimum value near $\beta = 0.96$ (minimum-ionizing particles), then undergoes what is called the relativistic rise, to level off at values of 2 - 3 MeV/g.cm^{-2} in most materials.

In practical calorimeters, the total energy loss $\Delta E/\Delta x$ may differ quite a bit from the value calculated from $<dE/dx>$. This is because of the relatively small number of collisions with atomic electrons, and of the very large fluctuations in energy transfer that may occur in such collisions. Therefore, the energy loss distribution will in general be peaked at values below the ones calculated from $<dE/dx>$ and have a long tail toward large energy losses, the so-called Landau tail[9]. Only for very substantial amounts of matter, at least 100 m of water equivalent, will the energy-loss distribution become approximately Gaussian.

Figure 5 shows some experimental results on muon energy loss, measured in the HELIOS calorimeter[10], which consists essentially of ~ 1 m of uranium. The figure clearly shows the asymmetric distribution of energy losses. It also shows that at increasing energies, and in particular at 200 GeV, higher-order QED processes such as bremsstrahlung start dominating over ionization losses, since the distribution of energy losses is shifted by a considerable amount to higher values.

2.2 Strong-interaction processes

The absorption of particles subject to the strong interaction (hadrons) in a block of matter proceeds in a way that is very similar, in many respects, to the one described for electromagnetically interacting particles, although in detail the particle-production mechanisms are substantially more complicated. When a high-energy hadron penetrates a block of matter, it will at some point interact with one of its nuclei. In this process, mesons are usually produced (π, K, etc.). Some other fraction of the initial particle energy is transferred to the nucleus. The excited nucleus will release this energy by emitting a certain number of nucleons and, at a later stage, low-energy γ's, and lose its kinetic (recoil) energy by ionization. The particles produced in this reaction (mesons, nucleons, γ's) may in turn lose their kinetic energy by ionization and/or induce new reactions, thus causing a shower to develop.

Some of the particles produced in this cascading process interact exclusively electromagnetically (e.g. π^0, η). Therefore, hadron showers contain in general a component that propagates electromagnetically. The fraction of the initial hadron energy converted into π^0's and η's may strongly vary from event to event, depending on the detailed processes occurring in the early phase of the shower development, i.e. the phase where production of these particles is energetically possible.

On average, approximately one third of the mesons produced in the first interaction will be π^0's. In the second generation of interactions, the remaining π^+, π^-, etc. may produce π^0's as well, if they are sufficiently energetic, and so on. And since production of π^0's by hadronically interacting mesons is an irreversible process, the average fraction of the initial hadron energy converted into π^0's increases (approximately logarithmically) with the energy.

Although the shower development by hadrons and electrons shows many similarities, there exist some characteristic differences which turn out to have crucial consequences.

2.2.1 Shower dimensions

The hadronic shower development is (for an important part) based on nuclear interactions and, therefore, the shower dimensions are governed by the *nuclear interaction length* λ_{int}. The nuclear interaction probability is determined by the fraction of a two-dimensional plane occupied by atomic nuclei; since the number of atoms per unit volume is to first order material-independent, λ_{int} will approximately scale with the nuclear radius, i.e. as $A^{1/3}$, when expressed in units of g/cm^2.

Existing experimental data indicates that the longitudinal and lateral profiles of hadronic showers scale roughly with λ_{int}. Figure 6 shows the results of measurements that give a good impression of the longitudinal and lateral development of 300 GeV showers in uranium[11]. The profiles look very similar to e.m. showers (figs. 3,4), albeit on a very different scale. It takes about 80 cm of uranium to contain the 300 GeV π^- showers at the 95% level, while 10 cm would be sufficient for electrons at the same energy.

The leakage as a function of the detector depth is shown in fig. 7, for hadron energies ranging from 5 to 210 GeV. It turns out that the detector size needed to contain, for instance, more than 99% of the shower energy, increases only very slightly with the energy, from $6\lambda_{int}$ at 5 GeV to $9\lambda_{int}$ at 210 GeV[12].

One may use the differences in characteristic energy deposit for particle identification. Since λ_{int} scales with $A^{1/3}$ and X_0 with A/Z^2, the separation between electromagnetically interacting particles (e, γ, π^0) and hadrons works best for high-Z materials, where the ratio λ_{int}/X_0 may reach values larger than 30 (see section 3.4).

2.2.2 Invisible energy

A second crucial difference between the shower development by high-energy electrons and by hadrons concerns the fact that, in the latter case, a certain fraction of the energy is dissipated in undetectable (*invisible*) form. Apart from neutrinos and high-energy muons, which may be generated in the hadronic shower-development process and which will generally escape the detector, we refer mainly to the energy needed to release nucleons from the nuclear field that keeps them

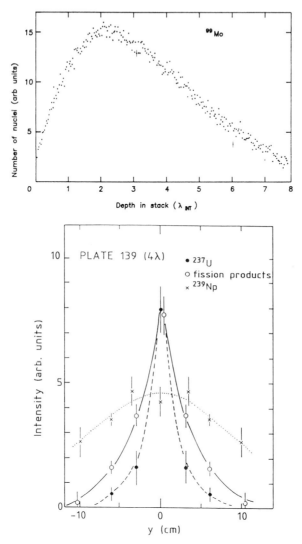

Fig. 6 Longitudinal (top diagramme) and lateral (lower diagramme) shower profiles for 300 GeV π^- interactions in a block of uranium, measured from the induced radioactivity. The ordinates indicate the number of radioactive decays of a particular nuclide, produced in the absorption process of the high-energy pion. Since the different nuclides are produced by different types of shower particles, such experimental data may yield valuable information on details of the shower development process (from ref. 11).

together. Some fraction of this nuclear binding energy loss may be recuperated when neutrons get captured by other nuclei. The protons, α's, and heavier nucleon aggregates released in nuclear reactions will, however, only lose their *kinetic energy*, by ionization.

The fraction of invisible energy can be quite substantial, up to 40% of the energy dissipated in non-e.m. form[13].

At low energies (< 2 GeV) the probability that charged hadrons lose their kinetic energy *without* causing nuclear interactions, i.e. by ionization alone, increases rapidly. In this case, as for muons and e.m. showers, there are no invisible-energy losses. As a consequence, hadron calorimeters suffer in general from signal non-linearities at low energy[14] (see section 3.2).

2.2.3 Non-relativistic shower particles

A third difference, which has important consequences for the calorimetric energy measurement of elementary particles, results from the fact that a large fraction of the energy deposited in *hadronic* showers is carried by (extremely) non-relativistic particles, i.e. protons and neutrons. We mention three consequences:

i) Many protons produced in the shower-development process have a specific ionization $<dE/dx>$ that is 10 to 100 times the minimum-ionizing value, depending on the Z of the traversed medium. As a consequence, the fraction of the energy of such particles detected by sampling calorimeters consisting of alternating layers of absorber and active material that usually have very different Z-values, may be considerably different from the fraction detected for minimum-ionizing particles. This is illustrated in fig. 8.

ii) Some frequently used active calorimeter media show a strongly non-linear behaviour in their response to densely ionizing particles. They suffer from saturation (scintillator)[15] or recombination effects (liquid argon[16], room-temperature liquids[17]). Such effects may suppress the response, i.e. the signal per unit deposited energy, by as much as a factor 5 for this shower component[18]. These effects are much smaller, or absent, when gases or silicon are used as the active calorimeter medium.

iii) Neutrons, which lose their kinetic energy *exclusively* through strong interactions may travel quite long distances before being finally absorbed. In calorimeters where the neutrons contribute significantly to the signal, this may lead to a considerable prolongation of the pulse duration for hadronic signals, compared with e.m. ones. Typical time constants for the neutron contribution to the calorimeter signal amount to 10 ns for the release of kinetic energy, and 0.5 μs for the γ's created in the thermal-neutron capture process. These phenomena may be exploited for particle identification, and in particular for e/π separation (see section 3.4.1).

2.2.4 The role of neutrons

Regarding calorimetric applications, perhaps the most crucial difference between e.m. and hadronic shower development comes from the fact that a considerable fraction of the energy is carried by *non-ionizing* particles, i.e. the soft (few MeV) neutrons from the nuclear evaporation processes.

Since these neutrons lose their kinetic energy exclusively through collisions with atomic nuclei, their contribution to the signal of *sampling* calorimeters is completely dependent on the nuclear peculiarities of the materials composing the calorimeter. It is well-known that in particular hydrogen is very efficient in slowing down neutrons.

It was shown both experimentally[11] and theoretically[13,19,20] that in calorimeters with hydrogenous active material, the neutrons generated in the shower development may deposit a large fraction of their kinetic energy in the active layers, while charged particles are only sampled at the few per cent level. This effect is an important tool for making so-called *compensating* calorimeters.

3. PERFORMANCE OF CALORIMETER SYSTEMS

In this section, we will discuss the performance of calorimeters, these instrumented blocks of dense matter in which the particles interact and get absorbed through the processes described in the previous section, yielding signals from which the particle properties (energy, direction, type) can be derived.

Historically, one may distinguish between calorimeters, according to the purpose that they serve, as e.m. and hadronic shower detectors. Nowadays, there is a growing tendency to combine both functions in one instrument.

Another distinction that may be made concerns their composition: homogeneous, fully sensitive devices as opposed to sampling calorimeters. The latter consist of a passive absorber with active material embedded into it, most frequently in the form of a sandwiched layer structure. In this way only a small fraction of the initial particle energy, ranging from 10^{-5} for gas calorimeters to a few percent for solid or liquid readout media, is deposited in the active layers.

Although additional fluctuations, affecting the energy resolution, are caused by the fact that only a fraction of the energy is deposited in the active material, the sampling technique is becoming more and more popular, particularly in accelerator-based experiments, for the following reasons:

i) Since very dense absorber materials can be used, calorimeters can be made extremely compact. Even at the highest energies envisaged today, 2 m of lead or uranium is sufficient to contain all showers at the 99% level (see fig. 7).

ii) At increasing energies, the energy resolution tends to be dominated by *systematic* effects; therefore, the effects of sampling fluctuations become less important.

iii) Contrary to fully sensitive devices, sampling calorimeters can be made *compensating*.

Before discussing in detail actual devices, we will first elaborate on the latter point, which is crucial for the performance of hadron calorimeters.

3.1 Compensation

3.1.1 The role of the e/h signal ratio

In a given calorimeter, hadron showers are detected with an energy resolution that is worse than for e.m. ones. This is mainly because, in hadronic showers,

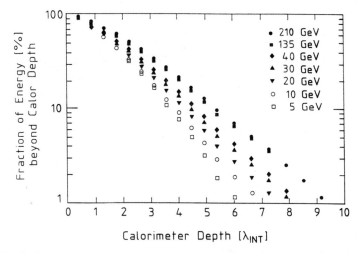

Fig. 7 The leakage as a function of the detector depth, for pions of 5 - 210 GeV, measured in a uranium/plastic-scintillator calorimeter. Data from ref. 23.

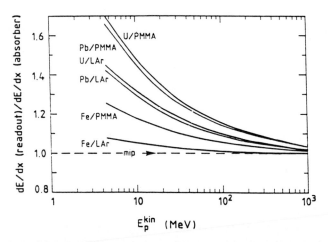

Fig. 8 The fraction of the ionization energy deposited in the active layers by non-interacting protons, in various calorimeter configurations. From ref. 14.

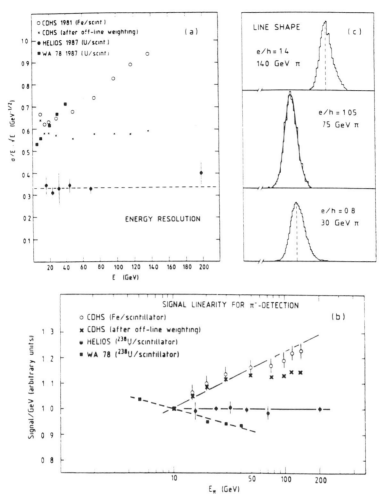

Fig. 9 Experimental observation of the consequences of $e/h \neq 1$. Results of measurements on pion absorption in undercompensating[22], compensating[10], and overcompensating[23] calorimeters. In diagramme a), the energy resolution $\sigma/E \cdot \sqrt{E}$ is given as a function of the pion energy, showing deviations from scaling for non-compensating devices. In diagramme b), the signal per GeV is plotted as a function of the pion energy, showing signal non-linearity for non-compensating detectors. The signal distribution for monoenergetic pions (the line shape, diagramme c) is only Gaussian for the compensating calorimeter.

fluctuations occur in the fraction of the initial energy carried by ionizing particles. Losses in nuclear binding energy (see section 2.2.2) may consume up to 40% of the incident energy, with large fluctuations about this average.

As a consequence, the signal distribution for monoenergetic pions is wider than for electrons at the same energy, and has in general a smaller mean value ($e/\pi > 1$). The calorimeter response to the e.m. (e) and non-e.m. (h) components of hadron showers shows a similar difference ($e/h > 1$). Since the event-to-event fluctuations in the fraction of the energy spent on π^0 production (f_{em}) are large and non-Gaussian, and since $<f_{em}>$ increases (logarithmically) with energy, the following effects have to be expected if $e/h \neq 1$:

 i) The signal distribution for monoenergetic hadrons is non-Gaussian.

 ii) The fluctuations in f_{em} give an additional contribution to the energy resolution.

 iii) The energy resolution σ/E does not improve as $E^{-1/2}$ with increasing energy.

 iv) The calorimeter signal is *not* proportional to the hadron energy (non-linearity).

 v) The measured e/π signal ratio is energy dependent.

Because of the latter effect, we prefer to use the energy-independent quantity e/h. In practice, the difference between e/h and e/π is small, and it vanishes at low energies and for e/h close to 1. All these effects have been experimentally observed[10,22,23] (fig. 9) and can be reproduced with a simple Monte Carlo.

At increasing energies, deviations from $e/h = 1$ (the *compensation condition*) rapidly become a dominating factor for the (lack of) calorimeter performance, e.g. for the energy resolution σ/E (fig. 9a). Signal non-linearities of $\sim 20\%$ over one order of magnitude in energy have been observed, both in overcompensating ($e/h < 1$) and undercompensating ($e/h > 1$) calorimeters (fig. 9b). But perhaps the most disturbing drawback of a non-compensating calorimeter, especially in an environment where high trigger selectivity is required, is the non-Gaussian response (fig. 9c), which may cause severe problems if one wants for example to trigger on (missing) transverse energy: it will be very difficult to unfold a steeply falling E_T distribution and a non-Gaussian response function. Moreover, severe trigger biases are likely to occur: if $e/h < 1$ (> 1) one will predominantly select events that contain little (a lot of) e.m. energy from π^0's.

There is general agreement that hadron calorimeters for future applications at high energy should be compensating. It should be emphasized that other sources of experimental uncertainty, such as calibration errors, will produce effects similar to the ones caused by deviations from $e/h = 1$. Therefore, it is not necessary that e/h be exactly 1. It has been estimated[13] that $e/h = 1 \pm 0.05$ is adequate to achieve energy resolutions at the 1% level.

3.1.2 Methods to achieve compensation

Because of the invisible-energy phenomenon, *i.e.* the nuclear binding energy losses typical for (the non-e.m. part of) hadronic showers, one might naïvely expect the e/h signal ratio to be larger than 1 for all calorimeters. This is, however a tremendous oversimplification. Based on our present understanding of hadron calorimetry[13,19,20], it may be expected that a large variety of very different structures can actually be made compensating. A wealth of available experimental data supports the framework of this understanding, and explicit predictions were experimentally confirmed.

The response of a sampling calorimeter to a showering particle is a complicated issue that depends on many details. This is particularly true for hadronic showers. It has become clear that showers can by no means be considered as a collection of minimum ionizing particles that distribute their energy to absorber and active layers according to $< dE/dx >$. The calorimeter signal is, to a very large extent, determined by very soft particles from the last stages of the shower development, simply because these particles are so numerous. Many observations support this statement.

Simulations of high-energy e.m. showers in lead or uranium sampling calorimeters show that about 40% of the energy is deposited through ionization by electrons softer than 1 MeV[13]. Measurements of pion signals in fine-sampling lead/plastic-scintillator calorimeters revealed that there is almost no correlation between the particles contributing to the signal of consecutive active layers[24,25]. This proves that the particles that dominate the signal travel on average only a very small fraction of a nuclear interaction length indeed.

For a correct evaluation of the e/h signal ratio of a given calorimeter, the last stages of the shower development must therefore be understood in detail, *i.e.* the processes at the nuclear and even the atomic level must be analysed. The particles that decisively determine the calorimeter response are *soft photons* in the case of e.m. showers, and *soft protons and neutrons* from nuclear reactions in non-e.m. showers. Since most of the protons contributing to the signal are highly non-relativistic, the saturation properties of the active material for densely ionizing particles are of crucial importance (see section 2.2.3).

There are many other factors that affect the signals from these shower components, and thus e/h. Among these, there are *material* properties, such as the Z values of the active and passive components, the hydrogen content of the active media (see section 2.2.4), the nuclear-level structure and the cross-section for thermal-neutron capture by the absorber; and there are the *detector* properties, such as the size, the signal integration time, the thickness of the active and passive layers, and the ratio of these thicknesses.

In order to achieve compensation, three different phenomena may be exploited:

i) The non-e.m. response may be selectively boosted by using depleted uranium (^{238}U) absorber plates. The fission processes induced in the non-e.m. part of the shower development yield extra energy, mainly in the form of soft γ's and neutrons[24]. This phenomenon also leads to the commonly used terminology, since the extra energy released in ^{238}U fission *compensates* for the nuclear binding energy losses.

ii) One may selectively suppress the e.m. response by making use of the peculiarities of the energy deposit by the soft-photon component of e.m. showers. Below 1 MeV, the photoelectric effect is an important energy loss mechanism. Since the cross-section is proportional to Z^5, soft photons will interact almost exclusively in the absorber layers of high-Z sampling calorimeters. They will only contribute to the signal if the interaction takes place sufficiently close to the boundary layer, so that the photoelectron can escape into the active material. This effect may lead to a considerable suppression of the em response[13]. It may be enhanced by shielding the active layers by thin sheets of passive low-Z material[18,21,26,27].

iii) The most important handle on e/h is provided by the neutron response, in particular for calorimeters with hydrogenous active material (see section 2.2.4). In this case, the fraction of the neutron's kinetic energy transferred to recoil protons in the active layers varies much more slowly with the relative amounts of passive and active material than does the fraction of the energy deposited by charged particles. Therefore, the relative contribution of neutrons to the calorimeter signal, and hence e/h, can be varied by changing the sampling fraction[13,19]. A small sampling fraction enhances the relative contribution of neutrons. It is estimated that in compensated lead- or uranium-scintillator calorimeters, neutrons make up $\sim 40\%$ of the non-e.m. signal, on average[20,21]. The lever arm on e/h provided by this mechanism may be considerable. It depends on the energy fraction carried by soft neutrons (favouring high-Z absorbers), on the hydrogen fraction in the active medium, and on the signal saturation for densely ionizing particles.

Apart from these methods, which aim at achieving $e/h = 1$ as an *intrinsic* detector property, a completely different approach has been applied, in order to reduce the mentioned disadvantages of an intrinsically non-compensating detector by means of off-line corrections to the measured data[22,28]. In this approach, which requires a very fine-grained detector, one tries to determine the π^0 content on a shower-by-shower basis, and a weighting scheme is used to correct for the different calorimeter responses to the π^0 and non-π^0 shower components.

An example of the results of calculations on e/h for uranium calorimeters is shown in fig. 10[13]. For hydrogenous readout materials (plastic scintillator, warm liquids) the e/h value sensitively depends on the relative amount of active material, and in any case configurations can be found with $e/h = 1$.

Experimental results clearly confirm the tendency predicted for plastic scintillator readout[10,23,29]. For non-hydrogenous readout, mechanism iii) does not apply. Here the neutron response, and hence the e/h ratio, can be affected through the signal integration time, taking more or less advantage of the considerable energy released in the form of γ's when thermal neutrons are captured by nuclei, a process that occurs at a time scale of 1 μs. Experimental results obtained so far seem to confirm the prediction that it will be hard to achieve *full* compensation with liquid-argon (LAr) readout[30,31,32]. In U/Si detectors, a full exploitation of mechanisms i) and ii) might yield a compensating calorimeter, since there are no saturation effects[18,26,27]. Detectors with gaseous readout media offer a convenient way to tune e/h to the desired value, i.e. through the hydrogen content of the gas mixture. This has been demonstrated experimentally by the L3 Collaboration[33].

The curves for TMP calorimeters given in fig. 10 are based on the assumption that the signal saturation in this liquid is equal either to LAr or to PMMA plastic scintillator. Preliminary experimental data indicate that the signal suppression in warm liquids is considerably larger, and that it depends, perhaps, on the electric-field strength and on the particle's angle with the field vector[34]. Figure 11 shows how sensitively the e/h signal ratio depends on the saturation properties. The UA1 Collaboration recently reported having measured an e/h value smaller than 1.1 with their first uranium-TMP calorimeter module[35].

In summary, compensation is *not* a phenomenon restricted to uranium calorimeters, nor is the use of uranium absorber a guarantee for achieving compensation. It has become clear that both the readout medium and the absorber material determine the e/h value. Compensation is easier to achieve with high-Z absorbers because of the large neutron production and the correspondingly large leverage on e/h. But even materials as light as iron allow compensation, if used in combination with, for instance, plastic scintillator, albeit with impractically thick absorber plates[13].

The neutron production in lead is considerably smaller than in uranium. In order to bring e/h to 1.0 for lead/scintillator detectors, the neutron signal has therefore to be more enhanced relative to charged particles than for uranium/scintillator calorimeters. As a consequence, the optimal sampling fraction is smaller for lead. The calculations[13] predicted e/h to become 1.0 for lead plates about 4 times as thick as the scintillator, while for uranium plates a thickness ratio of 1:1 is optimal. This prediction was experimentally confirmed by the ZEUS Collaboration[36]. They found $e/h = 1.05 \pm 0.04$, a hadronic energy resolution scaling with $E^{-1/2}$ over the energy range 3 - 75 GeV, and no deviations from a Gaussian line shape.

The mechanisms described above, which make compensating calorimeters possible, only apply to *sampling* calorimeters. They are based on the fact that only a small fraction of the shower energy is deposited in the active part of the calorimeter; thus, by carefully choosing the parameter values, one may equalize the response to the e.m. and non-e.m. shower components. This does *not* work for *homogeneous* devices, where, in the non-e.m. shower part, losses will inevitably occur that cannot be compensated for. Measurements performed so far with homogeneous hadron detectors support this conclusion, for what concerns the e/h ratio[37], as well as the resulting non-linearity, the non-Gaussian response, and the poor energy resolution[38].

3.2 Energy response and resolution

3.2.1 Fluctuations in the energy measurement

The detection of particle showers with calorimeters is based on *statistical* processes: the production of ionization charge, scintillation or Čerenkov photons, phonons, or electron-hole pairs in semiconductors. The energy resolution for particle detection is therefore determined, among other factors, by fluctuations in the number of primary, uncorrelated processes n. The width of the signal distribution σ_S for detection of monoenergetic particles with energy E will therefore relate to n as $\sigma_S/S \sim \sqrt{n}/n$, which leads for linear calorimeters to the familiar relation $\sigma_E/E = c/\sqrt{E}$.

It has become customary to give a value of c for expressing calorimetric energy resolutions, where E is given in units of GeV. Because of the statistical nature of calorimetry, the relative energy accuracy σ_E/E *improves* with increasing energy. This very attractive feature is one of the reasons why these instruments have become so popular.

Fluctuations in the number of primary processes constituting the calorimeter signal form the ultimate limit for the energy resolution. In most detectors, the energy resolution is dominated by other factors. We mention two exceptions. Firstly, there are the semiconductor nuclear-γ detectors, such as Ge, Ge(Li), and Si(Li) crystals. It takes very little energy to create one electron-hole pair in these crystals, only 2.9 eV in Ge. The signal of a 1 MeV γ fully absorbed in the detector will therefore consist of some 350,000 electrons. The fluctuations in this number lead to an energy resolution of $\sigma_E/E = 0.17\%$ (at 1 MeV!). Owing to correlations in the production of consecutive electron-hole pairs (the so-called Fano factor), the limit of the energy resolution given by fluctuations in the number of primary processes will be somewhat larger. In practice, resolutions of $\sim$ 2.0 keV at 1 MeV are indeed achieved with such detectors.

A second example are lead-glass e.m. shower counters. They are based on the detection of the Čerenkov light produced by the electrons and positrons from the shower. Particles travelling at a velocity lower than the velocity of light in the absorber will *not* emit this light, and therefore the lead-glass is only sensitive to electrons with a kinetic energy larger than $\sim$ 0.7 MeV. This means that at maximum only $1000/0.7 \sim 1400$ particles can produce Čerenkov light, per GeV of shower energy, and that the resolution σ_E/E cannot become better than $\sim 3\%$ at 1 GeV because of fluctuations in this number. The best lead-glass detector systems have reached $\sigma_E/E \approx 5\%$ for e.m. showers in the 1 - 20 GeV energy range[39].

More frequently, the energy resolution is determined by factors *other* than the fluctuations in the number of primary processes. These factors may concern statistical processes with a Gaussian probability distribution, or be of a different nature. In the latter case, their contribution to the energy resolution will cause deviations from the $E^{-1/2}$ scaling law. Such deviations are of course most apparent at high energies.

As an example, we mention homogeneous scintillation counters(e.g. NaI(Tl), CsI, BGO). Compared to what happens with the semiconductor crystals discussed before, the following complications arise:

i) The scintillation photons are not monoenergetic.

ii) Only a fraction of the photons reach the light-detecting element. The rest is either absorbed or refracted. These effects strongly depend on the detector geometry and on the position where the light is produced.

iii) The sensitivity of photocathodes or photodiodes depends on the wavelength.

Measurements with NaI(Tl) crystals on 8 keV X-rays yielded $\sigma_E/E \approx 15\%$. If we assume that this result is dominated by fluctuations in the primary processes, this means that on average ~ 40 photoelectrons are observed. Based on this result, one would then expect resolutions of $\sigma_E/E \approx 1.5\%$ at 1 MeV. Yet, the best resolutions obtained at this energy are only about 5%. For e.m. shower detection one finds resolutions $\sigma_E/E \approx 1\%$ at 1 GeV, while a factor 30 improvement should be expected when extrapolating the 1 MeV results with the $E^{-1/2}$ scaling law. For this reason, it is incorrect to express resolutions as $c/\sqrt{E}$ for such detectors.

Other factors which will cause deviations from $E^{-1/2}$ scaling are instrumental: noise and pedestal contributions to the signal, uncertainties coming from calibration and non-uniformities, or incomplete shower containment.

The energy resolution of *sampling* calorimeters is frequently dominated by the very fact that the shower is sampled[40]. The nature of these sampling fluctuations is purely statistical and, therefore, they contribute as $c/\sqrt{E}$ to the final energy resolution. A major contribution comes from fluctuations in the *number* of *different* shower particles contributing to the calorimeter signal. In some devices (e.g. gas or Si readout), also the fluctuations in the energy that the individual shower particles deposit in the active calorimeter layers have to be taken into account. In calorimeters with *dense* active material (plastic scintillator, LAr), the contribution of sampling fluctuations to the energy resolution tends to scale as $\sigma_{samp}/E = \sqrt{t_{abs}/E}$, for a particular combination of passive and active material, a fixed thickness of the active planes and a thickness t_{abs} of the passive planes. The sampling fluctuations depend also on the thickness t_{act} of the active planes. Photon conversions in these planes contribute a term which scales like $c\sqrt{1/t_{act}}$, for fixed t_{abs}[14]. The relative contribution of this term depends on the Z values of the active and passive calorimeter layers. For Fe/LAr, one finds that σ_{samp}/E scales approximately like $t_{act}^{-1/4}$, for fixed t_{abs}.

When detecting electrons and hadrons with the same calorimeter, sampling fluctuations for the latter particles are considerably larger. First of all, the number of *different* shower particles contributing to the hadronic signal is smaller, because

i) Individual shower particles may traverse many planes.

ii) The average energy deposited by individual particles in the active layers is larger (soft protons!)

Moreover, the spread in the dE/dx loss of individual shower particles in the active layers is much larger.

The contribution of sampling fluctuations to the energy resolution can be measured in a straightforward way, by comparing energy resolutions measured with different fractions of the active calorimeter channels[24,25].

In hadronic shower detection, two additional sources of fluctuation play a role, which have no equivalent for e.m. calorimeters, and which tend to dominate the energy resolution of practically all hadron calorimeters constructed up to now. Firstly, there are the effects of the non-Gaussian fluctuations in the π^0 shower component (section 3.1.1); these contribute a constant term to the energy resolution, which only vanishes for compensating detector structures.

Secondly, there are *intrinsic* fluctuations, in the fraction of the initial energy that is transformed into ionizing shower particles (the *visible* energy, see section

2.2.2). These form the ultimate limit for the energy resolution achievable with hadron calorimeters. In general one may therefore write (ignoring instrumental contributions such as shower leakage, calibration, etc.)

$$\sigma_{had}/E = \sqrt{\frac{c_{int}^2 + c_{samp}^2}{E}} + a$$

This formula shows that at high energy one will want to have a as small as possible (compensation); moreover, it is useless to make the sampling much finer than the limit set by the intrinsic fluctuations.

It turns out that calorimeters with hydrogenous readout are not only advantageous for achieving compensation: they may also yield considerably lower values for c_{int} than other detectors[13]. The intrinsic resolution is largely dominated by fluctuations in the nuclear binding energy losses. Since most of the released nucleons are neutrons in the case of high-Z target material, there is a correlation between this invisible energy and the kinetic energy carried away by neutrons. Efficient neutron detection therefore *reduces* the effect of the intrinsic fluctuations on the energy resolution.

Recently, the ZEUS Collaboration measured the intrinsic-resolution limit for compensating uranium- and lead-scintillator calorimeters to be $19\%/\sqrt{E}$ and $11\%/\sqrt{E}$, respectively[25]. This difference, which means that in principle better energy resolutions can be achieved with lead calorimeters than with uranium ones, can be explained as follows[41]. The extent to which the mechanism described above works depends on the *degree* of correlation between the nuclear binding energy losses and the kinetic neutron energy. This correlation is expected to be better in lead than in uranium, since in the latter case many of the neutrons come from fission processes. These fission neutrons are less strongly correlated to the nuclear binding energy losses.

3.2.2 Performance of electromagnetic calorimeters

The best energy resolutions for e.m. shower detection are obtained with homogeneously sensitive detectors and they remain the method of choice when ultimate performance is needed. While detectors based on NaI(Tl) have been in use for decades, delivering consistently energy resolutions of $\sigma/E \sim 0.02 E^{-1/4}$ [42] there are several recent developments providing alternative, sometimes even superior, performance characteristics. Particularly noteworthy are

i) The use of CsI(Tl), which offers similar performance but better mechanical properties compared with NaI(Tl)[43].

ii) The use of BGO, which allows the construction of very compact detectors due to its short radiation length[44].

iii) The use of BaF$_2$ crystals, which have a very fast (rise-time $\approx$ 500 ps) ultraviolet scintillation light component, and promise very high rate capability and good radiation resistance[45,46,47,48].

iv) Studies on homogeneously sensitive noble-liquid (Ar, Kr, Xe) e.m. calorimeters[49,50,51,52,53], which are evaluated to provide energy resolution in the NaI-range combined with ultimate radiation hardness.

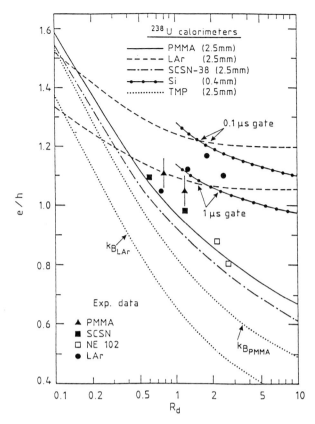

Fig. 10 The signal ratio e/h for uranium calorimeters employing different readout materials, as a function of the ratio of the thicknesses of absorber and readout layers. Results of experimental measurements are included. From ref. 13.

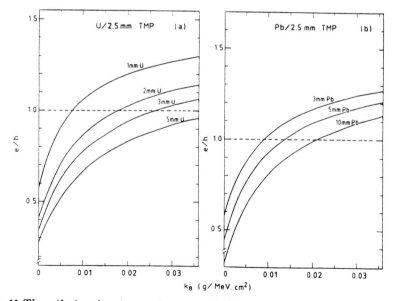

Fig. 11 The e/h signal ratio as a function of Birk's constant k_B, for TMP calorimeters with uranium (a) and lead (b) absorber. The liquid gaps are 2.5 mm wide. From ref. 18.

When σ/E drops below 1%, the resolution tends to become dominated by systematic effects such as calibration uncertainties. This is generally true for all calorimeters.

The energy resolution reported for lead-glass e.m. shower counters varies between $5\%/\sqrt{E}$[39] and $12\%/\sqrt{E}$, depending among others on the shower containment.

Achieving a sufficient degree of shower containment is usually no problem for the much cheaper sampling calorimeters. With calorimeters based on plastic-scintillator or LAr readout, energy resolutions are usually found to be in the range from $7\%/\sqrt{E}$[54] to $\sim 20\%/\sqrt{E}$. With gaseous readout media the energy resolution is usually worse, because of the additional effects of fluctuations in the energy that individual shower particles deposit in the active layers (path length, Landau fluctuations).

A very fine shower sampling can be obtained when scintillating plastic fibres are used as the active material. Energy resolutions as low as $8\%/\sqrt{E}$ (for 0.5 mm thick fibres) have been achieved with this technique[55].

Signal *linearity* is usually no problem for e.m. shower detection. Only calorimeters with gas gain readout operating in a digital mode (streamer, Geiger) may suffer from non-linearities at high energy, because of the increased density of particles in the core of the shower, leading to a saturation in the gas gain.

3.2.3 Performance of hadron calorimeters

As pointed out in section 3.1.1, the energy resolution σ/E does in general *not* scale as $E^{-1/2}$ for hadron calorimeters. Only for devices with e/h sufficiently close to 1.0 is such scaling observed down to values of $\sigma/E \sim 1\% - 2\%$, where instrumental effects start dominating the results.

The energy resolution of *homogeneous* detectors is dominated by their non-compensating nature. The value of σ/E does not improve below $\sim 10\%$, even at energies as high as 150 GeV[37,38].

Most *sampling* calorimeters currently employed as hadron detectors use iron as the absorber material, with active layers consisting of plastic scintillator, LAr, or wire chambers. None of these detectors has achieved energy resolutions better than $55\%/\sqrt{E}$ (at 10 GeV), while rapid deviations of $E^{-1/2}$ scaling occur at higher energies.

Efforts to determine the π^0 content on a shower-by-shower basis and to correct the signal by means of a weighting scheme[22] did result in a restoration of the $E^{-1/2}$ scaling for detection of single pions of known energy; these weigthing algorithms may, however, introduce signal non-linearities, and doubts remain about the applicability of such a scheme for detecting jets of unknown composition and energy[21,28].

With compensating calorimeters, considerably better energy resolutions have been obtained. For uranium/plastic-scintillator detectors, values of $\sim 35\%/\sqrt{E}$ were reported by the HELIOS[10] and ZEUS Collaborations[29], scaling with $E^{-1/2}$ up to ~ 200 GeV. ZEUS also measured $44\%/\sqrt{E}$ for a compensating Pb/plastic-scintillator sandwich detector (10 mm Pb/2.5 mm scintillator)[36]. Because of the dominating contribution of sampling fluctuations to the latter result, values of $30\%/\sqrt{E}$ or better may be expected for the fine-sampling compensating Pb/scintillating-fibre detector that is being developed at CERN[41] (see sect. 4).

The overcompensating U/scintillator calorimeter built by WA78 showed an energy resolution behaviour similar to the one measured for the undercompensating iron calorimeters mentioned before: $\sigma/E \sim 52\%/\sqrt{E}$ at 10 GeV, rapidly degrading at higher energies[23].

The deviations from the $E^{-1/2}$ scaling are less dramatic in U/LAr calorimeters, which have e/h values closer to 1. The energy resolution obtained by D0 amounts to about $60\%/\sqrt{E}$ at 10 GeV[30]; SLD found a similar value in their initial prototype studies[31]. HELIOS have recently found resolutions around $45\%/\sqrt{E}$[32].

First results obtained with the new UA1 uranium-TMP calorimeter modules show an energy resolution for pion detection of $45.7\%/\sqrt{E} + 7\%$[35].

The best absolute values for σ/E were reported by HELIOS, 1.9% for 3.2 TeV ^{16}O ions[10], 1.0% for 6.4 TeV ^{32}S ions[32]. Figure 12 shows the results of calorimetric measurements on the 3.2 TeV ^{16}O beam from the CERN SPS, dumped in their U/plastic-scintillator calorimeter. The good energy resolution allows a detailed study of the spectrum of the contaminating lower-mass ions.

The effects of non-compensation on the *signal linearity* were already discussed in section 3.1.1 and experimental results were shown in fig. 9b. Signal non-linearities of $\sim 20\%$ over one order of magnitude in hadron energy are commonplace in non-compensating calorimeters like the WA1 and WA78 ones shown in this figure.

The weighting procedure mentioned before did *not* eliminate these non-linearities, particularly below ~ 30 GeV. Therefore, a jet composed of 10 particles of 10 GeV will yield a signal that is considerably different from a 100 GeV pion signal and the resolution improvement is likely to be considerably smaller for jet detection.

At low energies, below ~ 2 GeV, a different kind of non-linearity occurs, also for compensating calorimeters. At these low energies, charged hadrons may lose their energy *without* undergoing strong interactions and the corresponding nuclear binding energy losses. In that case, they deposit their energy through ionization alone, like muons. The hadronic response h will go up, and the e/h ratio down. The response to hadrons thus *increases* at low energies. This is clearly confirmed by experimental data. Figure 13 shows the measured e/π signal ratio as a function of energy, for various calorimeter configurations. A clear decrease in this ratio, i.e. an *increase* in the hadronic response, is observed below ~ 2 GeV.

3.3 Position and angular resolution

The position and the angle of incidence of particles can be obtained with calorimeters, using measurements of the transverse and longitudinal shower distribution.

The localization of e.m. (σ_x^e) and hadronic (σ_x^h) showers is derived from the center of gravity of the transverse distributions, which are narrowest in the early part, before the shower maximum. Given adequate transverse granularity, the resolution σ_x^e or σ_x^h is determined by the signal/noise ratio and, therefore, improves with increasing particle energy E, approximately as $E^{-1/2}$.

In the narrow early part, more than 90% of an e.m. shower is contained in a cylinder of radius $r \approx 0.5 X_0$ and hence millimetre accuracies for σ_x^e are readily obtained, even with rather coarse (i.e. few $X_0 \times$ few X_0) granularity. As an example, mm-resolution is quoted for few-GeV showers in a lead-glass array of 3.5×3.5 cm^2 granularity[56], reaching submillimeter accuracy for 100 GeV showers[57,58]. Even higher spatial resolutions are obtained if a high-resolution detector, such as a Multiwire Proportional Chamber[59] or a Si-strip array[60], is inserted into the calorimeter at a depth of $\sim 5 X_0$. With such techniques, the position of a 100 GeV e.m. shower may be measured with an accuracy of ~ 100 μm!

Similar considerations apply to the localization of hadronic showers, which consist of a very narrow core surrounded by a *halo* of particles extending to several times the core diameter[11,61,62] (fig. 6). Measurements of the spatial resolution of the impact point[10,29,63,64] may be parameterized in the form

$$\sigma_x^h \,[\text{cm}] \leq 0.2\, \lambda_{eff}[\text{cm}]/\sqrt{E[\text{GeV}]}$$

In compact calorimeters, where the effective nuclear interaction length λ_{eff} may be as low as 20 cm, spatial resolutions in the range of a few cm at 1 GeV are achievable, reaching a few mm for $E \sim 100$ GeV. The resolution σ_x^h as a function of the transverse segment length $d(\lambda)$ has been evaluated[65] and a dependence of $\sigma_x^h \sim \exp(2d)$ may be derived, suggesting a limit on the useful transverse segmentation of $d \approx 0.1 \lambda$.

The measurement of the **angular** resolution for e.m. (σ_θ^e) and hadronic showers (σ_θ^h) has been carefully studied for several calorimeters used to investigate neutrino scattering[63,66,67,68] and, more recently, for 4π calorimeter facilities at future hadron colliders[60].

While the parametrization of the observed performance shows typically a $1/\sqrt{E}$ improvement of the angular resolution, the achieved values depend critically on the design of the apparatus. Typical results are quoted:

$\sigma_\theta^e[\text{mrad}] \leq 20/\sqrt{E[\text{GeV}]}$ (refs. 67,68).
$\sigma_\theta^e[\text{mrad}] \simeq 3.5 + 53/E[\text{GeV}]$ (ref. 66).

For carefully optimized hadronic angular resolution (achieved by choosing a material in which e.m. and hadronic showers have approximately the same spatial dimensions, i.e. $\lambda \approx 3 X_0$), a value $\sigma_\theta^h\,[\text{mrad}] \approx 160/\sqrt{E[\text{GeV}]} + 560/E[\text{GeV}]$ was reported[63,66,69].

3.4 Particle identification

Hadronic calorimeters provide identification of a class of particles that are not readily identified by other methods. In the following, I will discuss in some detail the identification of electrons, muons, and neutrinos.

3.4.1 Discrimination between electrons (photons) and hadrons

Most frequently, discrimination is based on the spatial differences in the shower profiles, accentuated in absorber materials with very different radiation and absorption lengths, i.e. at high Z (see fig. 14). High-Z materials (lead, tungsten, uranium) are most frequently used to optimize the electron-hadron discrimination.

The principal physics limitation to this technique is the charge-exchange reaction $\pi^- p \to \pi^0 n$ (or $\pi^+ n \to \pi^0 p$), which may simulate an e.m. shower. For few-GeV pions the cross-section for this reaction is at the 1% level of the total inelastic cross-section, and it decreases logarithmically with increasing energy[70]. Consequently, typical values for e/π discrimination are of the order of 10^{-2} in the 1 - 10 GeV range and 10^{-3} beyond 100 GeV[71,72,73,74,75]. A further improvement by a factor of 3 to 5 can be obtained if information on the transverse shower profile is available[76] (see also sect. 4.4).

The quoted rejection factors are based solely on the use of shower profile analysis. With additional momentum information, e.g. from magnetic momentum analysis, a further improvement in rejection of typically an order of magnitude is obtained.

A novel method, based on the differences in the time development of e.m. and hadronic showers, will be discussed in sect. 4.4.

One of the most challenging detection problems, and of increasing importance for modern particle-physics experimentation, is the identification of electrons close to the core of high-momentum jets. One proposed solution[60] is based on a detailed three-dimensional measurement of energy deposit, through which e.m. showers appear as relatively energetic deposits, concentrated in a very small volume compared with the extension of the hadronic showers from the majority of the jet particles.

A related technique will be used by the ZEUS Collaboration, who will embed in their hadron calorimeter at a depth of $4X_0$ and $7X_0$ a layer of 40000 silicon detectors, each one having an active area of several cm^2. Pulse-height measurements in these Si elements will allow them to discriminate between high-particle-density e.m. showers and the relatively wide hadron showers, permitting electron identification in the presence of hadrons[77,78].

3.4.2 Muon identification

Several calorimetric methods exist for discriminating between muons and hadrons, all based on the very large differences of energy deposit:

- *i*) Calorimeters with fine longitudinal segmentation: energetic muons are clearly recognized as isolated, minimum-ionizing tracks, possibly ranging far beyond the tracks from hadronic showers. This is the technique used in experiments on incident neutrinos, but also some of the highly granular collider detectors are expected to offer this identification potential.
- *ii*) Muon penetration through active or passive absorbers: hadrons are adequately absorbed, so that the 'punch-through' probability of pions ($P \sim$

$\exp[-d/\lambda]$) is sufficiently small. The absorber thickness d measured in *detectable* absorption lengths (λ), was found to agree closely with tabulated values for relatively thin absorbers[79,80,81] ($d \leq 10\lambda$), but shows significant deviations for absorber thicknesses $d > 10 - 15\lambda$[82,83].

The detailed rejection power against hadrons depends critically on the experimental precautions taken; it may be improved by

a) reducing the background from π and K decay in front of the calorimeter,

b) measuring the muon momentum after the calorimeter, for instance in magnetized iron[84,85] or a precision spectrometer[86],

c) correlating the direction or the momentum of muon candidates before and after the absorber.

Very good muon identification will continue to be emphasized when considering experiments for future hadron colliders. In these experiments, accurate muon-momentum measurements will be mandatory because they will exploit *missing-energy* signatures in the search for new physics phenomena.

3.4.3 Neutrino identification

Apparent *missing energy* or *missing momentum* has become a powerful technique to infer the presence of neutrinos among the collision products:

a) Missing energy relies on a measurement of the total energy using 4π calorimetric coverage for all particles (charged, neutral, muons). This can in practice be achieved for collisions at e^+e^- storage rings or in fixed-target experiments. Neutrino production is inferred whenever the measured energy is lower than the total available energy and incompatible with the resolution function of the detector. A total-energy measurement is *not* practical at hadron colliders, because a considerable fraction of the total

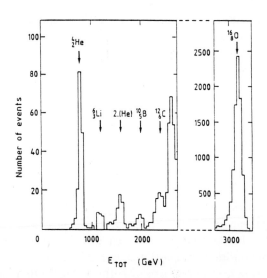

Fig. 12 The HELIOS calorimeter as a high-resolution spectrometer. Total energy measured in the calorimeter for minimum-bias events, showing the composition of the CERN heavy-ion beam. Data taken from ref. 10.

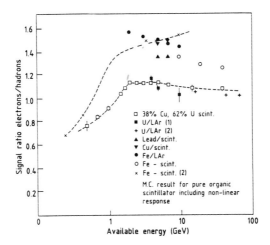

Fig. 13 The ratio of the e.m. to hadronic response as a function of energy, for various calorimeter configurations, showing the non-linearity for hadron detection at low energy.

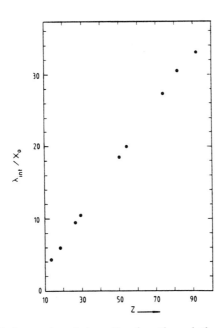

Fig. 14 The ratio of the nuclear interaction length and the radiation length as a function of Z.

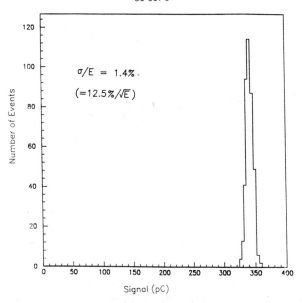

Fig. 15 The signal distribution for 80 GeV electrons measured with the SPACAL detector at 90° incidence.

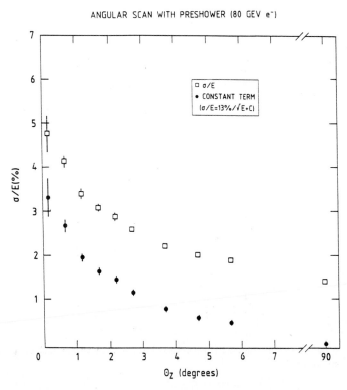

Fig. 16 The constant term in the e.m. energy resolution as a function of the angle θ_z between the particles and the fibre direction.

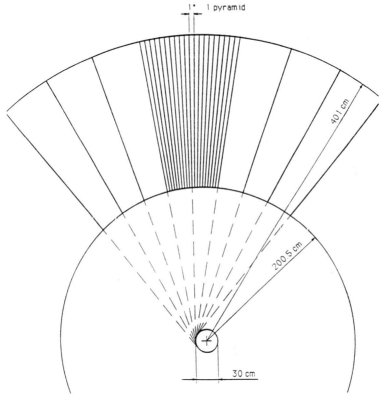

Fig. 17 A possible solution for the problems with electrons entering the detector at very small angles with the fibre direction. The modules are not pointing towards the interaction point but to a cylinder around the beam pipe. Lateral cross section of a possible LHC detector.

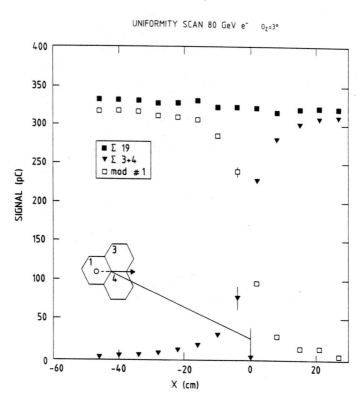

Fig. 18 The signal uniformity for e.m. shower detection. See text for details.

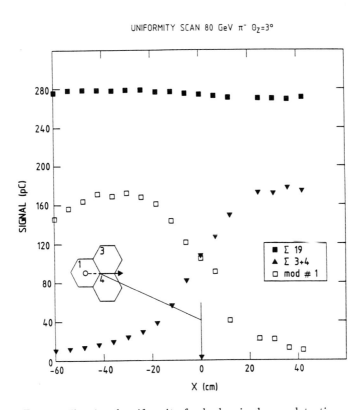

Fig. 19 The signal uniformity for hadronic shower detection.

energy is produced at angles too close to the colliding beams, and hence inaccessible for calorimetric measurements. In this case it is advantageous to implement

b) A *missing-transverse-momentum* measurement, where neutrino production is signalled by $\Sigma \mathbf{p}_{\perp,i} \neq 0$, incommensurate with detector resolution. The intrinsic quality of such measurements was estimated to be $\sigma(\mathbf{p}_{\perp,miss}/\mathbf{p}_{total})$ $\approx 0.3/\sqrt{\mathbf{p}_{total}}^{[87]}$, considerably better than the values achieved up to now: $\sigma(\mathbf{p}_{\perp,miss}/\mathbf{p}_{total}) \approx 0.7/\sqrt{\mathbf{p}_{total}}^{[88,89,90]}$ or $\sigma(\mathbf{p}_{\perp,miss} \approx 0.6 \mathbf{p}_\perp^{0.43} [\text{GeV}]^{[91]}$.

4. THE SPAGHETTI CALORIMETER PROJECT AT CERN

It is generally acknowledged that the advent of the new generation of particle accelerators, the multi-TeV high-luminosity *pp* colliders or Supercolliders (LHC at CERN, SSC in Texas), requires major improvements in the detector technology applied up till now, in order to allow for meaningful experiments at these machines. The event rates to be handled, the radiation levels to be sustained, and the required selectivity of the detector/data acquisition system are orders of magnitude beyond present experience. The quality of the physics will, therefore, much more depend on (and be limited by) the quality of the detectors than in previous generations of experiments.

Among the various detector components, the calorimeter is definitely one the most crucial elements. The calorimeter has to provide the first-level trigger information, it is the first filter for selecting the rare interesting events under extreme background conditions. Moreover, the calorimeter information is the basis for studying the production of jets and leptons, the likely messengers of new physics. It allows identifying them and measuring their properties with high precision.

In order to fulfill these tasks in the best possible way in the harsh conditions of an LHC/SSC experiment, the calorimeter should be

- As fast as possible,
- As hermetic as possible, and
- As radiation resistant as possible.

Apart from these 3 requirements which are specific for an LHC/SSC environment, the calorimeter should also meet the standards demanded from these devices in other applications:

- It should be compensating, i.e. give equal response to electromagnetic (e.m.) and hadronic showers.
- It should detect electrons, hadrons and missing energy (e.g. caused by neutrinos or supersymmetric particles) with a good energy resolution.
- The signals should be uniform, i.e. not depend on the impact point of the particles.
- It should be possible to distinguish e.m. and non-e.m. showers on the basis of the calorimeter information.

- It should be possible to calibrate the detector and to monitor its stability with a precision that is better than the energy resolution envisaged.

Although large calorimeters have been applied since about two decades in particle physics experiments, it is only since a few years that the factors which determine (the limitations in) their performance, and in particular the compensation mechanism, are sufficiently well understood to allow for dedicated improvement projects. The project that is the subject of this chapter was initiated after it was theoretically predicted[13], and subsequently proven experimentally[92], that a lead/plastic-scintillator sampling calorimeter can be made compensating, a feature which until that moment was believed to be a unique property of uranium devices. The project aimed for optimizing the technique of compensating lead/plastic-scintillator calorimetry by using large numbers of scintillating plastic fibres as active material (hence the name Spaghetti Calorimeter or SPACAL), rather than the traditionally used scintillator plates.

In order to make a lead/plastic calorimeter compensating, the detector should have a small sampling *fraction*, a factor 4 smaller than the typical value for current devices. Sampling fluctuations will, therefore, tend to dominate the energy resolution. This can be counterbalanced by using a high sampling *frequency*, hence the choice of fibres as active material. In addition, scintillating fibres produce extremely fast signals and allow constructing a very hermetic device, thus satisfying some of the essential conditions mentioned before.

In the past $2\frac{1}{2}$ years, most of the efforts of the SPACAL group went into optimizing the lead/scintillating fibre calorimeter technique. The aspects that were studied included techniques for incorporating large numbers of scintillating fibres in a block of lead, the optical properties of the fibres (e.g. maximizing the attenuation length) and the light collection uniformity. The results of the prototype studies were described in a number of articles[93,94,95].

Various generations of prototypes were constructed, starting with small e.m. modules of $\sim$ 10 kg, followed by hadronic modules of $\sim$ 100 kg, and by a 2-ton e.m./hadronic calorimeter. Recently our final detector, a 20-ton integrated e.m./hadronic calorimeter with 155 readout channels, was exposed for the first time to particle beams. After the technical problems involved were solved, the actual construction of this detector took about 4 months. Being obtained very recently, the hadron data that will be shown in the following are to be considered highly preliminary. They were obtained on-line, concern raw data, without the proper calibration factors, etc.

4.1 Electromagnetic shower detection

4.1.1 Energy resolution

In spite of its small sampling *fraction* (2.3%), the SPACAL detector provides good energy resolution for e.m. shower detection, because of the high sampling *frequency*. This is illustrated in fig. 15, where the signal distribution for 80 GeV electrons is shown. The resolution σ/E amounts to 1.4%, which corresponds to $12.5\%/\sqrt{E}$. This result was obtained for $\theta_z = 90°$, i.e. by shooting the particles perpendicular to the fibre direction, which is the most favourable situation. More relevant for operation at the LHC is the performance for very small angles θ_z.

We measured that the energy resolution scales as $\sigma/E = 13\%/\sqrt{E} + c(\theta_z)$, where c amounts to $\sim 1\%$ for $\theta_z = 3°$. For smaller angles, c increases further (fig. 16), and for $\theta_z < 1°$ deviations from a Gaussian line shape were observed. We demonstrated experimentally that these are due to particles entering the detector at a fibre position and depositing an anomalously large amount of energy (δ-rays) in this one fibre[95]. These results are in agreement with EGS4 predictions.

To avoid these phenomena, the LHC detector should be designed in such a way that electrons cannot enter it under angles less than e.g. $3°$. A possible way of doing this is shown in fig. 17.

4.1.2 Signal uniformity, hermeticity and linearity

Good energy resolution is only meaningful if the signal uniformity is good as well, i.e. if the variations of the mean signal with the impact point across a detector cell are small compared to the energy resolution. Several iterations of improvements were needed to achieve this goal[95]. Figure 18 shows results of the most sensitive measurement, a scan with an electron beam that was moved in small steps from the centre of a detector cell to the point were 3 modules join, and further into the crack between two modules. All individual modules were calibrated with an electron beam sent into their centre.

The total calorimeter signal measured in this crack scan is practically independent of the impact point of the particles. The same is true for a scan with pions (fig. 19), and for the measured energy resolution (fig. 20). We conclude that the particles practically do not notice that they are detected by a modular device and, therefore, that the calorimeter is excellent in terms of signal uniformity and hermeticity.

As can be seen from figure 21, the calorimeter is also very *linear* for e.m. shower detection. The signal per unit energy changes by less than 1% over the energy range 10 - 150 GeV.

4.2 Hadrons and jets

Being very recently obtained, the results in this subsection have to be considered highly preliminary, and further improvements from the off-line data analysis may be expected.

4.2.1 Compensation, shower containment

The original motivation for starting this project was the expectation that a lead/fibre calorimeter could be made compensating, provided that it had the proper sampling fraction. In fig. 22 the e/π signal ratio is shown as a function of energy. Measurements were done at 5, 10, 20, 40, 80 and 150 GeV. The ratio is indeed sufficiently close to 1.0 (typically 1.05) for making the undesirable consequences of non-compensation negligibly small[13].

Figure 23 displays the relation between the e/π signal ratio and the lateral size of the detector, a measurement made possible thanks to the fine lateral

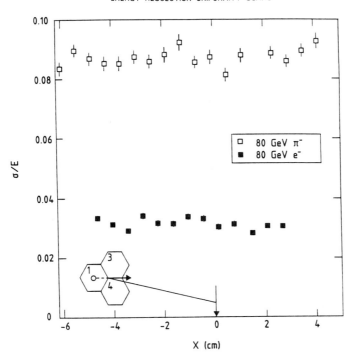

Fig. 20 The energy resolution as a function of the impact point of the particles.

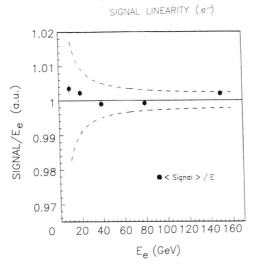

Fig. 21 The signal per unit energy measured for e.m. showers

granularity. The figure shows that the showers are very well contained in our detector and that in practice a region with a radius of ~ 30 cm around the impact point is probably sufficient for an accurate evaluation of the particle energy. The longitudinal leakage from this 10 λ_{int} deep detector was measured to be negligibly small at all the mentioned energies.

4.2.2 Energy resolution for hadrons and jets

The energy resolution was measured for negatively charged pions at energies ranging from 5 - 150 GeV. The results are shown in fig. 24. A least squares fit to the data yields: $\sigma/E = 28.9\%/\sqrt{E} + 2.6\%$. It is expected that especially the constant term will improve when the proper calibration factors are applied (the on-line gain equalization was not more accurate than $\sim 5\%$).

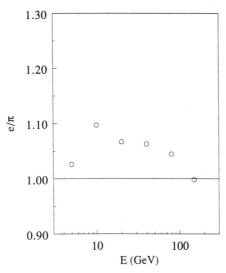

Fig. 22 The e/π signal ratio as a function of energy. Preliminary results.

Contrary to e.m. showers, the energy resolution for pions turns out to be *independent* of the angle of incidence θ_z (fig. 25), which can be understood since the process responsible for the deviations from a Gaussian lineshape in the case of e.m. showers has no equivalent for pions.

Figure 26 shows the hadronic energy resolution as a function of the lateral detector size. It illustrates that the improvement is indeed only marginal when a radius beyond $r \sim 30$ cm $(1.5\lambda_{int})$ is considered.

The detection of jets was simulated by placing a paraffine target upstream of the calorimeter. The pions were required to interact in this target, while the interaction products were detected with the calorimeter. Measurements were performed at 2 target distances, at 44 and 147 cm from the calorimeter front face, respectively. At the close target position, the energy resolution for detecting the reaction products was even better than for single pions (fig. 27), especially at the highest energies. For the far target position, the results were worse, which can be understood from the fact that secondaries may miss the detector in this case. By comparing the average pulse heights, the relation between the energy resolution and the lateral shower leakage can be obtained. Figure 28 confirms that the multiparticle resolution is indeed significantly better than the already impressive single-π result.

4.2.3 Position resolution

Thanks to the fine lateral granularity, the calorimeter also yields a good position resolution. The impact point of the particle was determined from the centre of gravity of the energy deposited in many calorimeter cells. The coordinates obtained in this way were compared with the ones derived from the wire chambers installed upstream of the detector. The distribution of the differences between these two sets of coordinates is shown in fig. 29, for 80 GeV π^- at $\theta_z = 0^0$. The position resolution (the σ of this distribution) amounts to ~ 5 mm. For angles $\theta \neq 0^0$ this distribution becomes asymmetric, since the π^- may penetrate the detector over a certain distance before interacting, and the position resolution deteriorates correspondingly.

For e.m. showers, where one detector cell may contain up to 95% of the shower energy, the position resolution depends on the impact point. Near cell boundaries the resolution is only a few mm, in the cell centres the precision is worse. The results would obviously improve if a smaller cell size were to be used.

Event displays of typical e.m. and hadronic showers are shown in fig. 30. This figure illustrates that although the energy of a pion shower is distributed over a large number of detector cells, a large fraction of it is deposited in the cell hit by the entering particle ($\sim 50\%$). Because of this feature, it is quite possible to *distinguish separate hadrons* at distances much smaller than the 30 cm needed for an accurate measurement of their energy. One may exploit this feature for measuring the detailed structure of jets, including the recognition of electrons from heavy-flavor decay (see sect. 2.4). An example of a multiparticle event, obtained with the interaction trigger, is shown in fig. 31. Also for this purpose, a finer granularity than the one used sofar, is likely to pay off in even better performance.

4.3 Muons

The Spaghetti Calorimeter can also contribute relevant information for the detection of muons, in particular in the case of a catastrophic energy loss in the calorimeter, which will frequently happen at high energies. The signals for minimum-ionizing particles are completely separated from the pedestal and also allow an accurate determination of the energy loss by individual muons traversing the full depth of the calorimeter (fig. 32).

4.4 Electron/pion separation

For the identification of *isolated* electrons, 2 methods were used. The first, classical method exploits the differences in the shower profiles between e.m. and hadronic showers, which are most pronounced for high-Z materials. Using a $2X_0$ thick preshower counter and the differences in lateral shower development, we obtained e/π separation at the 10^{-3} level for 10 GeV, improving to 2.10^{-4} at 80 GeV (fig. 33). The preshower counter, which absorbed on average $\sim 1\%$ of the e.m. shower energy, did *not* have a measurable effect on the energy resolution for e.m. shower detection.

The (large differences in) lateral shower characteristics also make it possible recognizing electrons *inside jets*. By superposition of electron and pion showers, we found that electrons impacting at only 4.3 cm from a pion of equal energy could be identified in 95% of the cases using a criterion that predicted a fake electron in less than 0.1% of the cases.

In the course of this project, a new way to identify electrons was exploited[94]. The method is based on small, but significant differences between the *time structures* of e.m. and hadronic shower signals. These differences, which are among others due to the different longitudinal shower development, can be exploited thanks to the high signal speed of this detector. The method is especially

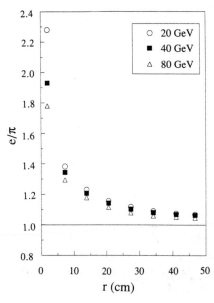

Fig. 23 The e/π signal ratio as a function of the radial detector size. Preliminary results.

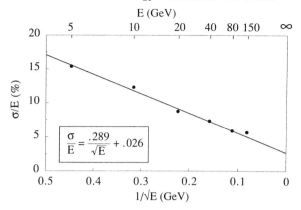

Fig. 24 The energy resolution for 5 - 150 GeV π^- particles. Preliminary results.

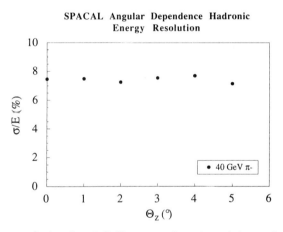

Fig. 25 The energy resolution for 40 GeV π^- as a function of the angle of incidence θ_z. Preliminary results.

interesting since it has the potential of offering an extremely fast electron trigger (no pattern recognition needed).

How powerful this method can be, is illustrated in fig. 34, which shows the distribution of the widths (measured at 20% of the amplitude) of e.m. and hadronic shower signals. The e.m. signals are all nearly identical ($\sigma \sim 130$ ps!), while practically all the hadronic signals are significantly broader. We achieved e/π separation at the 10^{-3} level (80 GeV) just using this method.

These results show that in spite of the absence of longitudinal segmentation, this detector performs very well in terms of e/π separation.

4.5 Signal speed

The lead/fibre calorimeter is definitely the fastest hadron calorimeter available today. The width of the signal is mainly determined (and limited) by the difference in the velocity at which the scintillation light travels inside the fibres ($c/1.6$, the index of refraction of the fibre core) and the velocity of the shower particles that generate the light. Due to the negligible cross section for thermal-neutron capture slow shower tails, characteristic for uranium calorimeters, are absent.

One example, illustrating the advantage of the fast detector response was already given in the previous subsection. Another advantage, crucial for LHC experiments, concerns the fact that one can use a very short gate width without deteriorating the calorimeter performance. Also, the effect of pile-up will be as small as can be. We measured the e/π signal ratio and the energy resolution as a function of the gate width. The results are given in figs. 35 and 36. It turns out that the calorimeter performance remains acceptable down to gate widths of the order of the LHC bunch crossing time.

5. CONCLUSIONS AND OUTLOOK

The use of calorimetric detection methods in high-energy physics has rapidly evolved from a technique employed for some rather specialized applications (neutrino scattering) to the prime experimental tool in modern experiments.

This evolution, inspired by the physics goals of the experiments concerned, has gone hand in hand with crucial developments in the calorimeter technology itself. The sophistication of the instrumentation, the understanding of the basic

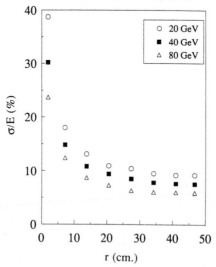

Fig. 26 The energy resolution for pions as a function of the lateral detector size. Preliminary results.

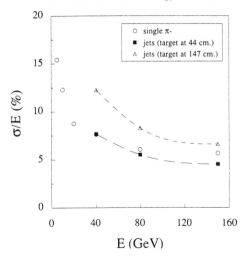

Fig. 27 The energy resolution for the reaction products of a pion interacting in a target upstream of the detector, compared to the single-π resolution, as a function of the pion energy. Preliminary results.

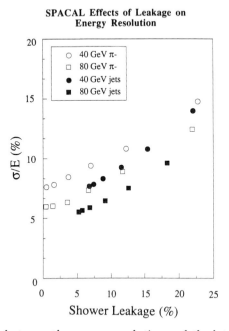

Fig. 28 The relation between the energy resolution and the lateral shower leakage, for single pions and for the reaction products of pions interacting in a target upstream of the calorimeter, at 40 and 80 GeV. Preliminary results.

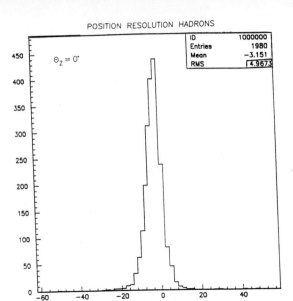

Fig. 29 Distribution of the reconstructed position of the impact point of 80 GeV π^- entering the calorimeter under 0^0 with the fibre axis. The horizontal axis has an arbitrary offset.

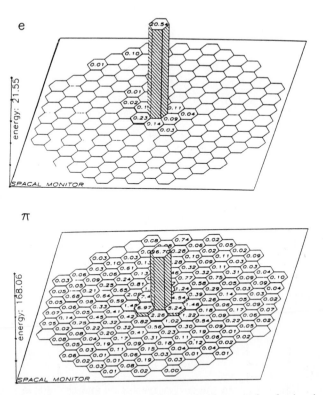

Fig. 30 Event displays of typical e.m and hadronic showers developing in the SPACAL calorimeter.

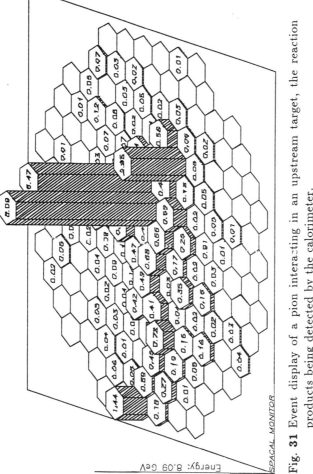

Fig. 31 Event display of a pion interacting in an upstream target, the reaction products being detected by the calorimeter.

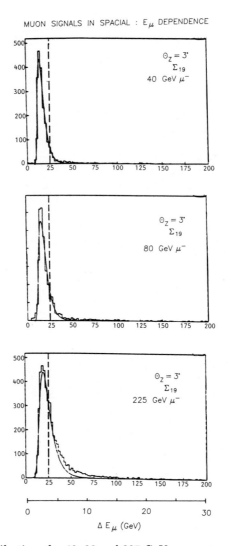

Fig. 32 Signal distributions for 40, 80 and 225 GeV muons traversing the Spaghetti Calorimeter.

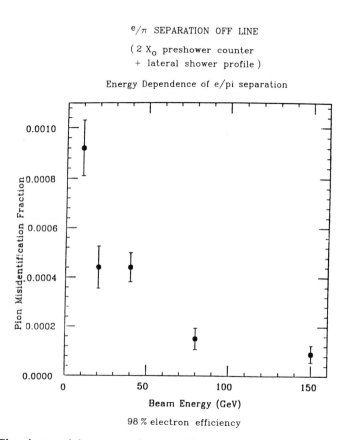

Fig. 33 The electron/pion separation as a function of energy, obtained from the differences in the shower profiles.

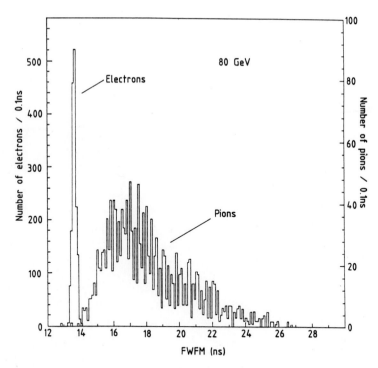

Fig. 34 Distribution of the widths, measured at 20% of the amplitude, of e.m. and hadronic shower signals at 80 GeV.

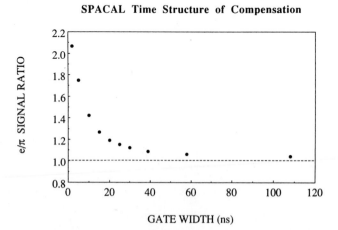

Fig. 35 The e/π signal ratio as a function of the gate width. Preliminary results.

principles of operation and of the limitations of the technique have all reached a very mature level, guaranteeing optimal exploitation possibilities.

Nevertheless, in view of the strongly increasing demands on calorimeter performance for future experiments, particularly at the proposed multi-TeV pp Colliders, further R&D is needed, which will have to focus on those calorimeter features which are likely to be essential in the next decade:

- hermeticity,
- energy resolution,
- rate capability,
- radiation resistance,
- electron/pion discrimination.

I see two trends that are likely to shape detectors for future experiments. Firstly, *compensating* calorimeters will become the standard. As was pointed out in section 3.1, the compensation requirement does *not* necessarily imply the use of uranium absorber. When hydrogenous readout material is used (plastic scintillator, room-temperature liquids, gas) it can be achieved with a variety of absorber materials of which lead is probably the most attractive one.

Secondly, I expect a development towards *integrated* calorimetry, i.e. combining in one instrument the functions of e.m. and hadronic shower measurement, electron and muon identification; the traditional subdivision in an e.m. calorimeter and a hadron calorimeter will disappear.

Research and development and prototype studies will be an even more essential part of the experimental programme than in the present generation of experiments. The time scales involved, the size and the cost of the calorimeters simply do not permit design errors.

I hope that the information provided in this review may serve as guidance and encouragement for those who want to contribute in one way or another to a further development of this very powerful and elegant experimental technique.

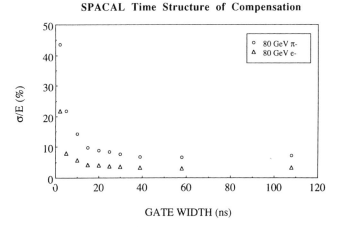

Fig. 36 The e.m. and hadronic energy resolution as a function of the gate width. Preliminary results,

REFERENCES

1. Y.S. Tsai, Rev. Mod. Phys. **46** (1974) 815.
2. E. Storm and H.I. Israel, Nucl. Data Tables **7** (1970) 565.
3. L. Pages et al., Atomic Data **4** (1972) 1.
4. W.R. Nelson, H. Hirayama and D.W.O. Rogers, The EGS4 Code System, Stanford, SLAC Report-165 (1985).
5. L. Landau and I. Pomeranchuk, Doklady Akad. Nauk. SSSR **92**, No. 3 (1953) 535.
6. A.B. Migdal, Phys. Rev. **103** (1956) 1811.
7. T. Yuda, Nucl. Instr. and Meth. **73** (1969) 301.
8. B. Rossi, High-Energy Particles (Prentice Hall, Englewood Cliffs, NJ, 1952), p. 16ff.
9. R. Kopp et al., Z. Phys. **C28** (1985)171.
10. T. Akesson et al., Nucl. Instr. Meth. **A262** (1987) 243.
11. C. Leroy et al., Nucl. Instr. and Meth. **A252** (1986) 4.
12. M.G. Catanesi et al., Nucl. Instr. and Meth. **A260** (1987) 43.
13. R. Wigmans, Nucl. Instr. and Meth. **A259** (1987) 389.
14. R. Wigmans, Energy Loss of Particles in Dense Matter - Calorimetry, Proc. of the ICFA School on Instrumentation in Elementary Particle Physics, Trieste, 1987, eds. C.W. Fabjan and J.E. Pilcher (World Scientific, Singapore, 1988).
15. See for example Y.K. Akimov, Scintillator Counters in High Energy Physics, Academic Press, 1965.
16. D.F. Anderson and D.C. Lamb, Nucl. Instr. and Meth. **A265** (1988) 440.
17. R.C. Munoz et al., J. Chem. Phys. **85** (1986) 1104.
18. R. Wigmans, Calorimetry at the SSC, Proc. of the Workshop on Experiments, Detectors and Experimental Areas for the Supercollider, Berkeley, 1987, eds. R. Donaldson and M.G.D. Gilchriese (World Scientific, Singapore, 1988), p.608.
19. H. Brückmann et al., Nucl. Instr. and Meth. **A263** (1988) 136.
20. J.E. Brau and T.A. Gabriel, Nucl. Instr. and Meth. **A238** (1985) 489.
21. R. Wigmans, Nucl. Instr. and Meth. **A265** (1988) 273.
22. H. Abramowicz et al., Nucl. Instr. and Meth. **180** (1981) 429.
23. M. de Vincenzi et al., Nucl. Instr. and Meth. **A243** (1986) 348.
24. C.W. Fabjan and W.J. Willis, in: Proc. of the Calorimeter Workshop, FNAL, Batavia, Ill., 1975, ed. M. Atač, p. 1; C.W. Fabjan et al., Nucl. Instr. and Meth. **141** (1977) 61.
25. H. Tiecke (The ZEUS Calorimeter Group), Nucl. Instr. and Meth. **A277** (1989) 42.

26. R. Wigmans, Signal equalization and energy resolution for uranium/silicon hadron calorimeters, Report NIKHEF Amsterdam, NIKHEF-H/87-13 (1987).

27. E. Borchi et al., Silicon sampling hadronic calorimetry: A tool for experiments at the next generation of colliders, preprint CERN-EP/89-28 (1989).

28. H1 Calorimeter Group, Performance of a Pb-Cu Liquid Argon Calorimeter with an Iron Streamer Tube Tail Catcher, preprint DESY 88-073, (1988).

29. G. d'Agostini et al., Nucl. Instr. and Meth. **A274** (1989) 134.

30. M. Abolins et al., Hadron and Electron Response of Uranium/Liquid Argon Calorimeter Modules for the D0 Detector, Brookhaven Report BNL-42336 (1989).

31. D. Hitlin, SLD liquid argon prototype tests, Proc. of the Workshop on Compensated Calorimetry, Pasadena, 1985, CALT-68-1305.

32. D. Gilzinger et al., The HELIOS Uranium Liquid Argon Calorimeter, in preparation.

33. Y. Galaktionov et al., Nucl. Instr. and Meth. **A251** (1986) 258.

34. M. Pripstein (WALIC Collaboration), Requirements for the Development of Warm Liquid Calorimetry, Proc. of the Workshop on Future Directions in Detector R&D for Experiments at pp Colliders, Snowmass, Co., 1988, and private communication.

35. E. Radermacher (UA1 Collaboration), First results from a UA1 Uranium-TMP calorimeter module, preprint CERN-EP/89-01 (1989).

36. E. Bernardi et al., Nucl. Instr. and Meth. **A262** (1987) 229.

37. E.B. Hughes et al., Nucl. Instr. and Meth. **75** (1969) 130.

38. A. Benvenuti et al., Nucl. Instr. and Meth. **125** (1975) 447.

39. R.M. Brown et al., IEEE Trans. Nucl. Sci. **NS-32** (1985) 736; P.W. Jeffreys et al., A Phototriode Instrumented Lead Glass Calorimeter for use in the Strong Magnetic Field of OPAL, Rutherford Lab report RAL-85-058 (1985).

40. U. Amaldi, Phys. Scripta **23** (1981) 409.

41. R. Wigmans, The Spaghetti Calorimeter Project at CERN, Proc. of the Workshop on Future Directions in Detector R&D for Experiments at pp Colliders, Snowmass, Co., 1988.

42. Y. Chan et al., IEEE Trans. Nucl. Sci. **NS-25** (1978) 333.

43. H. Grassmann et al., Nucl. Instr. and Meth. **228** (1985) 323.

44. J.A. Bakker et al., Study of the Energy Calibration of a High Resolution EM Calorimeter, CERN-EP/89-16 (1989).

45. M. Laval et al., Nucl. Instr. and Meth. **206** (1983) 169.

46. D.F. Anderson et al., Nucl. Instr. and Meth. **228** (1985) 33.

47. R. Bouclier et al., Nucl. Instr. and Meth. **A267** (1988) 69.

48. C.L. Woody and D.F. Anderson, Nucl. Instr. and Meth. **A265** (1988) 291.

49. K.L. Giboni et al., Nucl. Instr. and Meth. **225** (1984) 579.

50. T. Doke et al., Nucl. Instr. and Meth. **A237** (1985) 475.

51. E. Aprile et al., Nucl. Instr. and Meth. **A261** (1987) 519.

52. V.M. Aulchenko et al. (KEDR Collaboration), paper submitted to the 24th Int. Conf. on High-Energy Physics, Munich, 1988; see also D.G. Hitlin, Proc. of the 24th Int. Conf. on High-Energy Physics, Munich, 1988 (Springer, Berlin, 1989), p. 1187.

53. M. Chen et al., Nucl. Instr. and Meth. **A267** (1988) 43.

54. H. Burkhardt et al., Nucl. Instr. and Meth. **A268** (1988) 116.

55. P. Sonderegger, Nucl. Instr. and Meth. **A257** (1987) 523, and references therein.

56. G.A. Akopdjanov et al., Nucl. Instr. and Meth. **140** (1977) 441.

57. T. Kondo and K. Niwa, Electromagnetic shower size and containment at high energies, paper contributed to the Summer Study on the Design of the Superconducting Super Collider, Snowmass, Co. (1984).

58. I. Stumer and P. Yepes (HELIOS Collaboration), private communication (1989).

59. E. Gabathuler et al., Nucl. Instr. and Meth. **157** (1978) 47.

60. T. Akesson et al., Proc. Workshop on Physics at Future Accelerators, La Thuile and Geneva, 1987, ed. J. Mulvey, CERN 87-07, vol. I, p. 174 (1987).

61. A.L. Sessoms et al., Nucl. Instr. and Meth. **161** (1979) 371.

62. Y. Muraki et al., Radial and longitudinal behaviour of nuclear electromagnetic cascade showers induced by 300 GeV protons in lead and iron absorber, Univ. of Tokyo report ICR 117-84-6 (1984).

63. A.N. Diddens et al., Nucl. Instr. and Meth. **178** (1980) 27.

64. T. Akesson et al., Nucl. Instr. and Meth. **A241** (1985) 17.

65. F. Binon et al., Nucl. Instr. and Meth. **188** (1981) 507.

66. D. Bogert et al., IEEE Trans Nucl. Sci. **NS-29** (1982) 336.

67. J.P. DeWulf et al., Nucl. Instr. and Meth. **A252** (1986) 443.

68. C. DeWinter et al., Experimental results obtained from a low-Z fine-grained electromagnetic calorimeter, preprint CERN-EP/88-81 (1988).

69. I. Abt et al., Nucl. Instr. and Meth. **217** (1983) 377.

70. A.V. Barns et al., Phys. Rev. Lett. **37** (1970) 76. See also T. Ferbel in: Understanding the Fundamental Constituents of Matter, ed. A. Zichichi (Plenum Press, New York, NY, 1978).

71. J.A. Appel et al., Nucl. Instr. and Meth. **127** (1975) 495.

72. D. Hitlin et al., Nucl. Instr. and Meth. **137** (1976) 225.

73. R. Engelmann et al., Nucl. Instr. and Meth. **216** (1983) 45.

74. U. Micke et al., Nucl. Instr. and Meth. **221** (1984) 495.

75. C. DeWinter et al., An Electron-Hadron Separator for Digital Sampling Calorimeters, preprint CERN-EP/88-87 (1988).
76. J. Cobb et al., Nucl. Instr. and Meth. **158** (1979) 93.
77. J. Krüger (ed.), The ZEUS Detector, Status Report 1987, Report PRC 87-02, DESY (1987).
78. C. Gössling, Large Area Silicon Detectors, Proc. 24th Int. Conf. on High-Energy Physics, Munich, 1988 (Springer, Berlin, 1989), p. 1208.
79. L. Baum et al., Proc. Calorimeter Workshop, FNAL, Batavia, Ill., 1975, ed. M. Atač, p. 295.
80. A. Grant, Nucl. Instr. and Meth. **131** (1975) 167.
81. M. Holder et al., Nucl. Instr. and Meth. **151** (1978) 69.
82. R. Leuchs, Messung des hadronischen Untergrundes bei der Identifizierung von Myonen, Tech. Univ. Aachen, 1982; K. Eggert (UA1 Collaboration), private communication.
83. F.S. Merritt et al., Hadron Shower Punch Through for Incident Hadrons of Momentum 15, 25, 50, 100, 200 and 300 GeV/c, preprint Enrico Fermi Institute, ER 13065-41 (1985).
84. K. Eggert et al., Nucl. Instr. and Meth. **176** (1980) 217.
85. F. Abe et al., Nucl. Instr. and Meth. **271** (1988) 387.
86. Technical Proposal of the L3 Collaboration, CERN/LEPC/83-05 (1983).
87. W.J. Willis and K. Winter, in Physics with very high energy e^+e^- colliding beams, CERN 76-18 (1976), p. 131.
88. G. Arnison et al. (UA1 Collab.), Phys. Lett. **139B** (1984) 115.
89. P. Bagnaia et al. (UA2 Collab.), Z. Phys. **C24** (1984) 1.
90. P. Jenni (UA2 Collab.), Nucl. Phys. **B3** (Proc. Suppl.) (1988) 341.
91. L. Mandelli (UA2 Collab.), UA2 Results for the 1987 Run, preprint CERN-EP/88-182 (1988).
92. E. Bernardi et al., Nucl. Instr. and Meth. **A262** (1987) 229.
93. F.G. Hartjes and R. Wigmans, Nucl. Instr. and Meth. **A277** (1989) 379.
94. R. DeSalvo et al., Nucl. Instr. and Meth. **A279** (1989) 467.
95. D. Acosta et al., Nucl. Instr. and Meth. **A294** (1990) 193.

NOISE LIMITS IN DETECTOR CHARGE MEASUREMENTS

P.F. Manfredi* and V. Speziali

Dipartimento di Elettronica
Università di Pavia
Via Abbiategrasso 209
27100 Pavia, Italy

and Istituto Nazionale di Fisica Nucleare
Via Celoria 16
20133 Milano, Italy

*Visiting at Instrumentation Division
Brookhaven National Laboratory, Upton, N.Y. 11973

ABSTRACT

The noise in the front-end amplifiers sets a serious limitation in the accuracy of charge measurements with detectors that have no internal multiplication like gaseous, liquid and solid state ionisation chambers, solid state strip and pixel detectors and silicon drift chambers. The present paper, after discussing the limits set by noise and reviewing the noise properties of front-end components, analyzes the perspectives opened up by modern device and circuit technologies in low-noise detector applications. Particular attention will be devoted to the situations arising in elementary particle experiments of both accelerator and non-accelerator nature.

I INTRODUCTORY REMARKS

The accuracy in energy measurements, position sensing and event timing with most detectors of ionising events is related to the dispersion with which the preamplifier noise affects the charge released in the detector. This is especially true if the detector has no internal multiplication, so that the charge that it delivers is directly compared, to determine the signal-to-noise ratio, to the preamplifier equivalent noise charge ENC. The low-noise techniques associated with detectors are becoming a very important aspect in elementary particle experiments [1, 2, 3, 4, 5]. New, high energy and high luminosity colliders are fixing serious constraints in low noise design, in connection with the short intervals between bunch crossings, down to the 20 ns foreseen for SSC, and the accordingly short weighting functions that have to be adopted in signal processing [6, 7].

Besides, at the very high beam energies of future hadron colliders the thickness of the sensitive layers which determines the charge released in the detector by minimum ionizing particles, has, expectedly, to be reduced, in order to avoid extravagantly large production in the detector (8). This results in a reduction of the charge available, which degrades the signal-to-noise ratio.

One more problem connected with accelerator experiments has to be considered, that is, the noise increase in the front-end amplifiers which may occur because of their being exposed to ionizing radiation. In detector-preamplifier systems where the signal-to-noise ratio is intrinsically not very large, the increase in noise due to irradiation may bring to a point at which the event detectability is impaired. A low noise design, therefore, makes sense only if the minimum requirements in signal-to-noise ratio are met under the worst-case exposure conditions foreseeable during the expected front-end lifetime [9, 10].

Different, but by no means less difficult problems arise in low noise detection techniques for non-accelerator experiments. Here the limitations in the achievable signal-to-noise ratios are set by the small values of the charge to be detected and, frequently, by the large detector capacitances. To make the best use of the comparatively long weighting functions that are permitted by rates of events much lower than in the accelerator case, the effects of 1/f-noise must be kept under strict control [11].

In both cases of accelerator and non accelerator experiments, a considerable evolution in the complexity of the front-end systems has been recorded over the past few years. Until ten years ago, low noise techniques were applied to a few tens of signal processing channels that could be individually adjusted for optimum noise performances. Present and future experiments involve large scale front-end system with numbers of signal processing channels ranging from a few thousands to a million or even more. Low noise monolithc technologies become, therefore, of utmost relevance in the front-end design and the further requirement of a suitable insensitivity of the noise properties to the effects of irradiation becomes essential for front-end systems to be employed with high luminosity colliders.

The present paper aims at providing an insight into the low noise techniques associated with radiation detectors. It discusses the properties of those active devices that are of higher interest as front-end elements in detector applications and analyses how they would lend themselves to designing large scale, low noise monolithic circuits with a suitably low rate of degradation when exposed to ionising radiation.

In order to understand the extent to which an active device suits the requirements set by the actual detector application, it is advisable to review here the fundamental concepts about the accuracy limitations arising from the noise in a charge measuring system. Section II, III, IV in this paper are devoted to such a review. Section V discusses the noise properties of active devices by describing the results of measurements carried out with the most up-to-date and accurate techniques and illustrates some of the existing low noise front-end systems.

II NOISE SOURCES AND EQUIVALENT NOISE CHARGE

This section aims at specifying the noise-induced dispersion in the measurement of the charge delivered by a detector and discussing how it is affected by the noise sources in the detector and in the front-end.

For this purpose, a model has to be adopted to represent the detector and the analog channel which processes its signals to get from them the charge information.

The detector can be thought of as a capacitive source, which under the effect of an ionizing event delivers a current pulse i(t) that carries the charge to be measured.

The analog processor consists of two basic building blocks: a preamplifier and a filter. The processor as a whole implements on i(t) a filtering function which is expressed as a "weighting integral":

$$(1) \quad v_o(t_M) = \int_0^{t_M} i(x) \cdot W(x) \, dx$$

where $v_o(t_M)$, linearly related to Q is the output amplitude at the measurement instant t_M and W(t) is the processor weighting function. If the processor is of time invariant nature, (1) becomes the convolution integral

(2) $$v_o(t_M) = \int_0^{t_M} i(x) \cdot v_{o,\delta}(t_M - x) \, dx$$

where $v_{o,\delta}(t)$ is the output signal in response to a δ-shaped $i(t)$ carrying a unit charge. Comparison between (1) and (2) shows that in the time-invariant case the weighting function $W(t)$ is related to the δ-response $v_{o,\delta}(t)$ by

(3) $$W(t) = v_{o,\delta}(t_M - t).$$

The choice of $W(t)$ is oriented to enhance the signal-to-noise ratio as much as possible under the actual experiment constraints, that may require $W(t)$ to be shorter than a certain width dictated by the expected time distribution of events, to be flat-topped for an easier amplitude measurement or to be bipolar in shape to avoid excessive baseline shifts at high event rates.

As it is well known that processors of time variant and time-invariant nature having the same weighting function and producing the same $v_o(t_M)$ in response to a given $i(t)$ give the same signal-to-noise ratio, the model adopted here will be of time-invariant nature as this lends itself to a simpler description. On the base of the stated equivalence between time-invariant and time-variant systems, the latter case can be studied with the same model in which the δ-response is assumed to be connected through (3) to the weighting function $W(t)$ of the actual time-variant processor.

The model in question is shown in fig. 1.

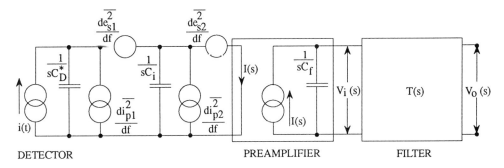

Fig. 1. Time-invariant model for the noise analysis of an analog channel for charge measurements.

As announced, it includes a preamplifier, assumed to be of the charge-sensitive type and described, accordingly, as a current integrator with transfer function $1/sC_f$ and a filter with transfer function $T(s)$, where s is the complex frequency.

The capacitance C_D^* results from the sum of the detector capacitance C_D, of the strays in the connection between detector and preamplifier and of the feedback capacitance C_f in the charge-sensitive loop.

C_i is the preamplifier input capacitance measured in the open-loop condition. It can be assumed to be equal to the capacitance which appears at the control electrode (base or gate) of the input device when its two other electrodes (source, drain or emitter, collector) are connected to ground through a low impedance path.

The model of fig. 1 includes four noise sources, described by their spectral densities, to take into account the noise contributions brought about by detector, preamplifier and related biasing networks.

The parallel current source $\overline{di^2_{p1}}/df$ can be adequately described as the sum of two terms, a white one and one with a linear frequency dependence:

(4) $$\frac{\overline{di_{p1}^2}}{df} = b_{w1} + b_{f1}|f|$$

The white term accounts for the shot noise in the detector leakage current I_L, for the thermal noise in the detector bias resistor R_D which exists in the case of ac coupling between detector and preamplifier and for the thermal noise in the feedback resistor R_f which provides the dc return in the preamplifier. Representing the noise sources with their mathematical spectral densities ($-\infty < f < +\infty$), b_{w1} is expressed as:

(5) $$b_{w1} = q\, I_L + 2kT\left(\frac{1}{R_D} + \frac{1}{R_f}\right)$$

where: q is the electron charge, $1.6 \cdot 10^{-19}$C
k is the Boltzmann's constant, $1.38 \cdot 10^{-23}$ J/°K
T is the absolute temperature

In eq. (4), the spectral density with a linear frequency dependence accounts for the noise arising from dielectric losses associated with the detector, as well as with connectors and coupling capacitors.

The series source $\overline{de_{s1}^2}/df$ accounts for the thermal noise in any energy dissipative element which may exist between detector and preamplifier. It may represent, therefore the noise arising from a series resistor r_D in the detector and the noise due to a spreading resistor $R_{GG'}$ in the base or gate of the preamplifier input device.

It terms of mathematical densities,

(6) $$\frac{\overline{de_{s1}^2}}{df} = 2kT(r_D + R_{GG'}) = a_{w1} \ .$$

As for any two-port, the preamplifier noise can be represented by a series voltage source and a parallel current source with a certain degree of correlation between them.

In most of detector applications, however, and with most of practical preamplifier circuits, the importance of this correlation is so small that no significant error is made in assuming them totally uncorrelated.

The parallel source with spectral density $\overline{di_{p2}^2}/df$ can be assumed to consist of two terms:

(4') $$\overline{di_{p2}^2}/df = b_{w2} + b_{f2} \cdot |f|.$$

The constant term b_{w2} represents the shot noise associated with the current at the control electrode of the input device: $b_{w2} = qI_B$ in a bipolar transistor operated at a base current I_B and $b_{w2} = qI_G$ in a junction field effect transistor with gate leakage current I_G.

The $b_{f2} \cdot |f|$ term accounts for the noise associated with the dielectric losses in the input device.

The series source with spectral density $\overline{de_{s2}^2}/df$ in fig. 1 accounts mostly for the noise in the standing current of the preamplifier input device, and can be adequately represented as the sum of a white term and a term with 1/f frequency dependence:

(7) $$\frac{\overline{de_{s2}^2}}{df} = a_{w2} + \frac{a_f}{|f|^\alpha}$$

The first term in (7) describes the shot noise associated with the collector current in a bipolar transistor or the channel thermal noise in a field-effect device. The spectral density a_{w2} can be expressed as:

(8) $\quad a_{w2} = 2kT \cdot \dfrac{\Gamma}{g_m}$

where g_m is the transconductance of the input device and Γ a coefficient whose values are:

$\Gamma = 0.5$ for bipolar transistors.
$\Gamma \cong 0.7$ for Si junction field effect transistors of the diffused type and for ordinary MOS.
$\Gamma > 1$ for short channel devices.

The second term in eq. (7) describes the noise behaviour of the device in the low-frequency range. The exponent α was found to have values in the range 0.5-1.6. A case which is very frequently encountered and which will be assumed here as a base for discussing the effects of low frequency noise in charge measurements is that of $\alpha = 1$, true 1/f-noise.

In a real preamplifier, the $\overline{de^2_{\mathcal{S}}}/df$ spectral density may be affected by contributions coming from devices other than the input one. In what follows it will be assumed that the transconductance of the input device is large enough and that the design of the rest of the circuit is suitably clever to make these contributions negligible.

The spectral density of the total noise current at the preamplifier input port is:

(9) $$\dfrac{\overline{di^2_N}}{df} = \omega^2 \left(C^*_D + C_i\right)^2 \cdot \left[a_{w2} + a_{w1} \cdot \dfrac{C^{*2}_D}{\left(C^*_D + C_i\right)^2} \right] + |\omega| \cdot \left[2\pi a_f \cdot \left(C^*_D + C_i\right)^2 + \dfrac{b_{f1} + b_{f2}}{2\pi} \right] + b_{w1} + b_{w2} \ .$$

From (9) the spectral density of the noise at the input of the filter of fig. 1 can be calculated to be:

(10) $$\dfrac{\overline{dv^2_{i,N}}}{df} = \dfrac{\left(C^*_D + C_i\right)^2}{C^2_f} \cdot \left[a_{w2} + a_{w1} \cdot \dfrac{C^{*2}_D}{\left(C^*_D + C_i\right)^2} \right] + \dfrac{1}{|\omega|} \cdot \dfrac{1}{C^2_f} \left[2\pi a_f \cdot \left(C^*_D + C_i\right)^2 + \dfrac{b_{f1} + b_{f2}}{2\pi} \right] + \dfrac{1}{\omega^2} \cdot \dfrac{1}{C^2_f} \cdot \left(b_{w1} + b_{w2}\right) \ .$$

The last two equations suggest the following remarks.

The noise source related to dielectric losses, which contributes to the $\overline{di^2_{p1}}/df$, $\overline{di^2_{p2}}/df$ densities of fig. 1 with the b_f|f| parts, appears in $\overline{dv^2_{i,N}}/df$ as a $1/|\omega|$ term that adds to the one describing the low-frequency noise in the preamplifier input device.

The series source with spectral density $\overline{de^2_{s1}}/df = a_{w1}$ in fig. 1, from an equivalent point of view, acts to increase the series white noise in the preamplifier input device of the quantity:

$a_{w1} \cdot m^2/(m+1)^2$ where $m = C^*_D/C_i$.

This quantity results from the spectral density a_{w1} weighted by a factor which gets close to 1 at $m \gg 1$ as which tends to zero at $m \ll 1$.

The spectral density of the total output noise can be obtained multiplying (9) by

$1/\omega^2 C_f^2 \cdot |T(j\omega)|^2$ or (10) by $|T(j\omega)|^2$.

The rms output noise voltage is calculated by integrating the output spectral densities obtained in either way on the $(-\infty, +\infty)$ frequency range.
Let h(t) be the reverse transform of $H(s) = T(s)/s$, that is:

(11) $\quad h(t) = L^{-1}[H(s)] = L^{-1}\left[\frac{T(s)}{s}\right]$.

The function h(t) is related to the δ-response of the processor of fig. 1 by:

(12) $\quad v_{o,\delta}(t) = \frac{1}{C_f} \cdot h(t)$.

The function h(t) is supposed to be normalized to unit peak amplitude, so that the output peak amplitudes in response to δ-impulses detector currents carrying a unit charge or a charge Q are respectively $1/C_f$ and Q/C_f.

The noise performances of an analog processor associated with a radiation detector are specified in terms of equivalent noise charge (ENC). This is the charge which delivered at the preamplifier input by an indefinitely short current pulse would produce an output peak amplitude equal to the actual rms value of the noise. Applying this definition, the following expressions are obtained for ENC²:

(13)
$$ENC^2 = A_w \cdot \left(C_D^* + C_i\right)^2 \cdot \int_{-\infty}^{\infty} |T(\omega)|^2 df + \left[2\pi a_f \left(C_D^* + C_i\right)^2 + \frac{B_f}{2\pi}\right] \cdot \int_{-\infty}^{\infty} \frac{|T(\omega)|^2}{|\omega|} df +$$
$$+ B_w \cdot \int_{-\infty}^{\infty} \frac{|T(\omega)|^2}{\omega^2} df$$

having put

$$A_w = a_{w2} + a_{w1} \frac{C_D^{*2}}{\left(C_D^* + C_i\right)^2}, \quad B_w = b_{w1} + b_{w2}, \quad B_f = b_{f1} + b_{f2}$$

To arrive at (13), the root mean square output noise is calculated from the total spectral density of the output noise integrated on the $(-\infty, +\infty)$ frequency range. The rms output noise is then put equal to ENC^2/C_f^2.

It can be now be observed that T(s) is actually a dimensionless function of the dimensionless variable sτ, where τ is a quantity which expresses the time scale of the δ-response of the processor or of its weighting function; τ can be either the time constant, peaking time or the width. It is worth then changing variable in the three integrals of eq. (13) by putting $\omega\tau = x$, by virtue of which ENC² can be rewritten as:

$$\text{ENC}^2 = A_w \cdot \frac{1}{\tau}\left(C_D^* + C_i\right)^2 \frac{1}{2\pi} \int_{-\infty}^{\infty} |T(x)|^2 dx + \left[2\pi a_f \left(C_D^* + C_i\right)^2 + \frac{B_f}{2\pi}\right] \frac{1}{2\pi} \int_{-\infty}^{\infty} \frac{|T(x)|^2}{|x|} dx +$$

(14)

$$+ B_w \cdot \tau \cdot \frac{1}{2\pi} \int_{-\infty}^{\infty} \frac{|T(x)|^2}{x^2} dx .$$

The three integrals in eq. (14) are pure numbers.
By labelling them as A_1, A_2, A_3, it can be written:

(15)
$$\begin{cases} A_1 = \frac{1}{2\pi} \cdot \int_{-\infty}^{\infty} |T(x)|^2 dx = \frac{1}{\pi} \cdot \int_0^{\infty} |T(x)|^2 dx = \frac{1}{\pi} \int_0^{\infty} x^2 \cdot \frac{|H(x)|^2}{\tau^2} dx \\[6pt] A_2 = \frac{1}{2\pi} \cdot \int_{-\infty}^{\infty} \frac{|T(x)|^2}{|x|} dx = \frac{1}{\pi} \cdot \int_0^{\infty} \frac{|T(x)|^2}{x} dx = \frac{1}{\pi} \cdot \int_0^{\infty} |x| \cdot \frac{|H(x)|^2}{\tau^2} dx \\[6pt] A_3 = \frac{1}{2\pi} \cdot \int_{-\infty}^{\infty} \frac{|T(x)|^2}{x^2} dx = \frac{1}{\pi} \cdot \int_0^{\infty} \frac{|T(x)|^2}{x^2} dx = \frac{1}{\pi} \cdot \int_0^{\infty} \frac{|H(x)|^2}{\tau^2} dx . \end{cases}$$

The last integrals in the expressions of A_1, A_2, A_3 have been written by remembering that:

$$|H(\omega)|^2 = \frac{|T(\omega)|^2}{\omega^2} = \tau^2 \cdot \frac{|T(x)|^2}{x^2}, \text{ whence } |T(x)|^2 = x^2 \cdot \frac{|H(x)|^2}{\tau^2}$$

A_2 can be considered as a shape coefficient of the filter for the noise which appears at its input as having an f^{-1} frequency dependence. Once the nature of τ has been specified, also A_1 and A_3 become shape coefficient of the filters for the noise terms appearing at its input as having respectively f^0 and f^{-2} frequency dependence.
ENC^2 can be finally written as:

(16) $$\text{ENC}^2 = A_w \cdot \left(C_D^* + C_i\right)^2 \cdot \frac{1}{\tau} \cdot A_1 + \left[2\pi a_f \left(C_D^* + C_i\right)^2 + \frac{B_f}{2\pi}\right] A_2 + B_w \cdot \tau \cdot A_3$$

which shows that ENC^2 is built up from a contribution, the one related to A_w, inversely proportional to τ, a contribution, the one related to B_w, proportional to τ and a contribution related to the 1/f-noise, independent of τ.
According to (16), ENC^2 attains a minimum at the value of τ:

(17) $$\tau_{\text{MIN}} = \left(\frac{A_w}{B_w}\right)^{1/2} \cdot \left(\frac{A_1}{A_3}\right)^{1/2} \cdot \left(C_D^* + C_i\right) = \left(\frac{A_1}{A_3}\right)^{1/2} \cdot \tau_c$$

independent of the amount of 1/f-noise (12). The quantity

$$\tau_c = \left(\frac{A_w}{B_w}\right)^{1/2} \cdot \left(C_D^* + C_i\right)$$

is called "noise corner time constant" and is determined by the spectral densities A_w and B_w and by the total capacitance at the preamplifier input, but is independent of the filter. The value of τ at the which the minimum in ENC^2 occurs, instead, depends on the filter through the shape coefficients A_1, A_3. By introducing (17) into (16), the value ENC$^2_{MIN}$ is obtained:

$$(18) \quad ENC^2_{MIN} = 2\left(A_w \cdot B_w\right)^{1/2} \left(A_1 A_3\right)^{1/2} \left(C_D^* + C_i\right) \left[1 + \frac{2\pi a_f \left(C_D^* + C_i\right)^2 + \frac{B_f}{2\pi}}{2\left(A_w \cdot B_w\right)^{1/2} \left(C_D^* + C_i\right)} \cdot \frac{A_2}{\left(A_1 A_3\right)^{1/2}}\right]$$

As it can be inferred from (16), at values of τ well below τ_{MIN}, ENC^2 is governed by the spectral density A_w, while at τ much larger than τ_{MIN} the importance of B_w becomes dominant in determining ENC^2. At values of τ in proximity of τ_{MIN}, ENC^2 may be significantly affected by 1/f-noise.

It has to be pointed out that in practical filters $(A_1/A_3)^{1/2}$ is close to unity, so that $\tau_{MIN} \approx \tau_c$. It is worth expressing τ_c through the actual noise parameters of detector and preamplifier. If the thermal noise from the bias resistors is neglected, which is a reasonable assumption if the detector is dc coupled to the preamplifier and this has a very large R_f or a dynamic charge reset, B_w can be expressed as: $B_w = q(I_L + I_G)$. Another hypothesis which will be made here is that a_{w1} be negligible compared to a_{w2}, which implies that both the detector series resistance and the input device spreading resistance be very small compared to $2kT \cdot \Gamma/g_m$. Such a condition is met in most of cases, the exceptions being that of a preamplifier employing a bipolar transistor or a very large JFET. With the two assumptions made before, τ_c can be written as:

$$(19) \quad \tau_c = \left[\frac{2\Gamma V_{th}}{\omega_T \cdot (I_L + I_G)}\right]^{1/2} \cdot \left(m^{1/2} + m^{-1/2}\right) \cdot C_D^{*1/2}$$

where V_{th} is the thermal voltage, $V_{th} = kT/q$ and

$$\omega_T = \frac{g_m}{C_i} = 2\pi f_T$$

f_T being the transition frequency of the device.

Going back to (16), some useful considerations can be made.

i) In detector applications at future hadron colliders, where τ will be restricted to a few tens of nanoseconds, ENC^2 will be governed by A_w. In such a case, neglecting the contributions arising from B_w and 1/f-noise, ENC^2 can be approximated as follows:

$$(20) \quad ENC^2 = 2kT\Gamma \cdot \frac{1}{\omega_T \tau} \cdot \left(m^{1/2} + m^{-1/2}\right) \cdot C_D^*$$

which stresses the importance of a large f_T in the input device to keep ENC^2 small at short values of τ.

ii) If the contribution related to B_w can be neglected, like in the case in which detector and preamplifier input stage are cooled, a common practice in X-ray spectrometry, ENC² is governed by A_w at short τ values and by 1/f-noise at long τ's. Let τ_F be the value of τ at which the ENC² contribution related to A_w equals the one due to 1/f-noise.

If, as it happens in practice, $2\pi a_f (C^*_D+C_i)^2 \gg b_f/2\pi$, τ_F can be written as:

(21) $$\tau_F = \frac{A_1}{A_2} \cdot \frac{kT\Gamma}{\pi g_m a_f} = \frac{A_1}{A_2} \cdot \frac{kT\Gamma}{\pi \omega_T \cdot H_f}.$$

The last expression points out that τ_F is determined for a given device, by two intrinsic parameters, independent of the transversal dimension W or of the number of units connected in parallel. They are ω_Ts the intrinsic parameter describing the white noise in drain or collector current and $H_f = a_f \cdot C_i$, the intrinsic parameter describing the 1/f-noise associated with that current. So, in the case in which the spectral density of the noise consists of the only two terms a_{w1} and a_f/f, ENC² is governed by the former at $\tau \ll \tau_F$, while the latter takes over at $\tau > \tau_F$ [11].

The ratio between the ENC² contribution coming from 1/f-noise, ENC²$_{1/f}$ and total ENC² can be expressed through the τ_F/τ ratio as:

(22) $$\frac{ENC^2_{1/f}}{ENC^2} = \frac{1}{1 + \tau_F/\tau}.$$

In a JFET, $H_f \approx 10^{-26}$J or less if the device is of the epitaxial-channel, diffused-gate type. For $\omega_T \approx 10^{-9}$ rad/s, τ_F would be of the order of 100 µs. The effect of 1/f-noise on ENC² would start becoming noticeable at $\tau \approx 10$ µs.

The limit set by 1/f-noise in the accuracy of a charge measurement is important also in the case of non accelerator experiments employing front-end devices at cryogenic temperatures. In these circumstances it is the 1/f-noise which makes an increase in τ, otherwise possible by virtue of the low counting rates, meaningless. Beyond a certain point indeed, an increase in τ does not yield a corresponding benefit in the noise dispersion, if ENC² is governed by 1/f-noise.

iii) Another interesting case as far as the relative impact of 1/f-noise on ENC² is concerned, is that of the microstrip front-end systems presently employed at LEP and using MOS as input devices. MOS have comparatively large 1/f-noise, $H_f \approx 10^{-24}$J and virtually zero I_G. Considering a detector with a leakage current in the 1 nA region and $C^*_D \approx 10$ pF, these values being frequently encountered in microstrip applications, it can be understood from (16) that again 1/f-noise contributes to ENC² in a significant way.

III FURTHER CONSIDERATIONS ON ENC²

It can be demonstrated that the knowledge of only two shape parameters of the filter, A_1 and A_3 allows an upper limit to be set for ENC² in a charge measuring channel of assigned noise sources [13].

For this purpose it is sufficient to employ Cauchy-Schwartz inequality:

(23) $$\left| \int_{-\infty}^{\infty} u(x) \cdot v(x) \, dx \right|^2 \leq \int_{-\infty}^{\infty} |u(x)|^2 \, dx \cdot \int_{-\infty}^{\infty} |v(x)|^2 \, dx$$

by assuming:

(24) $\quad u(x) = |T(x)|$, $\quad v(x) = \dfrac{|T(x)|}{|x|}$

or, alternatively:

(24) $\quad u(x) = |x| \cdot |H(x)|$, $\quad v(x) = |H(x)|$.

Either choice applied to (23) leads to the following result, bearing in mind expressions (15) of A_1, A_2, A_3:

(25) $\quad A_2 \leq (A_1 \cdot A_3)^{1/2}$.

For a given filter A_1 and A_3 can be evaluated in the frequency domain according to (15) or from the time-domain relationships:

(26) $\quad A_1 = \displaystyle\int_{-\infty}^{\infty} [h'(y)]^2 \, dy \quad A_3 = \int_{-\infty}^{\infty} [h(y)]^2 \, dy$

y being the normalized time, $y = t/\tau$ [2].

The time-domain calculation of A_2, instead, if feasible at all, may imply remarkable mathematical difficulties. According to (25), the time-domain analysis of the filter, which, in passing, becomes the only possible approach in the time-variant case, provides in a straightforward manner, an upper limit for A_2.

If (25) is now introduced into (18), the following inequality is obtained for ENC^2_{MIN}:

(27) $\quad ENC^2_{MIN} \leq 2(A_w \cdot B_w)^{1/2} \cdot (A_1 A_3)^{1/2} \cdot (C_D^* + C_i) \left[1 + \dfrac{2\pi a_f (C_D^* + C_i)^2 + \dfrac{B_f}{2\pi}}{2(A_w \cdot B_w)^{1/2} \cdot (C_D^* + C_i)} \right]$

The extent to which inequality (27) is met for some h(t) functions of frequent use is apparent from Table I.

The inequality (27) can be rewritten in a more expressive way. It is known from the theory of signal processing that the indefinite cusp,

TABLE I

	Shaping	h(t) Function	A_2	$\sqrt{A_1 A_3}$	$\dfrac{A_2}{\sqrt{A_1 A_3}}$	A_1	A_3	$\sqrt{\dfrac{A_1}{A_3}}$
1	indefinite cusp	$e^{t/\tau}$, $e^{-t/\tau}$	0.64 $\left(\dfrac{2}{\pi}\right)$	1	0.64	1	1	1
2	truncated cusp	$k = \tau'/\tau$, k=1	0.77	1.04	0.74	2.16	0.51	2.06
		k=2	0.70	1.01	0.69	1.31	0.78	1.30
		k=3	0.67	1	0.67	1.31	0.91	1.10
3	triangular	O τ 2τ	0.88 $\left(\dfrac{4}{\pi}\ln 2\right)$	1.15 $\left(\dfrac{2}{\sqrt{3}}\right)$	0.76	2	0.67 $\left(\dfrac{2}{3}\right)$	1.73
4	trapezoidal	O τ 2τ 3τ	1.38	1.83	0.76	2	1.67	1.09
5	piecewise parabolic	O τ 2τ	1.15	1.43	0.80	2.67	0.77	1.86
6	sinusoidal lobe	O τ 2τ	1.22	1.57	0.78	2.47	1	1.57
7	RC-CR	O τ	1.18	1.85	0.64	1.85	1.85	1
8	semigaussian (n = 4)	O 4τ	1.04	1.35	0.77	0.51	3.58	0.38
9	gaussian	$e^{-1/2(t/\tau)^2}$	1	1.26	0.79	0.89	1.77	0.71
10	clipped approximate integrator	O 0.5τ	0.85	1.34	0.63	2.54	0.71	1.89
11	bipolar triangular	O τ 2τ	2	2.31	0.87	4	1.33	1.73

$$h(t) = e^{-|t|/\tau_c} \text{ where } \tau_c = \left(\frac{A_w}{B_w}\right)^{1/2} \cdot \left(C_D^* + C_i\right)$$

describes the matched filter when only the two spectral densities A_w, B_w are present [2]. If the A_1, A_3 shape factors of the indefinite cusp given in Table I, A_1, $A_3 = 1$ are introduced into (18), the equivalent noise charge $ENCO_\infty^2$ relevant to a matched filter for the case of the A_w, B_w spectral densities is obtained:

$$(28) \quad ENCO_\infty^2 = 2(A_w B_w)^{1/2} \cdot \left(C_D^* + C_i\right)$$

If the same $h(t)$ is applied to 1/f-noise alone, the resulting equivalent noise charge would be, according to (16) and Table I:

$$(29) \quad ENC_{1/f}^2 = \left[2\pi \cdot a_f \cdot \left(C_D^* + C_i\right)^2 + \frac{B_f}{2\pi}\right] \cdot \frac{2}{\pi}$$

By virtue of (28), (29), inequality (27) can be rewritten as:

$$(30) \quad ENC_{MIN}^2 \le ENCO_\infty^2 \left[1 + \frac{\pi}{2} \cdot \frac{ENC_{1/f}^2}{ENCO_\infty^2}\right] \cdot (A_1 A_3)^{1/2}$$

Throughout the set of $h(t)$ functions considered in Table I, the ratio

$$\frac{A_2}{(A_1 A_3)^{1/2}}$$

does not undergo large excursions, for it ranges between 0.64 and 0.87. Yet, Table I includes the most widely employed $h(t)$ functions, either unipolar or bipolar.

It becomes reasonable, therefore, to represent (18) in the form:

$$(31) \quad ENC_{MIN}^2 = ENCO_\infty^2 \cdot \left[1 + \gamma \cdot \frac{ENC_{1/f}^2}{ENCO_\infty^2}\right] \cdot (A_1 A_3)^{1/2}$$

with γ ranging between 1 and 1.36.

By writing (31) as:

$$(32) \quad ENC_{MIN}^2 = \left[ENCO_\infty^2 + \gamma ENC_{1/f}^2\right] \cdot (A_1 A_3)^{1/2}$$

it is interesting to enquire where the limits to the two separate contributions, ENC_1^2 and ENC_2^2:

$$(33) \quad ENC_1^2 = ENCO_\infty^2 \cdot (A_1 A_3)^{1/2}$$

$$(34) \quad ENC_2^2 = \gamma ENC_{1/f}^2 \cdot (A_1 A_3)^{1/2}$$

come from.

Introducing the A_w, B_w expressions into (28), ENC^2_1 becomes:

$$(35) \quad ENC_1^2 = 2\left[2qkT\Gamma\frac{I_L+I_G}{\omega_T}\right]^{1/2} \cdot C_D^{*1/2} \cdot \left(m^{1/2}+m^{-1/2}\right) \cdot (A_1A_3)^{1/2}.$$

In the expression above, A_w has been assumed to consist of a_{w1} term alone.
If the input device is a JFET and the detector leakage current may, accordingly, be the dominant term in I_L+I_G, (35) can be commented by saying that ENC^2_1 is fixed by:

1) the ω_T of the input device;
2) the detector leakage current;
3) the detector capacitance;
4) the mismatch coefficient m, which states how far is C_i from C^*_D;
5) the $(A_1A_3)^{1/2}$ factor, which accounts for the worsening introduced by the actual filter as compared to the matched one.

As m can be made close to 1 by suitably tailoring the transversal dimension W of the device or by paralleling a suitable number of units, it can be concluded that ENC^2_1 is essentially determined by the detector characteristics, C^*_D and I_L, and by the transition frequency of the input device.

With reference to (35) it must not be surprising that ENC^2_1, which accounts for both A_w and B_w contribution is affected by the $(m^{1/2} + m^{-1/2})$ factor, which is a characteristic of ENC^2 contributions brought about by series noise sources, like the one with spectral density A_w and does not intervene in the case of parallel noise sources as the one with spectral density B_w is. The explanation is that ENC^2_1 in (35) was evaluated as the minimum, with respect to τ, of the first and third therm in the expression (16) of ENC^2. This makes the geometric mean $(A_wB_w)^{1/2}$ replace the individual spectral densities A_w, B_w with the consequence of $(C^*_D+C_i)$ appearing as a factor in (35).

The ENC^2_2 term given by (34) can be written, by introducing the characteristic parameter H_f of 1/f-noise as:

$$(36) \quad ENC_2^2 = A_2\left[2\pi H_f \cdot C_D^* \cdot \left(m^{1/2}+m^{-1/2}\right)^2 + \frac{B_f}{2\pi}\right] \cdot \frac{\pi}{2}$$

As shown by (36), the ENC^2 contribution brought about by series 1/f-noise, that is, the one describing the low-frequency noise behaviour in the input device current is proportional to H_f and C^*_D and to a function of m which attains a minimum for m = 1, when the condition of capacitive matching is met. Such a condition is so important to deserve a more detailed discussion, which will be done in the next section.

IV CAPACITIVE MATCHING BETWEEN DETECTOR AND PREAMPLIFIER

The following considerations refer to the process of optimizing in the expression (16) of ENC^2 the two terms arising from the series noise sources, the ones described respectively by A_w and a_f/f spectral densities. These two terms become prevailing at large values of C^*_D. It will be assumed, for the sake of simplicity, that $a_{w1} = 0$, so that the contribution brought about by series noise can be written as:

$$(37) \quad ENC_{SERIES}^2 = \left[2kT \cdot \frac{\Gamma}{g_m} \cdot \frac{A_1}{\tau} + 2\pi a_f A_2\right] \cdot \left(C_D^*+C_i\right)^2$$

From (37), the noise sensitivity to the detector capacitance, dENC/dC$_D$ is obtained as:

$$\frac{dENC_{SERIES}}{dC_D} = \left[2kT \cdot \frac{\Gamma}{gm} \cdot \frac{A_1}{\tau} + 2\pi a_f A_2\right]^{1/2}$$

Expressing white and 1/f-noise through the relevant intrinsic noise parameters, (37) becomes

(38) $$ENC^2_{SERIES} = \left[2kT \cdot \frac{\Gamma}{\omega_T} \cdot \frac{A_1}{\tau} + 2\pi H_f \cdot A_2\right] \cdot C_D^* \cdot \left(m^{1/2} + m^{-1/2}\right)^2$$

As their name implies, ω_T and H_f are related to the physical processes on which the device is based and on their channel lenght, but are invariant with respect to paralleling similar devices or, which is equivalent, to increasing the W gate dimension.
According to (38), for a given device type, which specifies ω_T and H_f, and at fixed τ and C^*_D, ENC^2_{SERIES} attains a minimum for m = 1, that is, when the device input capacitance C_i equals C^*_D. At large values of detector capacitances, when C_{STRAYS} and C_f are negligible compared to C^*_D, m = 1 means $C^*_D = C_i$, which corresponds to say that the detector is capacitively matched to the amplifying device. It is worth emphasizing that the considerations about capacitive matching done by tailoring the device size to the actual C^*_D are based on the assumption that ω_T remains constant. Which means that g_m changes proportionally to C^*_i as C^*_i is varied.
The condition of capacitive matching should always be met in the design of a low noise preamplifier (14).
In practice, this may not always be feasible, so that some controlled degree of mismatch may be deliberately accepted.
For instance, in microstrip vertex detectors, $C^*_D \approx 10pF$. Exact capacitive matching would require an input device of $\approx 2 \cdot 10^4$ μm^2 which in a monolithic approach to the front-end amplifier system would limit the density of channels on the chip.
At large C^*_D s, failure to meet the matching condition would result in unduly large ENC^2 values. This is especially true in calorimetry application as well as in energy measurements with thin silicon detectors of large area. These two fields have stimulated the development of large junction field-effect transistors, like NJ 3600 (W = 9·10^4 μm) and NJ 1800 (W = 4.5·10^{4}μm) that are able to match detector capacitances of respectively 400 and 200 pF [15, 16]. Parallel combinations of NJ3600 would make the capacitive matching feasible at accordingly larger C^*_Ds. The price to be paid with such an approach to capacitive matching is obviously the power dissipation in the front-end.
The use of a matching transformer between detector and preamplifier overcomes this limitation and, compared to the transformerless solution suggests the following additional considerations [17, 18].
There may be limitations in the use of a transformer if this has to be located in a magnetic field. Investigation about the possibility of shielding the matching transformer in order to extend their use to this operating condition is being carried on.
The transformer adds noise, basically the one related to core losses. Besides, it has to be used in association with a bipolar h(t). If this precaution is not taken, owing to the presence of a finite inductance in the transformer, the series 1/f-noise of the preamplifier would yield a divergent ENC^2 contribution.
The transformer has an advantage, though. By virtue of its core saturation, it acts as a natural surge protection, thereby preventing the preamplifier input device from being damaged by large voltage transients that may occur in the detector.

V NOISE PROPERTIES OF ACTIVE DEVICES

The noise properties of active devices have now to be reviewed, to explain how they affect the value of ENC^2 given by (16). This discussion will clarify the limits of each device in a specific application and will help to orient the choice toward the device which suits a certain experiment condition at best. Before going into detail about the noise properties of active devices, Table II will be presented, to summarise in a very concise way the fundamental limitations for the different device categories.

TABLE II

Device Type	A_w series white spectral density		1/f-noise H_f [J]	B_w parallel white spectral density $b_{w2} = qI_G$ [A^2/Hz]
	$a_{w1} = 2kTR_{BB'}$ [V^2/Hz]	$a_{w2} = 2k\Gamma T/\omega_T C_i$ [V^2/Hz]		
High Freq Bipolar Transistor	$10^{-19} \div 10^{-18}$	$10^{-22} \cdot 1/I_c$	$10^{-27} \div 10^{-26}$	$10^{-25} \div 10^{-24}$
Silicon junction field-effect transistor	Negligible in discrete devices	$3 \cdot 10^{-20} \div 5 \cdot 10^{-18}$	10^{-27}	$\approx 10^{-30}$
Silicon MOS field-effect transistor	Negligible in discrete devices	$10^{-19} \div 5 \cdot 10^{-18}$	N-channel enh. MOS 10^{-23} P-channel enh. MOS 10^{-25}	Virtually zero
GaAs MESFET	Negligible in discrete devices	$10^{-19} \div 10^{-18}$	10^{-23}	$10^{-27} \div 10^{-26}$

It appears from Table II as a first conclusion that the bipolar transistor, and the GaAs MESFET, featuring a large parallel noise are restricted in use to cases requiring low τ's, where otherwise they bring about the advantage of large transition frequencies. At values of τ well below 100 ns, which is the more useful range of both types of device in charge measurement, the bipolar transistor would still be limited by the white noise contribution arising from $R_{BB'}$ and the GaAs MESFET by the 1/f-noise. It is also true that for the latter, a substantial reduction in 1/f-noise can be obtained by operating the device at cryogenic temperatures.

The silicon JFET is still the most suitable device for charge measurement at values of τ from 50 ns to $\approx$ 10 µs, where it brings about the advantages of a small 1/f series noise and a small parallel white noise. Beyond 10 µs the JFET starts being limited at room temperature by its parallel noise. Such a limit is removed by operating the device at cryogenic temperatures [19].

Acceptance of MOS transistors in the category of devices suitable for charge measurements occurred when the need arose of realising monolithic preamplifier systems for collider experiments. Even before that, however, a depletion-type MOS, Philips BF992 featuring a good ω_T and a large g_m/I ratio provided a good solution for a microstrip vertex detector in the fixed target E687 experiment at Fermilab [20].

Enhancement-type MOSFET have some intrinsic limitations when employed in charge measurements. The N-channel type has a large 1/f-noise. The P-channel type, which has a nearly two order of magnitude lower spectral power density of 1/f-noise suffers from the ω_T limitations related to the low carrier mobility [11, 21, 22].

It is stressed in the first column of Table II that the considerations about the spreading resistances refer to the case of discrete devices.

It is almost impossible to quote data for devices belonging to monolithic circuits as they are strongly dependent on the technological process employed.

So, while in discrete JFETS and MOS the spreading gate resistance have been put equal to zero, as they actually are very small and because, as a difference from the bipolar transistor, there is no built-in reason why they should not be very small, things are much different in monolithic structures.

The devices quoted in Table II will now be considered individually from the point of view of their behaviour in affecting ENC^2.

V.1) BIPOLAR TRANSISTORS

If the series spectral power density of series noise in a bipolar transistor is a measured at high enough frequencies to make the 1/f-contribution negligible, but not too high to have the measurement affected by the cutoff frequencies, say, f in the 100 kHz range, A_w would be found to comply with the relationship:

$$(39) \quad A_w = 2kT\left(R_{BB'} + \frac{0.5}{g_m}\right) = 2kTR_{BB'} + \frac{(kT)^2}{qI_c}.$$

The first term in (39) describes the thermal noise in $R_{BB'}$, the second the shot noise in the collector current.

If the spectral power density of series noise referred to the input of the device is measured by means of a noise meter at different values of the standing collector current I_c and the results are plotted as functions of, $1/I_c$ he curve of fig. 4a) would be obtained, where the value of A_w as $1/I_c \to 0$ is $4kTR_{BB'}$. (Here, where experimental data are quoted, the physical unilateral spectral densities are used in place of the bilateral ones).

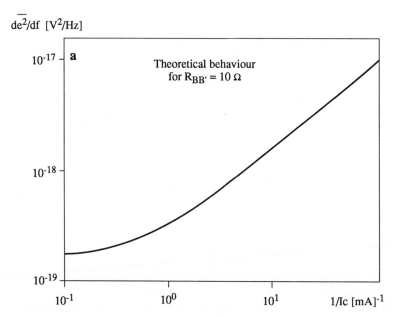

Fig. 2. a) Plot of theoretical spectral power density referred to input for the series white noise A_w in a bipolar transistor as a function of $1/I_c$.

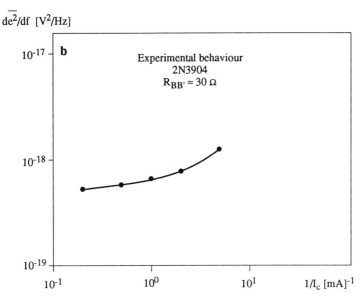

Fig. 2. b) Measured $A_w - 1/I_c$ dependence for the transistor 2N3904.

Microwave bipolar transistors have values of $R_{BB'}$ around $10 \div 15\ \Omega$. For instance, the Philips BFR92 transistor has an $R_{BB'}$ value, measured at 0.5 mA I_c, of 8 Ω.

A peculiar feature of the bipolar transistor employed as an input device in charge measurements comes from the fact that C_i, which can be reasonably approximated with the base diffusion capacitance C_B, depends in a predictable way on I_c. In a limited range of I_c values, where ω_T can be considered almost constant, C_i has a linear I_c dependence, as g_m does. This suggests that optimisation in ENC^2 be achieved rather than by tailoring W to the actual detector capacitance, simply by adjusting its standing collector current. This approach is feasible, as shown in fig. 3, where the theoretical expression of ENC^2, plotted as a function of C_B/C^*_D is compared with some experimental data [23].

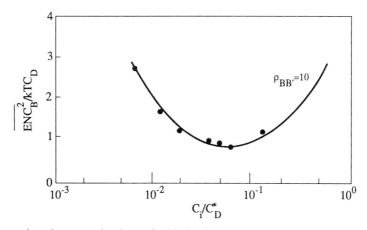

Fig. 3. Comparison between the theoretical behaviour (———) of the ENC as a function of C_i/C^*_D and some experimental points ($\bullet$); $\beta = 50$, $C^*_D = 100$ pF, $\tau = 180/\omega_T$ ($\rho_{BB'} = R_{BB'}\ \omega_T\ C_D$, $\beta = I_c/I_B$, where I_c is collector current and I_B is the base current).

The continuous curve in fig. 3 was obtained from (16) neglecting the 1/f-noise and assuming for h(t) the piecewise parabolic shape, shown at line 5 in Table I.

It is apparent from fig. 3 that, because of the large parallel noise, the minimum in ENC^2 does not occur in the capacitive matching situation, but at a smaller C_i/C^*_D ratio. This is not in conflict with the discussion on capacitive matching at section IV. That discussion took into account the series noise alone with the exception of the contribution due to $R_{BB'}$, while the present analysis refers to the minimisation of the entire ENC^2. The minimum value in ENC is given by:

$$(40) \quad ENC^{(M)} = 2kTC^*_D \cdot \frac{A_1}{\omega_T \tau} \left\{ \sqrt{1 + \frac{\omega_T^2 \tau^2}{\beta} \cdot \frac{A_3}{A_1}} + 1 + R_{BB'} \omega_T C^*_D \right\}.$$

The bipolar transistors have a comparatively small 1/f-noise. For instance, the already quoted BFR 92 has 6.5 nV/$\sqrt{Hz}$ at 10 Hz, with the device operated at 0.5 mA. This corresponds to $a_f \approx 4.5 \cdot 10^{-16} V^2$.

A considerable interest is being conveyed now on bipolar transistors as front-end elements in connection with the availability of semicustom processes based upon large f_T bipolar technologies and of the BICMOS technology. Some low-noise circuits have already been implemented and other projects are presently being carried on [24, 25]. Still some effort seems to be required in order to reduce the values of $R_{BB'}$, that vary largely from one case to another, but in some circumstances with minimum size devices can reach up to 1KΩ.

Among the results achieved so far, one seems to be particularly encouraging. A monolithic preamplifier featuring 1000 rms electrons at 5 pF - C^*_D and a $dENC/dC^*_D$ sensitivity of 39 e/pF with a weighting function of 15-ns basewidth has been realized at the University of Pennsylvania on the base of an AT & T bipolar process [26].

V.2 SILICON JUNCTION FIELD-EFFECT TRANSISTORS

Junction field-effect transistors (JFETs) have a small parallel noise,

$$\overline{di^2}/df \approx 10^{-30} A^2/Hz.$$

This makes them suitable for the amplification of signals from capacitive sources on a range of values of τ, which is broad already at room temperature and becomes broader with cryogenic operation.

This property, associated with their good behaviour in terms of 1/f series noise and the small noise contribution brought about by gate spreading resistances $R_{GG'}$, explains why Si JFETs are the most widely employed front-end devices in applications with detectors of ionizing radiation.

Besides, JFETs are intrinsically radiation tolerant and this makes them worth of consideration for uses in the environment of high luminosity colliders.

Three types of JFETs will be considered here:

a) Epitaxial channel JFETs with diffused gate
b) Implanted channe JFETs, adopted in monolithic design
c) JFETs realised on a high resistivity silicon, that are compatible with the detector fabrication process.

Type a) process is currently employed in discrete JFETs. As an example of type b), the JFETs compatible with CMOS process will be discussed. JFETs realised on a Si-detector material will be presented as an example of type c).

V.2.1 Type a) JFETs

Silicon N-channel JFETs with the channel realised in an epitaxial layer and the P+ gate diffused into it provide the best noise performances at values of τ down to about 50 ns. At lower τ's they get limited by their comparatively poor f_T characteristics, that have to be put in connection with the technology employed to realize the gate. Gate lengths below about 5 µm are difficult to obtain. At values of τ above $\approx$ 50 ns, however, the N-channel epitaxial JFET is a very suitable device for low noise applications by virtue of the two following properties. First, out of all field-effect devices, it features the smallest amount of series 1/f-noise. Secondly, its channel thermal noise complies quite well with the theoretical expression for the white spectral power density:

$$(41) \quad \frac{\overline{de^2}}{df} = 4kT \cdot \frac{0.7}{g_m} .$$

The following considerations about epitaxial-channel JFETs refer to specimens produced by INTERFET, about which the authors of the present paper are best knowledgeable. Examples of series noise spectra for JFETs of different W values are given in fig. 4.

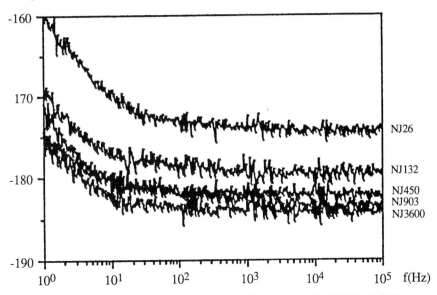

Fig. 4. Spectral density of series noise as a function of frequency for JFETs of different gate widths W.

The plots of fig. 4 clearly show the flat portion where white noise is the dominant contribution and the increasing density at decreasing frequency which reveals the presence of low-frequency noise. The curves of fig. 4 refer to selected units of the considered types.

Table III and IV give the low frequency noise parameters A_f and H_f, the measured density of the white noise and its theoretical value determined by measuring g_m and introducing it into (41), for JFETs of different gate widths W at two values of the standing current I_D. All experimental data of tables III and IV describe the average behaviour across batches of the considered types realized with low-resistivity channel material.

TABLE III
Results of noise measurement; $I_D = 2$ mA

Gate width W [μm]	$A_f [V^2]$	$H_f [J]$	White noise Theoretical [nV/Hz$^{1/2}$]	Experimental [nV/Hz$^{1/2}$]
3.3×10^3	19.00×10^{-17}	3.3×10^{-27}	0.97	1.10
1.2×10^4	6.8×10^{-17}	3.4×10^{-27}	0.75	0.78
2.2×10^4	4.3×10^{-17}	3.9×10^{-27}	0.64	0.68
4.5×10^4	2.2×10^{-17}	3.3×10^{-27}	0.61	0.63
9.0×10^4	0.9×10^{-17}	3.6×10^{-27}	0.55	0.55

TABLE IV
Results of noise measurements; $I_D = 10$ mA

Gate width W [μm]	$A_f [V^2]$	$H_f [J]$	White noise Theoretical [nV/Hz$^{1/2}$]	Experimental [nV/Hz$^{1/2}$]
3.3×10^3	13.0×10^{-17}	2.3×10^{-27}	0.70	0.84
1.2×10^4	3.5×10^{-17}	1.8×10^{-27}	0.51	0.56
2.2×10^4	2.2×10^{-17}	2.0×10^{-27}	0.40	0.49
4.5×10^4	1.2×10^{-17}	1.8×10^{-27}	0.37	0.42
9.0×10^4	0.7×10^{-17}	2.8×10^{-27}	0.31	0.36

As expected, Tables III and IV confirm that A_f decreases at increasing W, while $H_f = A_f \cdot C_i$ remains nearly constant, thereby justifying its name of intrinsic parameter describing 1/f-noise. At constant current, the white noise decreases at increasing W, this effect being related to the increase in transconductance which results from the reduced current density.

Tables III and IV also show the extent to which the measured white noise densities comply with the expected ones on the base of (41).

The agreement is satisfactory at $I_D = 2$ mA. At the larger I_D, more deviations are observed between experimental and theoretically predicted white noise densities. It seems

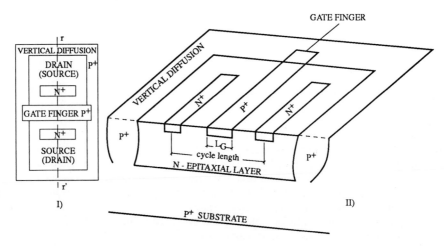

Fig. 5. Device structure of a single cell epitaxial-channel JFET. I) Top view; II) Cross section with r, r' plane.

reasonable to explain this with the remark that any noise associated with bulk resistors and gate spreading resistors affects the total series white noise more consistently at $I_D = 10$ mA, where the contribution due to the thermal noise in the channel is smaller, than at $I_D = 2$mA.

The structure of an epitaxial-channel JFET can be understood with reference to fig. 5, I) and II).

The JFET shown in figure is an N-channel unit, as usually JFETs employed in low noise amplifiers are. P-type devices are limited in these applications by the lower mobility of their majority carriers in the channel, that are holes.

To realize an N-channel JFET, the N-type epitaxial layer is grown on top of the P^+-type substrate. Vertical P^+diffusions all way through to the substrate define the active area of the device, fig. 5, I). Fig. 5 II) is the section of the JFET along the (r, r') vertical plane. The vertical diffusion and the substrate constitute the back-side gate.

The single-cell device of fig. 5 has a one-finger topside gate which actually touches the vertical diffusion and therefore is in electrical contact with the backside gate. The source and drain are N^+ diffused fingers on either side of the top-side gate. The distance between source and drain midpoints is the cycle length. The JFETs considered in this section have cycle lenghts which, depending on the device type vary between 15 μm and 25 μm. The transition frequency f_T is determined by the length of channel underneath the topside gate diffusion, L_G; f_T is inversely proportional to L^2_G.

A scaling down in the cycle length results in a reduced L_G and an increased f_T. The inverse proportionality to L^2_G has, as a consequence, that even a little reduction in the cycle length produces a substantial increase in f_T.

As already pointed out, epitaxial channel, diffused gate JFETs do not have outstanding f_T performances. Rarely f_T exceeds 500 MHz and this limitation is due both to the photolitographic process employed and to the lateral diffusion in the P^+ top gate region.

It would be worth, indeed, concentrating some effort toward the scaling down of the device cycle with the aim of increasing f_T and steps are being now taken in that direction. The goal, obviously would be the extension of JFETs to the 10 ns range of τ, as required by high luminosity hadron colliders.

Nowadays, N-channel JFETs for low noise applications are available with a large number of gate width values, so that satisfactory capacitive matching conditions can be achieved in a variety of detector applications. Actual devices are built-up by repeating the source-gate-drain cycles within the frame of the vertical diffusion as many times as needed to arrived at the required W.

V.2.2 IMPLANTED JFETs COMPATIBLE WITH A MONOLITHIC CMOS PROCESS

The JFETs described in this section are part of a monolithic process whose salient feature is the compatibility of four device types, N and P-channel JFETs, N and P-channel MOS on the same substrate [27, 28]. The JFET-CMOS process is available at Fraunhofer Institute It has been employed in the development of the preamplifier system for the microstrip vertex detector of Aleph experiment at LEP, in the framework of a cooperation between Fraunhofer Institute, Max Planck Institute and Pavia University. The configurations of the four compatible devices are shown in fig. 6, I, while the spectral densities of series noise voltage are plotted as functions of frequency in figure 6 II, a) through d).

It is apparent from fig. 6 II that the two types of JFETs are superior, noisewise to the MOS devices and that the P-channel MOS has much less 1/f-noise than the N-channel MOS, a fact which has been discussed in the literature. Out of the four devices belonging to the JFET-CMOS process the N-channel JFET, curve d) in fig. 6 II has the best noise properties. Yet, compared to a discrete, diffused-bype JFET, curve e) in fig. 6 II, it features a considerably larger low frequency noise. This can be understood from fig. 6 II, which clearly shows that the 1/f-noise for the diffused-type JFET as compared to the implanted-channel JFET is well below the factor 3 that the ratio in gate areas would imply if the H_f-coefficients were the same.

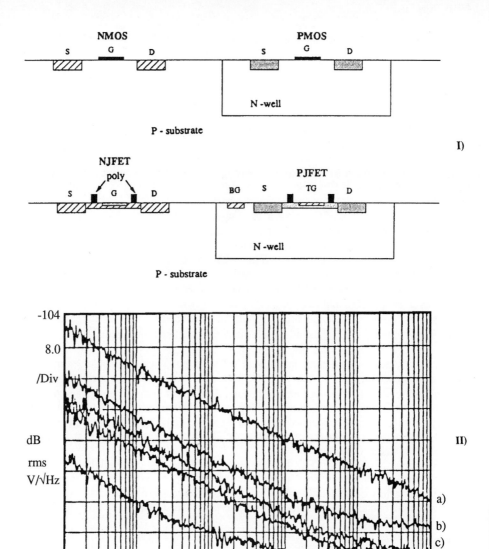

Fig. 6-I) Configurations of the four compatible devices in the JFET-CMOS process.
6-II) Spectral power densities of the series noise voltage for devices belonging to the JFET-CMOS compatible process (gate width W = 200 μm, length L = 4 μm). (a) N-channel MOS; (b) P-channel MOS; (c) P-channel JFET; (d) N-channel JFET; (e) The same for a discrete N-channel JFET, Interfet NJ16 gate size W = 400 μm, L = 6.5 μm. The spectral densities were measured with the devices operated at a standing current of 100 μA.

This difference seems to be put in relation to the channel structure in the two cases, which is implanted for the JFET belonging to the monolithic process and epitaxial in the diffused-type JFET.

The spectral density of series noise voltage as a function of frequency is plotted in fig. 7 for three N-channel devices of different gate width W, belonging the JFET-CMOS process. The measurements were carried out with the JFETs operating at the same drain current of 100 μA, which is a reasonable value to be assumed for the front-end component in a monolithic circuit with many preamplifier channels.

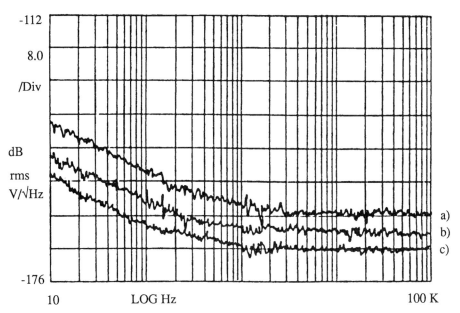

Fig. 7. Spectral power densities of the series noise voltage as a function of frequency for N-channel JFETs of different gate areas. (a) gate width W = 200 μm, length L = 4 μm; (b) W = 400 μm, L = 4 μm; (c) W = 1600 μm, L = 4 μm.

Fig. 7 shows how the noise voltage spectral density scales down with the device size at a fixed drain current. The N-JFET process to which fig. 7 refers cannot be compared, as far as white noise goes, with that of fig. 6. The difference in the noise behaviour between JFETs of fig. 6 and 7 has to be attributed to the larger gate spreading resistance in the devices of fig. 7.

Although the JFET-CMOS process makes four different device types available for front-end design, considerations of radiation sensitivity of noise suggest that the N-channel MOS be discarded. A design based exclusively on N and P channel JFETs may be taken into considerations. However, a front-end system requires several switching functions in both analog and digital sections and for them MOS are better suited than JFETs. The previous considerations led to the choice of N-JFET and P-MOS as basic design elements.

The N-JFET is employed as a preamplifier input device and wherever low noise characteristics are required. Design complementarity is guaranteed by the P-MOS, which, besides, lends itself to a simple realization of the switching functions [29].

As an application example of NJFET-PMOS design, a 64-channel preamplifier system for microstrip detectors has been developed. Each channel consists of a charge-sensitive preamplifier with a switch-actuated reset, a discrete-time filter implementing a trapezoidal weighting function and an analog memory which samples and holds the amplitude at the filter output [29].

The stored voltages, that are linearly related to the charges induced on the read out strips, are sequentially switched into a serializing output bus during the transfer phase.

In the design of the whole system, an issue of the utmost importance is the noise minimization in the charge-sensitive preamplifiers. The preamplifiers were at first realized as separate test structures and are now available as independent monolithic units with either a built-in, switch-actuated reset or with an externally connected reset resistor.

The circuit diagram of the charge-sensitive preamplifier is shown in fig. 8. In fig. 8 the input NJFET J1 and the P-MOS M2 constitute a complementary cascode, which has, as an active load the N-JFET cascode (J2, J3).

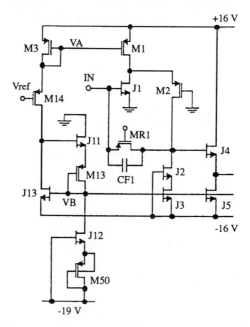

Fig. 8. Monolithic charge-sensitive preamplifier based on NJFET-PMOS technology.

The integrating capacitor CF1 is connected between the drain of M2 and the gate of J1. In the circuit version of fig. 8, the reset of the charge on CF1 is achieved through the MR1 switch. In the alternative version CF1 and MR1 are not present on the chip, so that the parallel combination of CF1 and a reset resistor can ben connected externally.

The noise behaviour of the monolithic charge-sensitive preamplifier is described in fig. 9, where the equivalent noise charge ENC is plotted as a function of the filter time constant τ at three different values of the capacitance C^*_D connected at the preamplifier input to simulate the detector. The ENC measurements were performed on a preamplifier with resistive charge reset. The preamplifier was associated to a spectroscopy amplifier with a semigaussian shaping of time constant τ.

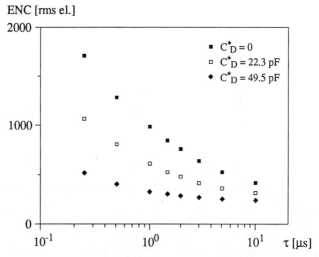

Fig. 9. ENC as a function of the time constant τ of the semigaussian spectroscopy amplifier amplifying and shaping the signal at the output of the preamplifier of fig. 8.

The main features of the monolithic NJFET-PMOS preamplifier are summarized in Table V.

It can therefore be concluded that in the case of comparatively small detector capacitances ($C^*_D \approx 10$ pF) as the microstrips' is, monolithic circuits employing JFETS as input elements and yielding satisfactory noise performances are feasible. These JFETs, that have an implanted channel are not as good, noisewise, as the diffused ones, where the channel is in an epitaxial layer. They lend themselves, however, to a more straightforward monolithic integration process and the design can benefit from the availability of other device types like the MOS, as it was demonstrated in this section.

TABLE V

Main characteristics of the NJFET-PMOS preamplifier

Supply voltages	± 6V
Power dissipation	1.6 mW
Risetime	120 ns ($C^*_D = 0$)
Risetime	250 ns ($C^*_D = 22$ pF)
Open loop gain	72 dB
Input device	N-JFET, W = 400 μm, L = 4 μm
Chip area	0.064 mm²

V.2.3 IMPLANTED JFETs COMPATIBLE WITH THE DETECTOR PROCESS

The need of realising JFETs compatible with the detector process arose in connection with the apperance of detector types with very low anode capacitance.

Examples of such detectors are the silicon drift chambers and the totally depleted CCDs, for which the output electrode capacitance can be as low as 100 fF or even less. Obviously, whatever connection to an external preamplifier chip would add a considerable amount of stray, thereby spoiling a potentially good signal-to-noise ratio [30, 31, 32].

This consideration stimulated the search for a solution to the problem of matching such a small detector capacitance, which necessarily implies that the preamplifier or at least its input element has to be integrated on the detector chip. The problem is not an easy one, for the detector is realised on a material whose resistivity is by some orders of magnitude larger than the one required for electron device implementation.

An implanted-type JFET compatible with the detector process has been developed as a basic element [33, 34, 35]. The cross section of the JFET is shown in fig. 10.

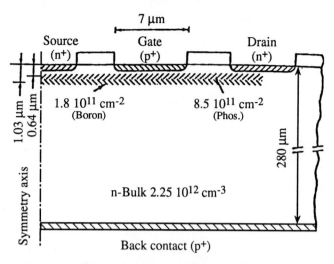

Fig. 10. Cross section of a JFET implanted on a high-resistivity, totally depleted substrate which houses also the detector sensitive region.

As apparent from fig. 10, the JFET channel is a deep phosphorus implant in the high resistivity substrate. A deeper boron implant separates the channel from the totally depleted region underneath.

JFETs with 7 μm gate lenght and different gate widths have been realized on the base of the structure shown in fig. 10, with a gate length of 7 μm.

Measurements carried out on units with -1.5 V pinch-off voltage and 108 μm gate width have shown values of the gate leakage current in the 10 pA range at room temperature, g_m/I_D ratios around 0.4 V^{-1} and output resistances of about 60 kΩ. Spectral analysis of the input noise voltage has shown that the density of channel thermal noise is close to the value expected on the base of the relationship $4 kT \cdot (0.7/g_m)$. As to the 1/f-noise, values of H_f around 10^{-25} J were found, which, though larger than for diffused type JFETs, have to be considered satisfactory in a device with implanted channel. The JFET with 7 μm gate length and 108 μm width has an input capacitance in the 100 fF range, suitable to match the capacitance appearing at the output electrode in a drift chamber or a totally depleted CCD.

A charge-sensitive preamplifier employing as active elements only JFETs of the type previously described and therefore suitable for integration on the detector material was also realized. The circuit includes also the integrating capacitor and a feedback resistor of large value. The monolithic preamplifier compatible with the detector process marks a significant step forward in the direction of integrating on the same chip detector and front-end electronics [35].

V.3 MOS TRANSISTORS

In the early days of research and development activities oriented to LEP experiments, it became clear that detectors with such a high density of read out electrodes like microstrips, employed in a collider experiment, require a monolithic front-end system. Such a system had to provide, besides low noise amplification and filtering, some kind of signal serializing into a single output bus, to substantially reduce the number of cables in the vertex region. At that time the most obvious approach to the design of a monolithic preamplifier system was the one based upon MOS devices. MOS are not intrinsecally low noise-devices. They present, indeed, a large amount of 1/f-noise which, as already pointed out, affects ENC to a significant extent in the experimental conditions of LEP. Conversely, MOS lend themselves to large integration densities and offer a good design versatility, which becomes an important feature in a front-end system where amplification, analog switching and storage and logic functions have to be combined on the same chip [36, 37, 38].

Investigation about the noise behaviour of MOS transistors in the specific application as front-end elements for radiation detectors has been carried-out by several authors.

Sansen has pointed out that an N-channel enhancement-type MOS has a thirty-to-fourty times larger power density of 1/f-noise than a P-channel enhancement MOS of equal area [21].

This fact has been confirmed by measurements carried out by other authors on different C-MOS processes. Values of the intrinsic coefficient $H_f = A_f \cdot C_i$ between 10^{-23} J and $2.5 \cdot 10^{-23}$ J for the N-MOS and between $5 \cdot 10^{-25}$ and $5 \cdot 10^{-26}$ for the P-MOS were measured [11, 22].

The P-MOS, therefore, may represent a suitable front-end device in those applications where ENC is significantly affected by 1/f-noise and the processing times are not so short that the intrinsic frequency limitations of the P-MOS, related to the low carrier mobility become important. So, the P-MOS behaves adequately well at times longer than a few hundreds of ns.

An example of low-noise design based upon P-channel MOS is provided by AMPLEX, a monolithic chip developed for processing signals from multielectrode detectors [39]. The input P-MOS has 5000 µm gate width and 3 µm gate lenght. Operating close to weak inversion, it features a transconductance of ≈ 1.4 mA/V at a drain current of 50 µA. The resulting ENC, obtained with a semigaussian shaping of 700 ns peaking time is ≈ 1000 electrons at an input capacitance C^*_D of 20 pF and the $dENC/dC^*_D$ sensitivity is ≈ 33 e/pF.

An alternative approach to improved noise performances in circuits based on MOS relies upon the use of depletion-type devices. The low-frequency noise behaviour in depletion-type MOS substantially differs from that of enhancement-type transistors, as shown in fig. 11, where the noise spectral densities for two N-MOS units of either type and equal gate area are compared [9, 40].

The enhancement-type MOS presents in the low-frequency region a spectral power density with a frequency dependence close to 1/f.

The depletion-type unit has a $1/f^{1.6}$ frequency dependence, not far from the $1/f^2$ spectrum which has to be expected according to Kandiah [41].

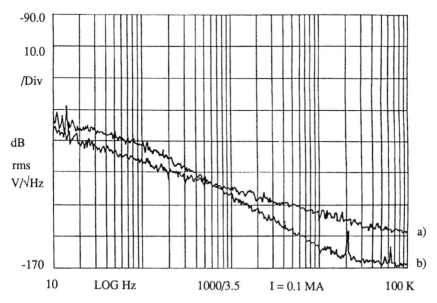

Fig. 11. Series noise spectral density as a function of frequency for two N-channel MOS having 1000 µm gate width and 3.5 µm length. (a) enhancement-type device; (b) depletion-type device

As it can be inferred from fig. 11, at time constants in the microsecond range, the depletion MOS has to have a lower ENC than the enhancement device.

This is the consideration which led to adopt depletion-type N-channel MOS as front-end devices in the MOS version of the monolithic chip for microstrip readout in Aleph experiment. The preamplifiers with the depletion MOS at the input were proven to yield ENC values at zero C^*_D not far from the JFET-PMOS circuit of fig. 8. Their ENC sensitivity to external capacitance, $dENC/dC^*_D$, is, however, definitely larger.

V.4 GaAs MESFETs

The GaAs MESFETs have a high transition frequency, from a few GHz to tens of GHz, which makes them suitable for charge amplification at processing times in the ns range. At room temperature they have a large amount of 1/f-noise. Values of the characteristic coefficient H_f as large as 10^{-23} J have been measured.

To get an idea about how ENC is affected by the channel thermal noise and by 1/f-noise in a GaAs MESFET, reference can be made to fig. 12.

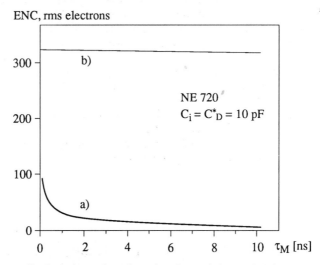

Fig. 12. ENC contributions brought about by channel thermal noise, a) and channel 1/f-noise; b) in a GaAs MESFET. ENC is plotted as a function of the peaking time τ_M of a piecewise quadratic h(t), case 5 in TABLE I.

The curves of fig. 12 are theoretical evaluations carried out from the knowledge of the experimentally determined spectral densities of white and 1/f-noise. For the calculations, a reference h(t) of piecewise quadratic shape with peaking time τ_M was assumed. The case of a 10 pF-C^*_D and an input device of suitable gate width W to match the actual C^*_D has been considered. Fig. 12 clearly points out that the dominant ENC limitation is related to the 1/f-noise in the drain current.

Despite its large 1/f-noise the GaAs MESFET may provide a solution to the problem of amplifying signals from microstrip detectors at operating times down to 10 ns, as demonstrated experimentally [42]. Investigation has been carried out about the noise behaviour in GaAs MESFET at cryogenic temperatures [43, 44]. An interesting point which emerged from this analysis is the reduction of 1/f-noise with temperature. For instance, reducing the temeprature from 300 °K to 77 °K produces a two orders of magnitude reduction in H_f. Such a property suggersts that the GaAs MESFET may compete with the Si JFET in applications with cryogenic detectors, for instance liquid Argon calorimeters [45].

The GaAs MESFET does not have an optimum temperature for cryogenic operation, as the Si JFET does. Besides, it has a larger transition frequency, which makes it more suitable than Si JFET in operation at short processing times.

It is worth quoting the following result.

A hybrid preamplifier made of GaAs MESFETS, with 10 paralleled elements at the input, each of them operating at 0.6 mA-I_D features at 77 °K with a semigaussian shaping of 1 ms time-constant, ≈ 100 electrons ENC at zero C^*_D and 1.1 electrons/pF dENC/dC^*_D.

VI EFFECTS OF IONIZING RADIATION ON THE NOISE IN ACTIVE DEVICES AND CIRCUITS

In collider experiments the front-end electronics may be very close to the beam pipe and be exposed, therefore, to ionising radiation. The noise in devices and circuits would increase under the effect of absorbed radiation. In front-end systems that have an intrinsically poor signal-to-noise ratio, the situation may come to the point at which the event detectability is impaired [46].

The problem will be particularly serious in experiments at high luminosity colliders. Any front-end design must, accordingly, concentrate adequate attention on the problem of radiation sensitivity of the noise and on the measures that have to be taken to reduce it. In what follows, some experimental results aiming at describing the effects of radiation on the noise in active devices and circuits will be presented.

Some of these results refer to effects that have been noticed at moderately low doses and were obtained in the framework of tests carried out with the LEP conditions in mind.

Measurements at much larger doses were carried out at Brookhaven National Laboratory with the aim of reproducing the situation that will occur during a front-end lifetime at SSC [47].

The radiation-sensitivity tests oriented to LEP applications were carried out on MOS, JFETs and monolithic circuits based on the JFET-CMOS process. It emerged from them, that in ordinary MOS the spectral density of the noise in the low-frequency region is significantly affected by radiation. Up to 12 dB increase in noise was observed in an N-channel, enhancement type MOS exposed to 25 krad integrated dose of γ-rays from a ^{60}Co source, as shown in fig. 13.

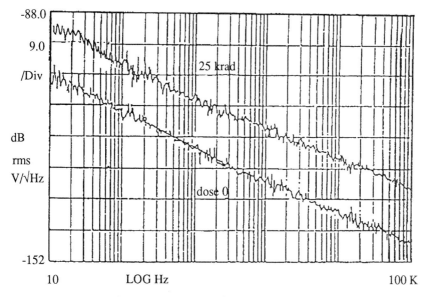

Fig. 13. Noise spectrum of an N-channel MOS with 120 μm gate width and 5 μm length before and after irradiation.

Noise in JFETs was found to be significantly less radiation-sensitive. Fig. 14 shows, as an example, the variation which takes place in the noise spectrum of an N-JFET part of a JFET-CMOS process exposed to 1 Mrad dose of γ-rays from a ^{60}Co source.

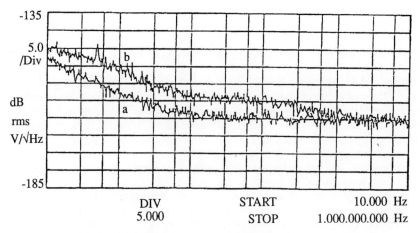

Fig. 14. Noise spectrum in an N-channel JFET part of a monolithic process.
a) before irradiation
b) after absorption of a 1 Mrad dose of ^{60}Co γ-rays.

Fig. 14 shows that the high-frequency portion of the spectrum, the channel thermal noise, is practically unaffected. Below 10 kH$_z$, a change in the shape of the spectrum is recorded, with an increase in the density of about 5 dB.

The effects described in fig. 14 are in agreement with those observed by Stephen [48].

It is also interesting to compare the radiation sensitivity of ENC in a CMOS preamplifier and in a JFET-CMOS preamplifier employing a JFET at the input, figs. 15 and 16.

For the measurements of figs. 15 and 16 the preamplifier output signal was amplified and shaped by a semigaussian amplifier with panel-selectable time-constant τ.

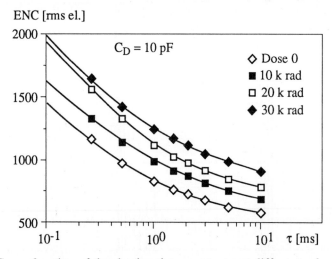

Fig. 15. ENC as a function of the shaping time-constant τ at different values of absorbed dose.

The preamplifiers to which figs. 15 and 16 refer are intended for microstrip readout applications and the measurements were, accordingly, performed at $C^*_D = 10$ pF, which is a typical case in the microstrip case.

As apparent in fig. 15, the ENC in the CMOS preamplifier is noticeably affected by irradiation. For the JFET-CMOS circuit the effect of absorbed radiation is restricted to the region of larger τ values, and has to be attributed to the increase in 1/f-noise and in the parallel noise associated with gate-leakage current.

All the tests discussed so far were carried out with devices powered and set to work in the nominal bias conditions.

Experiments in the conditions set by high luminosity hadron collider will require that the front-end be able to stand much larger doses. Typical values expected during the front-end lifetime are 12 Mrad integrated dose of γ-rays and 10^{14} N/cm^2.

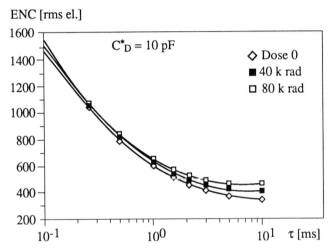

Fig. 16. ENC as a function of the shaping time-constant τ at different values of absorbed dose.

Investigation about the radiation effects on noise for hybrid preamplifiers employing a discrete, diffused-type JFET at the input was recently carried out at the Instrumentation division of Brookhaven National Laboratory, where several conditions in terms of operating temperature, powered on unpowered circuits and delays between end of irradiation and measurement were considered [47].

Some results are summarized in fig. 17. They refer to 12 Mrad integrated dose of ^{60}CO γ-rays. Irradiation in the actual case was carried out with power applied and the circuit was kept powered for about two hours after irradiation and prior to measurement.

As shown in fig. 17, irradiation mainly affects the ENC values at larger τ's. This behaviour is in substantial agreement with that presented in fig. 16, which was attributed to the combined radiation sensitivity effects in 1/f-noise and in the parallel noise associated with the gate leakage current.

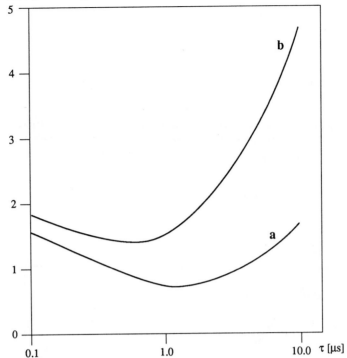

Fig. 17. ENC as a function of shaping time-constant τ for a hybrid preamplifier employing a discrete, diffused-type NJFET at the input.
a) pre-irradiation
b) post-irradiation

The quoted examples point out the growing interest which is now being conveyed into the analysis of radiation effects on the noise in front-end devices and circuits. Such a field of investigation is at its very beginning and deserves more attention and effort. The reliability of front-end systems for future hadron colliders is strongly dependent on the results of such an effort.

CONCLUSIONS

The limits set by front-end noise in the measurement of the charge released by a radiation detector have been theoretically investigated on the base of a simple model of the analog processing channel. Once the noise behaviour of the channel has been defined in terms of its equivalent noise charge (ENC), the ENC dependence on the detector characteristics, on the experiment constraints and on the noise parameters of the front-end devices has been analyzed. Several important conclusions were drawn about the limitations that occur in detector charge measurements and the measures that have to be taken to reduce the dispersion due to noise. Among these measures, the correct choice of the front-end amplifying device, based on the knowledge of the noise behaviour and noise sensitivity to absorbed radiation in active components is of fundamental importance.

ACKNOWLEDGEMENTS

Part of this paper is based upon the results of a research program on suboptimal filters of 1/f-noise carried out with E. Gatti and M. Sampietro, whose contribution is gratefully acknowledged.

Valuable information about several items analyzed in this paper comes from discussions the authors had with H. Kraner, V. Radeka, S. Rescia.

Some experimental results quoted in this paper come from collaboration programs with W. Buttler, G. Lutz, M. Hoye, L. Rehn, D. Roberts.

Discussions with D. Camin about the noise behaviour of GaAs MESFETs are acknowledged.

REFERENCES

1 - V. Radeka, Ann. Rev. Nucl. Part. Sci. 38 (1988) 217.

2 - E. Gatti, P.F. Manfredi, Processing the Signals from Silicon Detectors in Elementary Particles Physics. Rivista del Nuovo Cimento, 9 (1986).

3 - V. Radeka, Nucl. Instr. and Meth. A 226 (1984) 209.

4 - P.F. Manfredi, F. Ragusa, Nucl . Instrum. and Meth. A 235 (1985) 345.

5 - V. Radeka, S. Rescia, Nucl. Instrum. and Meth. A 256 (1988) 228.

6 - D.G. Cassel, Report of the Central Tracking Group, Proc. of 1986 Summer Study on the Physics of SSC. Snowmass, Colorado, (1986).

7 - A. Asner et al., Proceeedings of ECFA. CERN Workshop on the Large Hadron Collider in the LEP Tunnel. Lausanne, Genéve 1984, Vol. I, 49.

8 - M. Giorgi, Paper presented at the 1st National Meeting on Perspectives in Electronics for Experiments at Future Accelerators. Isola del Giglio (Italy), May 26, 27 (1988).

9 - W. Buttler et al., Nucl. Instrum. and Meth. A 288 (1990) 140.

10 - P.F. Manfredi, V. Speziali, Nucl. Instrum. and Meth. A 279 (1989) 152.

11 - G. Lutz et al., Nucl. Instrum. and Meth. A 277 (1989) 194.

12 - E. Baldinger, W. Franzen, Advances in Electronics and Electron Physics 8, (1956) 225.

13 - E. Gatti, P.F. Manfredi, M. Sampietro, V. Speziali, Suboptimal Filtering of 1/f-noise in Detector Charge Measurements to be published.

14 - V. Radeka, Position-Sensitive Semiconductor Detectors. Proceedings of a Workshop Held at Feimilab. T. Ferbel Editor (1981) 21.

15 - M. Bertolaccini et al., Nucl. Instr. and Meth. A 264 (1988) 399.

16 - M. Demicheli et al., Nucl. Instrum. and Meth. A 289 (1990) 418.

17 - V. Radeka, Summary of Noise Relations for Liquid Argon Ion Chambers Calorimeter Notes. Brookhaven National Laboratory, 9 (1973).

18 - E. Gatti, P.F. Manfredi, D. Marioli, Nucl. Instr. and Meth. 193 (1982) 539.

19 - P.F. Manfredi et al. Nucl. Instr. Meth. A 274 (1989) 477.

20 - J.P. Avondo et al., Nucl. Instr. and Meth. A 241 (1985) 107.

21 - W. Sansen, Nucl. Instr. and Meth. A 253 (1987) 427.

22 - P. Jarron, CMOS Front-End Electronics for Silicon Detectors: the Present and the Trends for the Future. Presented at 4th PISA Meeting on Advanced Detectors. La Biodola (Italyt), May 21-25, (1989).

23 - E. Gatti, A. Hrisoho, P.F. Manfredi, IEEE Trans. Nucl. Sci. NS 30 (1) (1983) 319.

24 - D. Dorfan, Nucl. Instr. and Meth. A 279 (1989) 186.

25 - G. Gola, G. Pessina, P.G. Rancoita, Fast Front-End Electronics for Experiments Using Silicon Calorimeters at SSC/LHC Colliders. Due to appear in Nucl. Instr. and Meth.

26 - H.H. Williams, M. Newcomer, Private Communication.

27 - H. Vogt, Nucl. Instr. and Meth. A 253 (1987) 439.

28 - G. Lutz et al., Nucl. Instr. and Meth. A 264 (1988) 391.

29 - W. Buttler et al., A Microstrip Preamplifier System Based on a Low Noise Radiation Hard Innovative Technology. Presented at 4th Topical Seminar on Experimental Apparatus San Miniato, May 28-June 1st, (1990).

30 - E. Gatti et al., Nucl. Instr. and Meth. 226 (1984) 129.

31 - P. Rehak, E. Gatti, Nucl. Inst. and Meth. A 289 (1990) 410.

32 - L. Strüder et al., Nucl. Instr. and Meth. A 288 (1990) 227.

33 - V. Radeka et al., IEEE Trans. Nucl. Sci. NS 35 (1) (1988) 155.

34 - V. Radeka et al., IEEE. El. Dev. Letters 10, 2 (1989) 91.

35 - P. Rehak et al., Nucl. Instr. and Meth. A 288 (1990) 168.

36 - T. Walker et al., Nucl. Instr. and Meth. 226 (1984) 200.

37 - R. Hoffmann et al., Nucl. Instr. and Meth. 226 (1984) 196.

38 - S. Kleinfelder et al., IEEE Trans. Nucl. Sci. NS 35 (1) (1988) 171.

39 - E. Beuville et al., Nucl. Instr. and Meth. A 288 (1990) 157.

40 - G. Lutz et al., Nucl. Instr. and Meth. A 263 (1988) 163.

41 - K. Kandiah, Nucl. Instr. and Meth. A 288 (1990) 150.

42 - D. Marioli, P.F. Manfredi, P. Massetti, Nucl. Instr. and Meth. A 269 (1988) 109.

43 - D.V. Camin, Nucl. Instr. and Meth. A 277 (1989) 204.

44 - A. Alessandrello et al., Low Noise GaAs Charge-Sensitive Preamplifier for Low Temperature Particle Detectors. Presented at 1989 Nucl. Sci, Symposium, San Francisco, 17. 19 Jan. (1990). To be published in IEEE Trans. Nucl. Sci.

45 - D.V. Camin, Private Communication.

46 - V. Ferrari, P.F. Manfredi, V. Speziali, Gamma radiation effects on the electric properties of electronic components of monolithic circuits. First Trilateral Meeting on Radiation Chemistry and Physics of Solids. Pavia, November 7-10, (1988).

47 - Kraner et al., Radiation Damage Studies on Hybrid Preamplifiers. Preliminary Report, Brookhaven National Laboratory, March (1990).

48 - J.H. Stephen, IEEE Trans. Nucl. Sci. NS. 33, 6 (1986) 1465.

EXERCISES ON NOISE AND SIGNAL PROCESSING
(The solution of the following exercises requires as
a background the paper NOISE LIMITS IN DETECTOR
CHARGE MEASUREMENTS by P.F. Manfredi and V. Speziali)

I TIME-INVARIANT SYSTEM

Determine the response to the δ-impulse detector current of the processing system sketched in fig. 1, which consists of a charge sensitive preamplifier followed by a shaper. DL is an ideal delay line with transmission time τ and characteristic impedance R_o.

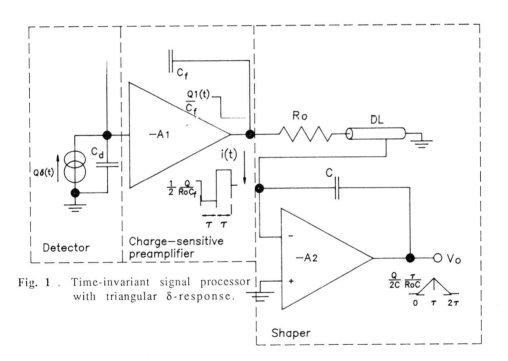

Fig. 1 . Time-invariant signal processor with triangular δ-response.

Solution

The system sketched in fig. 1 is of time-invariant nature. The δ-impulse-like detector current, carrying a charge Q is transformed by the charge-sensitive preamplifier into a voltage step with an amplitude Q/C_f. The step through the resistor R_o, is applied to the transmitting end of the delay line DL. Pay attention to the configuration of the delay line. Its inner conductor is shorted to ground at the receiving end, while the outer conductor is connected to the virtual ground of the integrator designed around the operational amplifier A_2. The current signal i(t) appearing on the outer conductor of the delay line in response to the voltage step which develops at the output of the charge-sensitive preamplifier has the bipolar shape sketched in fig. 1. The lobes, of opposite sign, have amplitudes $Q/(2R_0 C_f)$ and duration τ, equal to the transmission time of the delay line. The bipolar i(t) signal flowing across the capacitor C of the integrator produces the triangular signal V_o shown in fig. 1, which has basewidth 2τ and a peak amplitude $Q\tau/(2R_0 C C_f)$.

The solution described in fig. 1 provides a very effective, though simple method of realizing, with a time-invariant approach, a triangular weighting function.

II TIME-VARIANT SYSTEM

In the system sketched in fig. 2, prior to the detector signal, S_1 is open and S_2 is closed. Assume the detector signal occurring at t=0. At $t=0^-$ S_1 closes and S_2 opens and this configuration of the switches is maintained throughout a time interval of duration τ following the arrival of the detector signal. At $t=\tau$ S_1 opens. S_1 and S_2 remain both open during the next time interval $\tau < t < \tau+\tau'$. At $t=\tau+\tau'$ S_2 closes. Determine the output amplitude V_o in correspondence to a charge Q carried by the detector current pulse and the weighting function of the system.

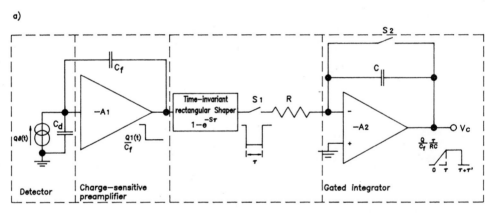

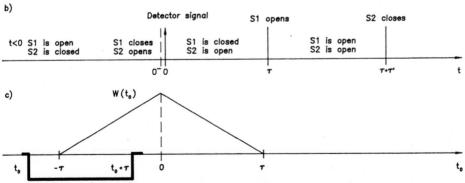

Fig. 2.   a) Time-variant signal processor.
b) Sequence of switching operations referred to the instant at which the detector signal occurs.
c) Building-up of the weighting function W(to).

Solution

The system is of time-variant nature. The detector current signal $Q\delta(t)$ occurs at t=0, when S_1 is closed and S_2 is open, so that the gated integrator consisting of the operational amplifier A_2, switches S_1 and S_2, resistor R and capacitor C is enabled to integrate the signal appearing at the output of the time-invariant rectangular shaper with transfer function $1-e^{-s\tau}$, where s is Laplace's complex variable.
The detector signal produces at the output of the charge-sensitive preamplifier a negative voltage step with an amplitude Q/C_f. The step is clipped to a duration τ by the time-invariant rectangular shaper. The resulting negative rectangle across the series connection of the closed S_1 switch and resistor R forces a costant current of magnitude Q/RC_f to flow to the virtual ground of the gated integrator.

At time $t=\tau$, when the rectangular signal ends and S_1 opens, the voltage V_o reaches the value $Q\tau/(R_oCC_f)$. Such a value, which carries the information about the charge Q to be measured, remains costant in the $(\tau,\tau+\tau')$ interval. During this interval the voltage V_o is sampled and the charge measurement is completed. At $t=\tau+\tau'$ S_2 is closed, thereby resetting the charge on C to zero and making the system ready for the next measurement.
To determine the weighting function W(t) of the system, a detector signal consisting of a δ-impulse carrying a unit charge and occurring at the variable instant to has to be employed. The reference instant t_o has to be slided on the time-axis, from $-\infty$ to $+\infty$. At any given t_o the amplitude V_o appearing at the output of the gated integrator as a result of the unit charge injected at the preamplifier input by the δ-impulse occurring at that given t_o must be recorded. This is the value of the weighting function at that t_o. It has to be pointed out that, in order to determine W(t) for this system, as for any time-variant system, the knowledge of the instant at which the event to be measured takes place is essential.
The shape of W(t) is determined with reference to fig. 2 c).
If $-\infty <t_o<-\tau$, the signal at the output of the time-invariant rectangular shaper recovers to zero before the gated integrator is enabled and therefore V_o remains zero. That is, $W(t_o)=0$ for any to between $-\infty$ and $-\tau$.
If $-\tau<t_o<\tau$ the signal at the output of the rectangular shaper is different from zero during part of the active interval of gated integrator. As t_o varies between $-\tau$ and 0, $W(t_o)$ increases and reaches a maximum whose value is $\tau/(C_fRC)$ for $t_o=0$, that is, when the rectangular signal at the output of the time-invariant shaper exactly overlaps the time window during which the integrator is enabled. If t_o is increased beyond t=0, $W(t_o)$ decreases and reaches zero at $t_o=\tau$. $W(t_o)$ then remains zero for $\tau<t_o<+\infty$.
The weighting function of the system in fig. 2 a) is therefore triangular, with basewidth 2τ and peak amplitude $\tau/(C_fRC)$.

III DISCRETE-TIME SYSTEM

In experiments with accelerators of the collider type, where the detector signals are confined within the bunch crossings and therefore occur at predictable instants, a possible signal-processing method is the one described in figs. 3 a), b).
According to figs 3 a), b) the baseline $V_o(t)$ at the preamplifier output is sampled at equally spaced instants $t_1, t_2, \ldots\ldots t_n$ prior to the signal arrival. The detector signal induces a step-like transition of amplitude Q/C_f in the baseline at the preamplifier output. The varied baseline is sampled again at equally spaced instant $t_1', t_2', \ldots\ldots t_n'$ after the detector signal has occurred.
The information about the charge to be measured is obtained from the difference:

(1) $$\sum_{k=1}^{n} V_o(t_k') - \sum_{j=1}^{n} V_o(t_j)$$

Determine the relevant weighting function.

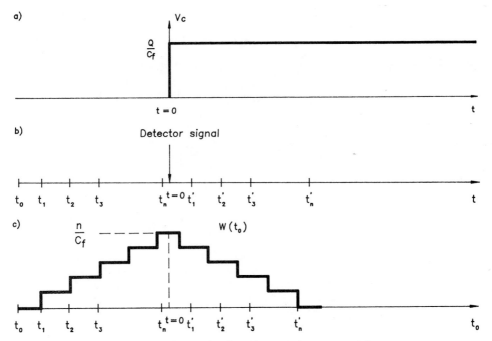

Fig. 3. a) Step induced on the baseline at the preamplifier output by a detector signal occurring at t=0.
b) Sequence of instants at which the baseline at the preamplifier output is sampled before and after the occurrence of the event.
c) Building-up of the weighting function $W(t_o)$.

Solution

Suppose that a unit charge is injected by a δ-impulse at the preamplifier input at t_o [fig. 3 c)]. This impulse induces a step of amplitude $1/C_f$ at the preamplifier output and the discontinuity in this step occurs at $t=t_o$. If $-\infty < t_o < t_1$, the two sums in eq.(1) are equal and their difference is zero, yielding $W(t_o)=0$.
$W(t_o)$ starts getting different from zero as soon as to exceeds t_1. For $t_1 < t_o < t_n$ $W(t_o)$ has a staircase shape which increases in amplitude as to increases, according to the relationship:

(2) $\quad W(t_o) = \dfrac{j}{C_f} \quad$ for $t_j < t_o < t_{j+1}$

$\quad W(t_o) = \dfrac{n}{C_f} \quad$ for $t_n < t_o < 0$

$W(t_o)$ starts decreasing as soon as to exceeds t_1' and tends to zero with a descending staircase shape. $W(t_o)=0$ for $t_n' < t_o < +\infty$.
The $W(t_o)$ function plotted in fig. 3 c) is the discrete-time implementation of a triangular, better say trapezoidal, weighting function.
In the real case, because of the gain-bandwidth limitations in the preamplifier, the $V_o(t)$ response of fig. 3 a), rather than a step, is an exponentially rising signal of finite time-constant and the transitions in the staircase shaped edges of $W(t_o)$ are accordingly smoothed.

The described behavior in the preamplifier response has, as a beneficial effect, that of preventing infinitely large values in the derivative of $W(t_0)$, that may lead to divergent ENC contributions from the preamplifier series noise.

SILICON JUNCTION FIELD-EFFECT TRANSISTORS IN LOW-NOISE CIRCUITS:

RESEARCH IN PROGRESS AND PERSPECTIVES

P.F. Manfredi* and V. Speziali

Dipartimento di Elettronica
Università di Pavia
Via Abbiategrasso 209
27100 Pavia, Italy

and Istituto Nazionale di Fisica Nucleare
Via Celoria 16
20133 Milano, Italy

*Visiting at Instrumentation Division
Brookhaven National Laboratory, Upton, N.Y., 11973

ABSTRACT

The interest for Silicon junction field-effect transistors as front-end devices in preamplifiers for radiation detectors is very well alive by virtue of their very good noise characteristics and of their small noise-sensitivity to the absorbed radiation. This paper discusses some recent achievements in the area of preamplifiers based on Si JFETs.

I INTRODUCTION

Over the past three decades, the silicon junction field-effect transistor (Si JFET) has been the most widely employed input device in preamplifiers for radiation detectors. Its small parallel noise, which becomes negligible at cryogenic temperatures, the low amount of 1/f-noise, actually the lowest among field-effect devices and the very reliable way in which the channel thermal noise depends on the device transconductance conveyed onto the Si JFET such a widespread favour to confine all other devices to very specific and narrow areas.

The search for alternative solutions in the design of front-end preamplifiers was stimulated by particular problems that arose in elementary particle physics at different times. One of these problems was related to the need of narrowing the weighting functions, as required by some detector applications in high rate fixed-target experiments. Such a need could not be met by the existing SiJFET, that have rather limited characteristics of gain-bandwidth product and accordingly the domain of applications where short weighting functions, less than 100 ns, were required, was taken over by discrete microwave bipolar transistors. A second problem which in the very first instance led to turn down the Si JFET and to switch the interest to the CMOS by accepting, as a price to be paid, a sizable deterioration in noise, was brought about by the microvertex detectors employed in collider experiments. These applications set the requirement of a low-noise monolithic front-end, which could not be met, at the very beginning, by the Si JFET. Monolithic integration of low-noise circuits based on Si JFETs is now feasible, more research is in progress and some results will be disclosed in this paper.

Removing the existing frequency limitations of SiJFET is another issue of relevant importance.

It may be wondered what makes the effort aiming to overcome the previously discussed limitations of Si JFETs worthwhile. An answer comes from the consideration that Si JFETs are receiving due attention in view of the design of front-end systems for experiments at future hadron colliders, where the good noise characteristics of JFETs, associated to the small sensitivity of noise to the absorbed radiation are highly attractive.

Recent activities, some of which now in progress, involving JFETs are going to be described. Concisely, large JFETs have been developed to make transformerless matching possible with detector capacitances up to several hundreds of pF. Tetrode structures of JFETs, where the channel conductance is controlled by two independent gates have been designed. Monolithic integration of circuits employing epitaxial-channel JFETs and dielectric isolation techniques is now in progress. Monolithic preamplifier systems based upon implanted-type JFETs and PMOS have been developed. A new implanted JFET, where the gate length has been reduced to 1.5 µm to enhance the gain-bandwidth product will also be described.

II LARGE JFETS FOR TRANSFORMERLESS CAPACITIVE MATCHING

The development of large JFETs was suggested by the need of actuating the capacitive matching or, at least keeping the mismatch as small as possible, in some calorimetric applications involving detector capacitances up to 1 nF or more, when the presence of a strong magnetic fields limits the use of matching transformers [1].

As a result of a collaboration between Pavia University, Milano INFN and INTERFET, two units were developed. Both devices are large interdigitated structures based upon an epitaxial channel process, with the gate diffused onto it. One of them, called 3600, has a gate width W of 3600 mils, that is, $9 \cdot 10^4$ µm. The topview configuration of the lateral p$^+$ diffusion is shown in fig. 1 along with a complete channel cycle, not in scale with the rest of the figure.

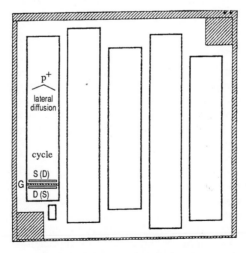

Fig. 1. Structure of 3600 JFET

The second unit, labelled 1800 is one half of 3600 in gate width. The values of input capacitances are about 400 pF for 3600 and 180 pF for 1800. Both units were designed to have a large g_m/I_D ratio, in order to achieve a satisfactory noise behaviour at a comparatively low power dissipation. Such a feature, which makes them suitables also as front-end elements in monolithic preamplifiers, where the power dissipated on the chip has to be constrained below given levels, is apparent in fig. 2.

Figure 2 shows the transconductance g_m and the g_m/I_D ratio as functions of drain current I_D for 3600 and 1800 JFETs. According to fig. 2, the 3600 JFET at a standing current of 2 mA features a 43.8 mA/V transconductance and a 21.7 V^{-1} g_m/I_D ratio. At the same current, transconductance and g_m/I_D ratio are respectively 33.8 mA/V and 16.8 V^{-1} for the 1800 JFET.

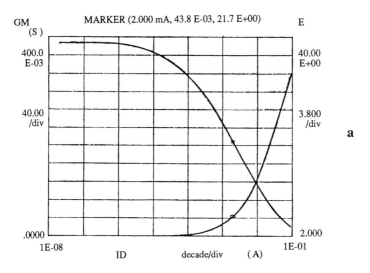

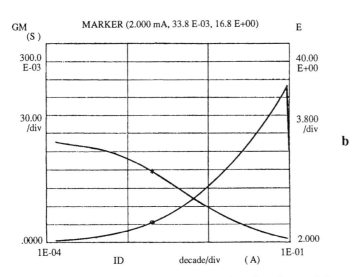

Fig. 2. Transconductance g_m and g_m/I_D ratio plotted as functions of the standing drain current I_D for 3600, a) and 1800, b) JFETs

III JFETS WITH TETRODE STRUCTURE

The TETRODE denomination refers to a JFET configuration where the channel conductance can be controlled by two independent gate electrodes. The structural difference between an ordinary (TRIODE) JFET and a TETRODE can be understood with reference to fig. 3.

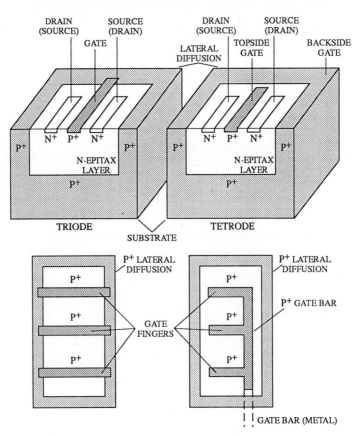

Fig. 3. Cross sections and gate configurations for TRIODE (left) and TETRODE (right) JFETS

Figure 3 shows the cross section and the gate top view for a triode and a tetrode JFET. In both types of device the N epitaxial layer where the channel is located, is surrounded by the P+ region constituted by the substrate and the lateral diffusion that penetrate into it. In the triode the gate fingers overlap the lateral diffusions. The gate is, therefore, made of fingers, substrate and lateral diffusions, as all these elements are electrically connected togheter. The extension of the gate region and the high doping employed to implement it explain why in the triode-type JFETs the gate spreading resistances are so small in value, that negligible contributions are added to the channel thermal noise to determine the whole series thermal noise referred to JFET input.

In the tetrode JFET the fingers are arranged in a comb-shaped structure which constitutes the topside gate and ends up to a metal connection isolated from the lateral diffusions.

The backside gate, made of substrate and lateral diffusion is not connected to the topside-gate.

The tetrode configuration offers, as a major advantage, the design versatility related to the possibility of controlling the device working point by acting on two independent electrodes; such a versatility can be usefully exploited in the design of low-noise preamplifiers [2]. As a limitation, it may suffer from the thermal noise due to the spreading resistance in the topside gate, whose comb-shaped structure is usually made of comparatively thin diffusion layers. To avoid this effect, the comb-shaped topside gate should be metallized, which is not such a straightforward procedure. For instance, in the 1800 C tetrode, an epitaxial-channel device developed as a cooperation between Pavia University, Milan INFN and INTERFET, the configuration shown in fig. 4 has been adopted.

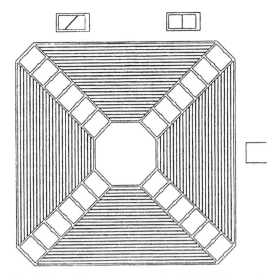

Fig. 4. Topside gate structure in the 1800 C tetrode

In the 1800 CJFET the gate fingers run between metallized bars at 90° angles. The fingers are not metallized and this seems to be responsible for some additional thermal noise over the one related to the channel current. Effort aiming at a complete gate metallization is now being produced.

IV MONOLITHIC PREAMPLIFIER BASED ON DIFFUSED-GATE, EPITAXIAL CHANNEL JFETS

The JFETs considered in sections II and III are discrete devices with epitaxial channel, diffused-gate. Over the implanted-channel JFETs, they have less 1/f-noise and a better compliance of the channel thermal noise to the theoretical expression for the spectral power density, $\overline{de^2}/df = 4kT \cdot 0.7/g_m$. A collaboration programme, involving Brookhaven National Laboratory, Pavia University, Milan INFN and INTERFET has been initiated with the aim of developing a monolithic charge-sensitive preamplifier employing only epitaxial-channel, diffused-gate JFETs.

The goal is that of retaining in the integrated circuits the characteristics of noise and radiation sensitivity of noise that are typical in this type of component as a discrete unit.

For this purpose, a charge-sensitive preamplifier employing no other element but N-channel JFETs has been designed.

Its performances have been investigated by means of SPICE simulation and of a breadboard circuit realization based upon discrete devices with the same geometric characteristics as the ones that will be integrated. The breadboard preamplifier has been also very useful in the analysis of the noise performances and of the relative contributions brought about to the input equivalent noise charge (ENC) by the different parts of the circuit [3].

The circuit diagram of the N-JFET preamplifier is shown in fig. 5. The circuit makes use of devices with two difference values of the drain-to-source distance (channel cycle) according to whether more importance is attributed to the g_m/C_i ratio or to the output resistance. Devices labelled SC (short channel) in fig. 11 have 15 µm-cycle and are employed on the signal path, like J_1, J_2, that constitute the input cascode, and the source followers J_7, J_9. Devices labelled LC (long channel) have 25 µm-cycle and implement the current sources (J_3, J_4), (J_5, J_6) (J_8), for which a large output resistance is of fundamental importance in providing a large dc gain.

In the circuit of fig. 5 the input cascode (J_1, J_2) acts as a transconductance amplifier which has, as an active load on the drain of J_2, the current source (J_3, J_4, J_7). J_7 serves also as an intermediate buffer which transmits the signal to J_9 and provides the dc return through the eight-diode level shifter and the externally connected large resistor R_f. The capacitive feedback takes place through C_f, which is an off-the chip component. Fig. 11 shows the values of the gate width W for all the JFETs except J_1, whose dimension has to be tailored to the specific value of the detector capacitance C_D in order to ensure the desired degree of mismatch.

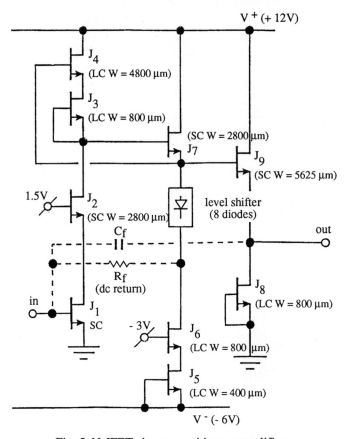

Fig. 5. N-JFET charge-sensitive preamplifier

The layout of the preamplifier is shown in fig. 6, in the case of a J_1 FET with W=11.250 μm.

Monolithic integration of the circuit of fig. 5 is being carried out on the base of the three different processes sketched in fig. 6, where only two JFETs per wafer are shown for the sake of simplicity (4).

Process a) in fig. 7 employs two epitaxial layers, P and N type, grown on top of an N-type substrate and V-groove isolation.

Case b) is a dielectric isolation process with a polysilicon substrate and SiO_2-isolated tubs that house the individual JFETs.

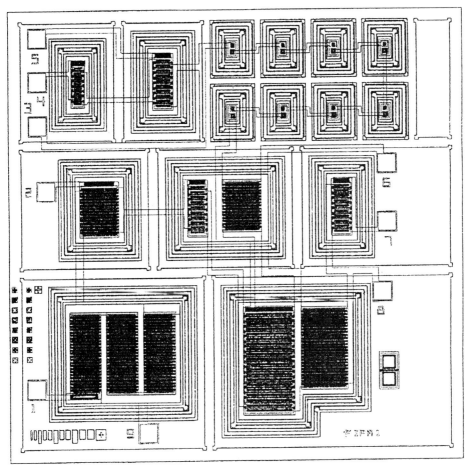

Fig. 6. Layout of the N-JFET preamplifier

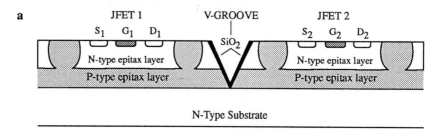
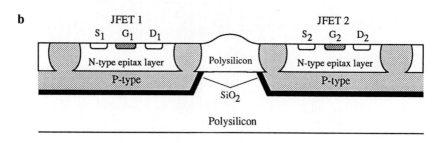
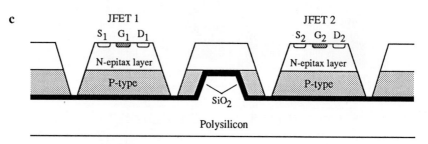

Fig. 7. Monolithic integration processes employed for the realization of the N-JFET preamplifier
  a) Double epitaxial layer, V-groove isolation.
  b) Dielectric isolation, planar process.
  c) Mesa process.

In process (c) the JFETs are realized on dielectrically isolated mesas.

The preamplifier of fig. 5 has been integrated along with several test structures according to the three processes shown in fig. 7 and investigation on the already delivered devices and circuits is now in progress. A very preliminary result relevant to the integration process a) of fig. 7 is presented in fig. 8.

Fig. 8 compares the input JFET J_1 of the monolithic preamplifier with a discrete component of the same gate width W and gate cycle, from the point of view of g_m versus I_D and g_m/I_D versus I_D characteristics. As shown in fig. 8, the device part of the monolithic preamplifier tracks the discrete one very well, which means that in the integration process, the transconductance characteristics of the discrete device have been retained.

However, the JFET in the monolithic circuit was found to have a comparatively large gate spreading resistance, which is not present in the discrete unit. The reason for this has been understood and measures to remove this problem are presently being taken.

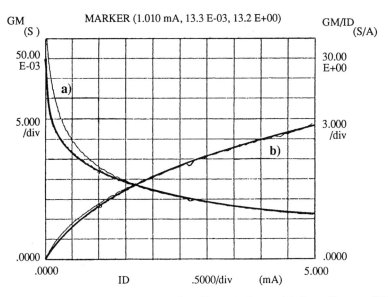

Fig. 8. Transconductance gm (curve a), and g_m/I_D-ratio (curve b), for a discrete JFET (thin line) and a similar unit, part of the monolithic preamplifier (thick line).

V NJFET-PMOS PREAMPLIFIER SYSTEM FOR MICROSTRIP DETECTORS

A monolithic process where JFETs are made available along with CMOS has been employed to realize a 64-channel preamplifier system for microstrip detectors. This project has been carried out in the framework of a collaboration between Pavia University, Milan INFN, Max Planck Institut of Munich and Fraunhofer Institute of Duisburg. Considerations of noise and radiation tolerance have suggested that the system can be designed by making use of two device types: the N-channel JFET and the P-channel MOS. The JFET is employed as a preamplifier input stage and, more generally, as a low-noise element. The P-MOS is used as a complementary element to the N-JFET and, besides, upon it rely all switching functions on the chip. The N-JFET employed in the monolithic preamplifier system have an implanted channel, because of which they have inferior noise performances as compared to the epitaxial channel devices discussed in sections II, III, IV. Nonetheless, they are good enough to provide a very valuable solution to the problem of implementing a monolithic, multichannel preamplifier system for radiation detectors [5, 6, 7].

The preamplifier system has 64 analog channels, each of them consisting of a charge-sensitive preamplifier with a switch actuated reset and a discrete-time filter which implements an almost triangular weighting function.

The filter consists in a set of capacitors and switches and a gated integrator which performs also the function of analog storage.

The preamplifier system is primarily intended for operation with a collider-type accelerator. The sampling filter action is initiated some time before the bunch crossing and keeps going some time after it. At the end of the filter action, an analog amplitude proportional to the charge appearing at the preamplifier input as a result of the event possibly present in the bunch crossing, is stored in the integrator.

During the readout phase all the stored analog amplitudes are sequentially switched onto a serializing bus and made available at the output.

The salient features of the preamplifier system, named JAMEX 64 are summarized in Table I.

The detailed circuit diagram of the analog channel is shown in fig. 9.

TABLE I

Multichannel preamplifier system Jamex 64
Main characteristics

No. of channels	64
Input pitch	100 μm
Chip size	6.4 · 4.9 mm^2
Supply voltages	±6 V - 9 V
Power dissipation	400 mW
Charge sensitivity	15 mV/fC

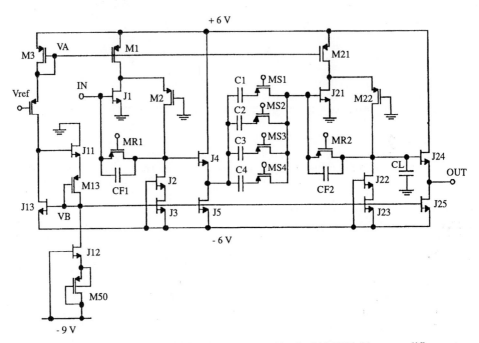

Fig. 9. Detailed circuit diagram of the analog channel in the JAMEX 64 preamplifier system

The input charge-sensitive-loop in fig. 9 is made of a complementary cascode J_1(NJFET), M_2 (PMOS) and related current sources M_1 and J_2, J_3.

The P-MOS MR1 in parallel to the feedback capacitor C_{F1} is the reset switch. The JFET J_4 fed by the current source J_5 is a buffer working at the same current density as J_1, so that the dc level at the preamplifier output is near zero. The four capacitors (C_1 through C_4), the four switches (MS_1 through MS_4) and the integrator designed around the J21, M22, M21, J22, J23 amplifier implement the switched capacitor, discrete-time filter. The analog information at the filter output is stored on C_{F2}, and made available, through the buffer J24, to the output serializing multiplexer.

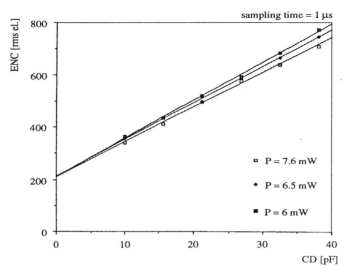

Fig. 10. Equivalent Noise Charge as a function of detector capacitance C_D for the analog channel of JAMEX 64 operated at three different power levels.

The analog channels have the average noise behaviour shown in fig. 10, where the equivalent noise charge ENC is plotted as a function of an external capacitance C_D simulating the detector. The plots refer to three different values of the power dissipated in the channel.

VI SHORT CHANNEL JFET

An N-channel JFET with 1.5 μm gate length has been realized in the framework of the same collaboration as for JAMEX 64. The short channel JFET is a CMOS compatible, implanted-channel device which is intended for replacement of the existing 4 μm gate-length JFET in the upgraded version of JAMEX 64.

The transconductance characteristics of the 1.5 μm gate-length JFET are described by fig. 11.

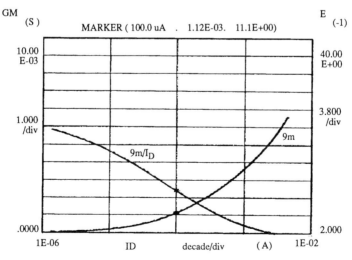

Fig. 11. Transconductance g_m and g_m/I_D ratio as functions of drain current I_D for a 1.5 μm gate length JFET with W = 400 μm.

The device of fig. 11, operated at 100 μA drain current, which is the working point presently assumed for the input JFETs in the preamplifiers of JAMEX 64, features a transconductance of 11 mA/V, with a g_m/I_D ratio of about 11.

The noise spectra of two 1.5 μm gate-length JFETs are plotted in fig. 12, to show how white and 1/f-noise scale down with the device size at a fixed drain current. The value of I_D employed for the measurements of fig. 12 is suitably large to characterize the noise behaviour of the JFETs in a more favourable operating condition.

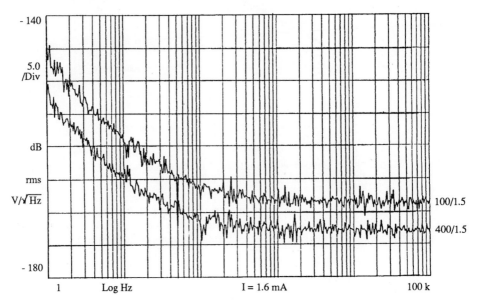

Fig. 12. Noise spectra of JFETs with 1.5 μm gate length.
a) W = 100 μm
b) W = 400 μm

According to fig. 12, the device with W = 400 μm at 1.6 mA drain current, has about 3 nV/√Hz spectral voltage density in the white noise region and less than 6nV/√Hz at 10 Hz. The low-frequency noise behaviour seems to be particularly good for a device of such a short gate length.

The transition-frequency charateristics of the 1.5 μm gate-length JFET have not yet been thoroughly investigated. Device simulations predict for this JFET a transition frequency in excess of 2 GHz. This result, if confirmed by experimental evidence, would put the 1.5 μm gate-length unit at the top of presently available silicon JFETs.

VI   CONCLUSIONS

In progress research activities aiming at extending the applications of JFETs as front-end elements in preamplifiers for radiation detectors and at improving their performances have been analyzed. Effort is presently being channelled to develop new discrete units, like large JFETs and tetrodes, to make the monolithic integration of epitaxial-channel JFETs possible, to improve the frequency behaviour of implanted-channel JFETs in monolithic processes and to upgrade the already existing monolithic preamplifier systems. Epitaxial-channel JFETs that, as discrete units present the best noise behaviour, still suffer from two limitations. First, they don't lend themselves to a well mastered integration process. Second, they don't offer yet adequately high transition frequencies.

ACKNOWLEDGEMENTS

The authors acknowledge the contribution of V. Radeka and S. Rescia in the programs based on epitaxial channel JFETs. Implementation of the same programs has been made possible by the technical skill of INTERFET management and staff, particularly by M. Hoye, L. Rehn, D. Roberts.

The activities on JFET-CMOS monolithic circuits and short-channel JFETs are being carried out with W. Buttler, G. Lutz, H. Vogt to whom the authors are indebted.

REFERENCES

1 - M. Bertolaccini et al. Nucl. Industr. and Meth. A 264 (1988) 399.

2 - S. Rescia, private communication.

3 - M. Demicheli et al. Nucl. Instr. and Meth. A 289 (1990) 418.

4 - L.A. Rehn-D.E. Roberts, Monolithic JFET preamplifiers for Ionization Chamber Calorimeters. Paper presented at International Industrial Symposium on Supercolliders. Miami Beach (Fla), March 14-16 (1990).

5 - H. Vogt - Nuclear Instrum and Methods, A 253 (1987) 434.

6 - W. Buttler et al. - Nucl. Instr. and Meth., A 279 (1989) 204.

7 - W. Buttler et al. - Nucl. Instr. and Meth., A 288 (1990) 140.

Index

Accelerator instabilities, 296-302
Accelerators, (see particle accelerators)
Asymptotic freedom, 38

Baryogenesis, 67-69
$B^0 - \overline{B}^0$ mixing, 170-174
Beamstrahlung, 316-317
β-function, 288-291
Betatron motion, (see transverse motion)
Bottom quark, 158-175
BRST symmetry, 28-30, 37

Cabibbo-Kobayashi-Maskawa matrix, 124-125
Calorimetry, 325-381
    absorption mechanisms, 327-339
    compensation, 339-345, 362-364
    energy response, 345-351, 361-362, 364-365
    particle identification, 352-360, 365-367
    position response, 351-352, 365
Capacitive matching, 393-396
CDF detector, 83-87
Charm mesons, 262-272
Chromaticity, 306
Collider luminosities, 75
Collider physics, 71-196
Compensation, 339-345
Cosmological constant, 52-53
Crab crossing, 319-321

Dijet production, 91-98
    angular distribution, 94-98
    mass spectrum, 94-95
Dimensional regularization, 9-16
Dirac field theories, 21-23
Direct photons, 105-108
Disruption, 313-316

Dynamic aperture, 309-312
Einstein-de Sitter model, 53-54, 59-60
Electromagnetic showers, 327-334, 348-350, 361-362
Electron/hadron response, 339-345
Electroweak standard model
    bottom quark, 158-175
    experimental tests, 126-144
    theory, 39-45, 122-126
    top quark, 144-158
Emittance, 291-293, 296-298
Energy motion, (see longitudinal motion)
Equivalent noise charge (ENC), 386-396
Expanding universes, 50-61

Gauge bosons, (see W and Z)
    additional, 179-182
Gauge theories, 1-47
Glashow-Weinberg-Salam, (see electroweak model)
Gluonium, 254-259
Goldstone field, 21, 31

Hadronic showers, 335, 345, 350-351, 362-365
Heavy leptons, 182
Heavy quarks, 144-175, 259-272
Higgs field
    mechanism, 30-31
    searches, 176-179
Hill's equation, 287-288
Hubble parameter, 53-55, 59

Jet production, 87-105

K^* spectroscopy, 215-244

Lepton universality, 136-138
Local symmetry, 23-24
Longitudinal motion, 283-286

Meson spectrum, 197-282
   angular decay, 202-210
   classification, 211-214
   quantum numbers, 199-201
Microwave background, 66
Multi-jet production, 98-104

Noether's theorem, 17-18
Noise properties of active devices
   bipolar transistors, 396-398
   effect of ionization, 409-412
   GaAs MESFETs, 408-409
   MOS transistors, 406-408
   silicon junction FETs, 398-406
Noise sources, 382-389
Non-Abelian symmetries, 32-38
Nucleosynthesis, 64-66

Octupole field, 306-309
Olber's paradox, 60-61
Onia, 259-262

Particle accelerators, 283-325
   beam-beam interaction, 312-321
   nonlinear effects, 302-312
   single particle motion, 283-293
   wake fields, 293-302
Parton model, 87-91

Quantum chromodynamics (QCD), 36-38
   experimental tests, 87-122
     electromagnetic, 105-109
     electroweak, 108-122
Quantum electrodynamics (QED), 25-30

Redshift, 57-58

Scalar field theory, 4-21
   complex, 16-21, 30-31
   self interacting, 5-8
Sextupole field, 304
Silicon JFETs, 417-428
   monolithic preamplifiers, 421-427
   noise properties, 398-406
   tetrode structure, 419-421
Soft collisions, 76-78
SPACAL detector, 360-375
Spontaneous symmetry breaking, 19-21, 30-31
Substructure, 182-185
Supersymmetric (SUSY) models, 185-189
Strangeonium ($s\bar{s}$) mesons, 244-253
Synchrotron tune, 286

Top quark searches, 144-158
Transverse motion, 286-291
Triple gauge coupling, 176

UA2 detector, 83-87

V-A current, 40

Wakefields, 294-296
W bosons, 108-122, 133-136, 138-144
Weinberg angle, 44-45, 122-123

Z bosons, 108-122, 131-133, 135-136, 138-144